Advances in Industrial Control

More information about this series at http://www.springer.com/series/1412

Matthew Ellis · Jinfeng Liu
Panagiotis D. Christofides

Economic Model Predictive Control

Theory, Formulations and Chemical Process Applications

Springer

Matthew Ellis
Department of Chemical and Biomolecular Engineering
University of California, Los Angeles
Los Angeles, CA
USA

Panagiotis D. Christofides
Department of Chemical and Biomolecular Engineering
University of California, Los Angeles
Los Angeles, CA
USA

Jinfeng Liu
Department of Chemical and Materials Engineering
University of Alberta
Edmonton, AB
Canada

ISSN 1430-9491 ISSN 2193-1577 (electronic)
Advances in Industrial Control
ISBN 978-3-319-41107-1 ISBN 978-3-319-41108-8 (eBook)
DOI 10.1007/978-3-319-41108-8

Library of Congress Control Number: 2016944902

Printed on acid-free paper

This Springer imprint is published by Springer Nature
The registered company is Springer International Publishing AG Switzerland

Series Editors' Foreword

The series *Advances in Industrial Control* aims to report and encourage technology transfer in control engineering. The rapid development of control technology has an impact on all areas of the control discipline. New theory, new controllers, actuators, sensors, new industrial processes, computer methods, new applications, new design philosophies…, new challenges. Much of this development work resides in industrial reports, feasibility study papers, and reports of advanced collaborative projects. The series offers an opportunity for researchers to present an extended exposition of such new work in all aspects of industrial control for wider and rapid dissemination.

The model predictive control (MPC) design philosophy may be used for a low-level loop controller, replacing regulating loop control systems based on the proportional integral differential (PID) controllers. It also has the flexibility to work at a higher level as a tracking control supervisor to follow a reference trajectory by providing the reference inputs to low-level loop controllers, typically of the PID variety. These types of industrial formulations for MPC usually involve a quadratic cost functional, a (linear) dynamic process model, and a set of physical system constraints on inputs and outputs. This class of MPC problems has the advantage of being readily numerically solvable and implemented in real time. Clearly these loop controller and reference-tracking uses of MPC fit very nicely with the lower level/supervisory-level architectures of the traditional process control hierarchy based on process function and sampling-time-separation arguments.

Since this paradigm has met with considerable success in the process industries, the obvious question to ask is where next with industrial developments of MPC. The academic control community has certainly been busy extending the theory of MPC into the field of nonlinear predictive control. Indeed, our sister series

Advanced Textbooks in Control and Signal Processing was recently fortunate to publish *Model Predictive Control* by Basil Kouvaritakis and Mark Cannon (ISBN 978-1-319-24851-6, 2016) in this growing field of nonlinear MPC.

However, an alternative and pragmatic way forward is to follow the example of researchers Matthew Ellis, Jinfeng Liu, and Panagiotis D. Christofides and look again at the process trends and requirements of industry. These researchers are mainly involved with the chemical process industries and what they find is an increasing focus on "dynamic market-driven operations which include more efficient and nimble operations". Their solution is a new re-formulation and re-interpretation of the MPC method termed economic model predictive control (EMPC) that is reported in this *Advances in Industrial Control* monograph, *Economic Model Predictive Control: Theory, Formulations and Chemical Process Applications*.

Their aim is to exploit the ever-increasing power of computing technology to enhance the traditional control hierarchy by extending and re-interpreting the MPC method in several respects. This approach may be applied to an upper-level control more concerned with management functions and scheduling, or to an intermediate level involving the multivariable loop controls. The cost functional is selected to capture some of the economic objectives of the process. The dynamic process model is extended to represent both economic and physical variables in the process. The physical process constraint set remains largely unchanged in interpretation and formulation, but a new set of "economic" process constraints is appended to the MPC problem description. The resulting formulation involves nonlinear mathematical representations and constrained nonlinear system optimization.

The researchers put their ideas and solutions to a test with a set of applications from the chemical process industries and there are three challenges in this work:

i. how to capture mathematically the "economic dimension" of a process in the construction of the cost functional, the process model, and the constraints;
ii. how to provide an EMPC control theory to guarantee essential control properties such as closed-loop stability; and
iii. how to develop the numerical computational algorithms that will allow the application of the desired control actions in real-time operation.

The EMPC method is a challenge for control theory analysis, and the resulting nonlinear optimization problems are difficult to solve and implement. The monograph presents new results in these areas that are original to the authors and they demonstrate their results with detailed chemical process simulation examples. The success of MPC in the process industries has been largely due to the economic benefits provided using linear system and quadratic cost problems. However, further improvements will require more accurate nonlinear plant models and cost measures which this text explores. The *Advances in Industrial Control* monograph series was originally created for the promotion of new methods for industrial applications. The level of originality and the new research results presented in this monograph meet this series' aim and make an excellent contribution to *Advances in Industrial Control*.

Michael J. Grimble
Michael A. Johnson
Industrial Control Centre
University of Strathclyde
Glasgow, Scotland, UK

Preface

Traditionally, economic optimization and control of chemical processes have been addressed with a hierarchical approach. In the upper layer, a static economic optimization problem is solved to compute an optimal process steady state. The optimal steady state is sent down to the lower feedback control layer to force the process to operate at the optimal steady state. In the context of the lower feedback control layer, model predictive control (MPC) has become a ubiquitous advanced control methodology used in the chemical process industry owing to its ability to control multiple input, multiple output process/systems while accounting for constraints and performance criteria. Recent pressure to make chemical processes operate more efficiently, cost effectively, and reliably has motivated process control researchers to analyze a more general MPC framework that merges economic optimization with process control. In particular, economic MPC (EMPC), which incorporates an economically motivated stage cost function in its formulation, has attracted significant attention and research over the last 10 years. The rigorous design of EMPC systems that operate processes in an economically optimal fashion while maintaining stability of the closed-loop system is challenging as traditional notions of stability may not apply to the closed-loop system under EMPC.

This book covers several rigorous methods for the design of EMPC systems for chemical processes, which are typically described by nonlinear dynamic models. The book opens with a brief introduction and motivation of EMPC and a background on nonlinear systems, control and optimization. An overview of the various EMPC methods proposed in the literature is provided. Subsequently, an EMPC scheme designed via Lyapunov-based techniques, which is the main focus of this book, is described in detail with rigorous analysis provided on its feasibility, closed-loop stability and performance properties. Next, the design of state-estimation-based EMPC schemes is considered for nonlinear systems. Then, several two-layer EMPC frameworks are presented that address computational efficiency and industrially relevant control designs. The book closes with additional EMPC designs that address computational efficiency and real-time implementation.

Throughout the book, the EMPC methods are applied to chemical process examples to demonstrate their effectiveness and performance.

The book requires some knowledge of nonlinear systems and nonlinear control theory. Because EMPC requires the repeated solution of a nonlinear optimization problem, a basic knowledge of nonlinear optimization/programming may be helpful in understanding the concepts. This book is intended for researchers, graduate students, and process control engineers.

We would like to acknowledge Dr. Mohsen Heidarinejad, Dr. Xianzhong Chen, Dr. Liangfeng Lao, Helen Durand, Tim Anderson, Dawson Tu, and Anas Alanqar all at UCLA who have contributed substantially to the research efforts and results included in this book. We would like to thank them for their hard work and contributions. We would also like to thank our many other collaborators and colleagues who contributed in some way to this project.

In particular, we would like to thank our colleagues at UCLA and the University of Alberta, and the United States National Science Foundation and the Department of Energy for financial support. Finally, we would like to express our deepest gratitude to our families for their dedication, encouragement, patience, and support over the course of this project. We dedicate this book to them.

Los Angeles, CA, USA — Matthew Ellis
Edmonton, AB, Canada — Jinfeng Liu
Los Angeles, CA, USA — Panagiotis D. Christofides

Contents

List of Figures

List of Tables

Chapter 1
Introduction

1.1 Motivation

Optimal operation and control of dynamic systems and processes has been a subject of significant research for many years. Within the chemical process industries, operating a process for any substantial length of time at globally optimal operating conditions with respect to some meaningful economic-oriented performance criterion is almost certainly impossible. Room for process operational performance improvement will always exist.

One methodology for improving process performance while achieving operational targets and constraints is to employ the solution of optimal control problems (OCPs) on-line. In other words, control actions for the manipulated inputs of a process are computed by formulating and solving on-line a dynamic optimization problem. With the available computing power of modern computers, solving complex dynamic optimization problems, which may take the form of large-scale, nonlinear, and non-convex optimization problems, on-line is becoming an increasingly viable option. The resulting controller design may be capable of improving the closed-loop dynamic and steady-state performance relative to other controller designs.

The process performance of a chemical process typically refers to the process economics and encapsulates many objectives: profitability, efficiency, variability, capacity, sustainability, etc. As a result of continuously changing process economics, e.g., variable feedstock, changing energy prices, variable customer demand, process operation objectives and strategies need to be frequently updated to account for these changes. Traditionally, a hierarchical strategy for planning/scheduling, optimization, and control has been employed in the chemical process industries. A block diagram of the hierarchical strategy is shown in Fig. 1.1 (adapted from [1]). Although the block diagram provides an overview of the main components, it is a simplified view of modern planning/scheduling, optimization, and control systems employed in the chemical process industry in the sense that each layer may be comprised of many distributed and hierarchical computing units. The underlying design principle of the hierarchical strategy invokes time-scale separation arguments between the

M. Ellis et al., *Economic Model Predictive Control*,
Advances in Industrial Control, DOI 10.1007/978-3-319-41108-8_1

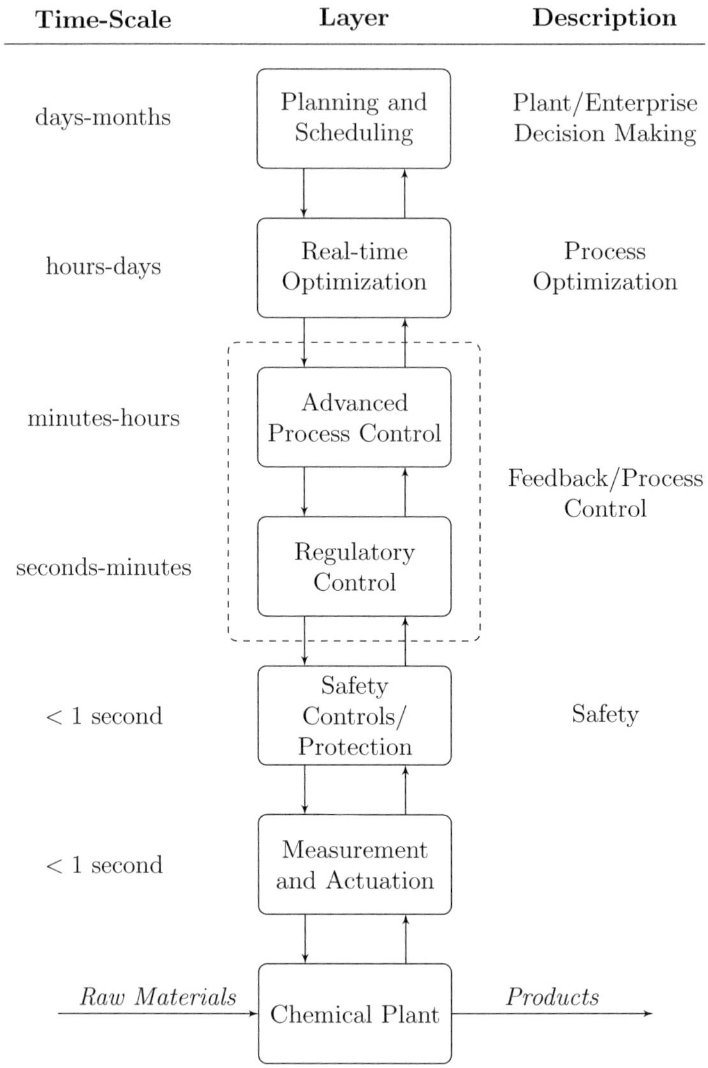

Fig. 1.1 The traditional hierarchical paradigm employed in the chemical process industries for planning/scheduling, optimization, and control of chemical plants (adapted from [1])

execution/evolution of each layer (Fig. 1.1). In the highest level of the hierarchy, enterprise-wide and/or plant-wide planning and scheduling decisions are made on the order of days-months. These decisions are made on the basis of multiple operating processes even multiple operating plants, and are out-of-scope of the present monograph.

In the next layers of the hierarchy of Fig. 1.1, economic optimization and control of chemical processes is addressed in the multi-layer hierarchical architecture, e.g., [2, 3]. The upper-layer, called real-time optimization (RTO), is responsible for process optimization. Within the RTO layer, a metric, defining the operating profit or operating cost, is optimized with respect to an up-to-date and rigorous steady-state process model to compute the optimal process steady-state. The computed steady-state is sent to the feedback process control systems, which consists of the supervisory control and regulatory control layers. The process control system steers the process to operate at the steady-state using the manipulated inputs of the process. The process control system must work to reject disturbances and ideally guide the trajectory of the process dynamics along an optimal path to the steady-state.

The advanced or supervisory process control layer of Fig. 1.1 consists of control algorithms that are used to account for process constraints, coupling of process variables and processing units, and operating performance. In the advanced process control layer, model predictive control (MPC), a control strategy based on optimal control concepts, has been widely implemented in the chemical process industry. MPC uses a dynamic model of the process in an optimization problem to predict the future evolution of the process over a finite-time horizon and to determine the optimal input trajectory with respect to a performance index. Furthermore, MPC can account for the process constraints and multi-variable interactions in the optimization problem. Thus, it has the ability to optimally control constrained multiple-input multiple-output nonlinear systems. Handling constraints and multivariate interactions are the two key advantages of MPC relative other control designs.

The standard MPC approaches are the regulating and tracking formulations that employ a quadratic performance index. Regulating MPC is used to force the process to the (economically) optimal steady-state, while tracking MPC is used to force the process to track a pre-specified reference trajectory. The quadratic performance index is a measure of the predicted squared weighted error of the states and inputs from their corresponding steady-state or target reference values. To date, there are hundreds, if not thousands, of papers on regulating/tracking MPC addressing many issues. A complete review of the MPC literature is beyond the scope of this book. For reviews and books on regulating/tracking MPC for the process industries, the interested reader is referred to [4–15].

The regulatory control layer is composed of mostly single-input single-output control loops like proportional-integral-derivative (PID) control loops that work to implement the control actions computed by the supervisory control layer; that is, it ensures that the control actuators achieve the control action requested by the MPC layer. Often, the dynamics of the regulatory control layer and control actuators are neglected in the dynamic model used in the MPC layer owing to time-scale separation arguments.

As previously mentioned, the overall control architecture of Fig. 1.1 invokes intuitive time-scale separation arguments between the various layers. For instance, RTO is executed at a rate of hours-days, while the feedback control layers compute control actions for the process at a rate of seconds-minutes-hours [1]. Though this paradigm has been successful, we are witnessing the growing need for dynamic market-driven operations which include more efficient and nimble process operation [16–19]. To enable next-generation or "Smart" operations/manufacturing, novel control methodologies capable of handling dynamic economic optimization of process operations should be designed and investigated. More specifically, there is a need to develop theory, algorithms, and implementation strategies to tightly integrate the layers of Fig. 1.1. The benefits of such work may usher in a new era of dynamic (off steady-state and demand and market-driven) process operations.

In an attempt to integrate economic process optimization and process control as well as realize the possible process performance improvement achieved by consistently dynamic, transient, or time-varying operation, i.e., not forcing the process to operate at a pre-specified steady-state, economic MPC (EMPC) has been proposed which incorporates a general cost function or performance index in its formulation. The cost function may be a direct or indirect reflection of the process economics. However, a by-product of this modification is that EMPC may operate a system in a possibly time-varying fashion to optimize the process economics and may not operate the system at a specified steady-state or target. The notion of time-varying operation will be carefully analyzed throughout this monograph. The rigorous design of EMPC systems that operate large-scale processes in a dynamically optimal fashion while maintaining safe and stable operation of the closed-loop process system is challenging as traditional notions of stability, e.g., asymptotic stability of a steady-state, may not apply to the closed-loop system/process under EMPC. While the concept of using general cost function in MPC has been suggested numerous times in the literature, e.g., [20–23], closed-loop stability and performance under EMPC has only recently been considered and rigorously proved for various EMPC formulations [24–27].

1.2 Tracking Versus Economic Model Predictive Control: A High-Level Overview

A high-level overview of the key differences between tracking and economic MPC is provided. A mathematical discussion of tracking and economic MPC, which requires a formal definition of notation and preliminary results, is delayed until the subsequent chapters. Also, the term tracking MPC is used throughout this monograph to refer to both regulating MPC or MPC formulated to force the process to operate at a given steady-state and tracking MPC or MPC formulated to force the process track a given reference trajectory. Model predictive control, whether tracking or economic, is a feedback control methodology where the control actions that are applied to the closed-loop process/system are computed by repeatedly solving a nonlinear constrained optimization problem on-line. The main components of MPC are:

1. A mathematical model of the process/system to predict the future evolution of the process/system over a time interval called the prediction horizon.
2. A performance index or cost functional that maps the process/system (state and input) trajectories over the prediction horizon to a real number that is a measure of the tracking or economic performance. The cost functional is the objective function of the optimization problem.

3. Constraints on the process/system including restrictions on the control inputs, system states and other considerations, e.g., stability and performance constraints.
4. A receding horizon implementation (described further below).

To make the optimization problem of MPC a finite-dimensional one, the prediction horizon of MPC is typically selected to be finite and the input trajectory over the prediction horizon, which is the decision of the optimization problem, is parameterized by a finite number of variables.

The receding horizon implementation is the strategy that involves repeatedly solving the optimization problem on-line to compute the control actions. Specifically, real-time (continuous-time) is partitioned into discrete time steps called sampling times where the time between two consecutive sampling times is called the sampling period. At each sampling time, the MPC optimization problem is initialized with a state measurement or estimate. The MPC optimization problem is solved to compute the optimal input trajectory over the prediction horizon. The control action(s) computed over the first sampling period of the prediction horizon is/are applied to the closed-loop process/system. At the next sampling time, the MPC problem is resolved after receiving an updated state measurement/estimate. The algorithm is repeated over the length of operation. The receding horizon implementation is important because it introduces feedback to compensate for disturbances, modeling errors, and other forms of uncertainty. Moreover, the receding horizon implementation allows for a better approximation of the solution of the corresponding infinite-horizon optimal control problem, i.e., the MPC problem in the limit as the prediction horizon tends to infinity. The infinite-horizon solution, assuming the solution exists, arguably gives the best solution as chemical processes are typically operated over long periods of time without a natural termination or shutdown time.

Tracking MPC optimization problem takes the following general form:

Optimize:	Tracking cost functional	
Subject to:	Dynamic model initialized with state measurement/estimate	(1.1)
	State/input constraints	
	Stability constraints	

while economic MPC problem takes the following general form:

Optimize:	Economic cost functional	
Subject to:	Dynamic model initialized with state measurement/estimate	
	State/input constraints	(1.2)
	Economic-oriented constraints	
	Stability constraints	

The main difference between tracking MPC of Eq. 1.1 and economic MPC of Eq. 1.2 is that the tracking MPC problem is formulated with a tracking cost functional, while the economic MPC problem is formulated with an economic cost functional. The tracking cost functional usually uses a quadratic stage cost that penalizes the deviation of state and inputs from their corresponding steady-state, target, or reference values. However, the EMPC cost functional may potentially use any general stage cost that reflects the process/system economics. Since the idea of EMPC is to compute control actions that directly account for the economic performance, economic-oriented constraints may also be added. For example, the economic-oriented constraints may limit the average amount of raw material that may be fed to the process/system or may ensure that the total production of the desired product over a specified length of operation meets demand.

1.3 Chemical Processes and Time-Varying Operation

In this section, a few chemical process examples that will be used to study the closed-loop properties of EMPC are presented to motivate the need for unsteady-state process operation to improve economic performance. As discussed in the introduction, steady-state operation is typically adopted in chemical process industries. In this operating paradigm, the control system is used to force a chemical process to a pre-specified steady-state and maintain operation at this steady-state thereafter until the desired operating steady-state is changed. However, steady-state operation may not necessarily be the best operation strategy with respect to the process economics. In fact, the chemical process control literature is rich with chemical process examples that demonstrate performance improvement with respect to specific cost metrics with dynamic process operation. In particular, many studies have analyzed the economic benefit of periodically operated reactors, e.g., [28–47], and the numerous references therein. To help identify systems that achieve a performance benefit from periodic operation, several techniques have been proposed including frequency response techniques and the application of the maximum principle [28, 36, 48–51]. Periodic control strategies have also been developed for several applications, for instance, [35, 39, 42–44].

While the periodic operating strategies listed above do demonstrate economic performance improvement, in the case of forced periodic operation, i.e., periodic operation induced by periodic switching of manipulated inputs, the periodic operating policies described in many previous works have been identified through a low-order control parameterization, e.g., a bang-bang input profile and in an open-loop fashion. Owing to recent advances in dynamic optimization (numerical solution strategies or direct methods), it is possible that these chemical process examples previously considered in the context of periodic operation may achieve further economic performance improvement under EMPC. Moreover, EMPC may systematically determine, in real-time, the optimal operating strategy based on the current economic factors while meeting operating constraints. When accounting for time-varying economic factors, e.g., real-time energy pricing, and time-varying disturbance, it is certainly possible that more complex operating strategies beyond steady-state and periodic operation are economically optimal. Developing EMPC schemes that dictate such complex operating strategies, which are generally referred to as time-varying operating strategies, has motivated much of the work contained in this book. Two chemical process examples that benefit from time-varying operation are provided below.

1.3.1 Catalytic Oxidation of Ethylene

Consider a benchmark chemical reactor example depicted in Fig. 1.2 that has previously studied in the context of forced periodic operation [38, 39]. Within the reactor, ethylene oxide (C_2H_4O) is produced from the catalytic oxidation of ethylene with air. Ethylene oxide is an important raw material within the chemical industry because it is used for the synthesis of ethylene glycol which is subsequently used to produce many materials. The reactor is modeled as a non-isothermal continuous stirred-tank reactor (CSTR) with a coolant jacket to remove heat from the reactor. Two combustion reactions occur that consume both the reactant and the product, respectively. The reactions are given by

$$\begin{aligned} C_2H_4 + \frac{1}{2}O_2 &\xrightarrow{r_1} C_2H_4O \\ C_2H_4 + 3O_2 &\xrightarrow{r_2} 2CO_2 + 2H_2O \\ C_2H_4O + \frac{5}{2}O_2 &\xrightarrow{r_3} 2CO_2 + 2H_2O \end{aligned}$$

where r_i, $i = 1, 2, 3$ is the reaction rate of the ith reaction, and the reaction rate expressions are

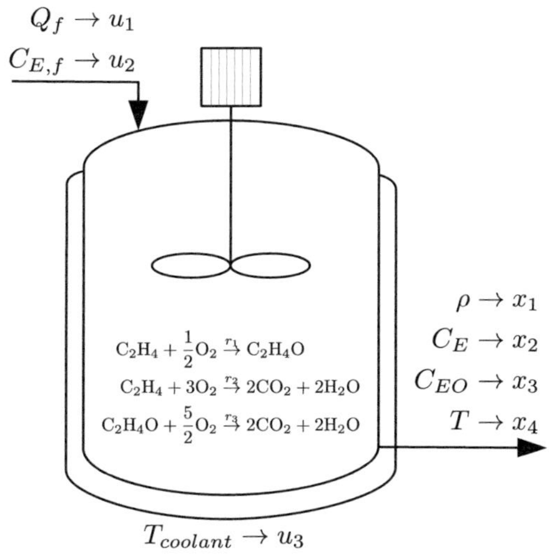

Fig. 1.2 Diagram of the catalytic reactor that produces ethylene oxide from ethylene

$$r_1 = k_1 \exp\left(\frac{-E_1}{RT}\right) P_E^{0.5} \tag{1.3}$$

$$r_2 = k_2 \exp\left(\frac{-E_2}{RT}\right) P_E^{0.25} \tag{1.4}$$

$$r_3 = k_3 \exp\left(\frac{-E_3}{RT}\right) P_{EO}^{0.5} \tag{1.5}$$

where k_i and E_i, $i = 1, 2, 3$ are the reaction rate constant and activation energy for the ith reaction, respectively, T is the temperature, R is the gas constant, and P_j is the partial pressure of the jth component in the reactor ($j = E, EO$ denotes ethylene and ethylene oxide, respectively). The reaction rate expressions are from [52] where catalytic oxidation of ethylene using an unmodified, commercial catalyst was studied over the temperature range 523–573 K. To model the gaseous mixture within the reactor, ideal gas is assumed and the concentration of the jth component within the reactor, denoted by C_j, is

$$C_j = \frac{P_j}{RT}. \tag{1.6}$$

A model describing the dynamic behavior of the reactor is derived through first principles under standard modeling assumptions, e.g., ideal gas and constant heat capacity. The dimensionless states are

$$x_1 = \rho/\rho_{\text{ref}}, \quad x_2 = C_E/C_{\text{ref}}, \quad x_3 = C_{EO}/C_{\text{ref}}, \quad x_4 = T/T_{\text{ref}}$$

where ρ/ρ_{ref} is the dimensionless vapor density in the reactor, C_E/C_{ref} is the dimensionless ethylene concentration in the reactor, C_{EO}/C_{ref} is the dimensionless ethylene oxide concentration in the reactor, and T/T_{ref} is the dimensionless reactor temperature. The manipulated inputs are

$$u_1 = Q_f/Q_{\text{ref}}, \quad u_2 = C_{E,f}/C_{\text{ref}}, \quad u_3 = T_c/T_{\text{ref}}$$

where Q_f/Q_{ref} is the dimensionless volumetric flow rate of the feed, $C_{E,f}/C_{\text{ref}}$ is the dimensionless ethylene concentration of the feed, and T_c/T_{ref} is the dimensionless coolant temperature. The model of the reactor is given by the following set of nonlinear ordinary differential equations:

$$\frac{dx_1}{dt} = u_1(1 - x_1x_4) \tag{1.7}$$

$$\frac{dx_2}{dt} = u_1(u_2 - x_2x_4) - A_1\bar{r}_1(x_2, x_4) - A_2\bar{r}_2(x_2, x_4) \tag{1.8}$$

$$\frac{dx_3}{dt} = -u_1x_3x_4 + A_1\bar{r}_1(x_2, x_4) - A_3\bar{r}_3(x_3, x_4) \tag{1.9}$$

$$\begin{aligned}\frac{dx_4}{dt} &= \frac{u_1}{x_1}(1 - x_4) + \frac{B_1}{x_1}\bar{r}_1(x_2, x_4) + \frac{B_2}{x_1}\bar{r}_2(x_2, x_4)\\ &\quad + \frac{B_3}{x_1}\bar{r}_3(x_3, x_4) - \frac{B_4}{x_1}(x_4 - u_3)\end{aligned} \tag{1.10}$$

where

$$\bar{r}_1(x_2, x_4) = \exp(\gamma_1/x_4)(x_2x_4)^{1/2} \tag{1.11}$$

$$\bar{r}_2(x_2, x_4) = \exp(\gamma_2/x_4)(x_2x_4)^{1/4} \tag{1.12}$$

$$\bar{r}_3(x_3, x_4) = \exp(\gamma_3/x_4)(x_3x_4)^{1/2} \tag{1.13}$$

and the parameters are given in Table 1.1 from [38, 39]. The admissible input values is considered to be bounded in the following sets:

$$\begin{aligned}u_1 &\in [0.0704, 0.7042],\\ u_2 &\in [0.2465, 2.4648],\\ u_3 &\in [0.6, 1.1].\end{aligned}$$

Following [38], the profitability of the reactor is assumed to scale with the yield of ethylene oxide. Therefore, to optimize the profitability or economics of the reactor, one seeks to maximize the time-averaged yield of ethylene oxide. The time-averaged yield of ethylene oxide over an operating time t_f is given by

Table 1.1 Dimensionless process model parameters of the ethylene oxidation reactor model

Parameter	Value	Parameter	Value
A_1	92.80	B_3	2170.57
A_2	12.66	B_4	7.02
A_3	2412.71	γ_1	–8.13
B_1	7.32	γ_2	–7.12
B_2	10.39	γ_3	–11.07

The parameters are from [38]

$$Y = \frac{\frac{1}{t_f}\int_0^{t_f} x_3(\tau)x_4(\tau)u_1(\tau)\,d\tau}{\frac{1}{t_f}\int_0^{t_f} u_1(\tau)u_2(\tau)\,d\tau}. \tag{1.14}$$

Since two combustion reactions occur in the reactor that consume both the desired product and the reactant, the yield is a measure of the amount of ethylene oxide leaving the reactor relative to the amount of ethylene fed into the reactor and thus, a good measure for the economic performance of the reactor.

For practical reasons, one may want to optimize the yield while also ensuring that the time-averaged amount of ethylene that is fed to the reactor be fixed, i.e., determine the method to distribute a constant time-averaged amount of ethylene to the reactor that maximizes the yield. Limiting the time-averaged amount of ethylene that may be fed to the reactor is described by the following constraint:

$$\frac{1}{t_f}\int_0^{t_f} u_1(\tau)u_2(\tau)\,d\tau = \dot{M}_E \tag{1.15}$$

where $\dot{M}_E$ is a given time-averaged dimensionless molar flow rate of ethylene that may be fed to the reactor. If $u_{1,\min}u_{2,\min} < \dot{M}_E < u_{1,\max}u_{2,\max}$, the constraint of Eq. 1.15 prevents one from simply considering feeding in the minimum or maximum flow rate of ethylene to the reactor for all time. Within the context of EMPC, the constraint of Eq. 1.15 gives rise to a class of economically motivated constraints which take the form of integral or average constraints. In stark contrast to traditional or conventional control methodologies, e.g., proportional-integral-derivative control or tracking MPC, economically motivated constraints may be directly incorporated into EMPC. As previously mentioned this example has been previously studied within the context of periodic operation, e.g., [38], and closed-loop operation of this reactor under EMPC is considered in later chapters.

1.3.2 Continuously-Stirred Tank Reactor with Second-Order Reaction

A well-known chemical engineering example that demonstrates performance improvement through time-varying operation is a continuously stirred tank reactor (CSTR) where a second-order reaction occurs. Specifically, consider a non-isothermal CSTR where an elementary second-order reaction takes place that converts the reactant A to the desired product B like that depicted in Fig. 1.3. The reactant is fed to the reactor through a feedstock stream with concentration C_{A0}, volumetric flow rate F, and temperature T_0. The CSTR contents are assumed to be spatially uniform, i.e., the contents are well-mixed. Also, the CSTR is assumed to have a static liquid hold-up. A jacket provides/removes heat to/from the reactor at a

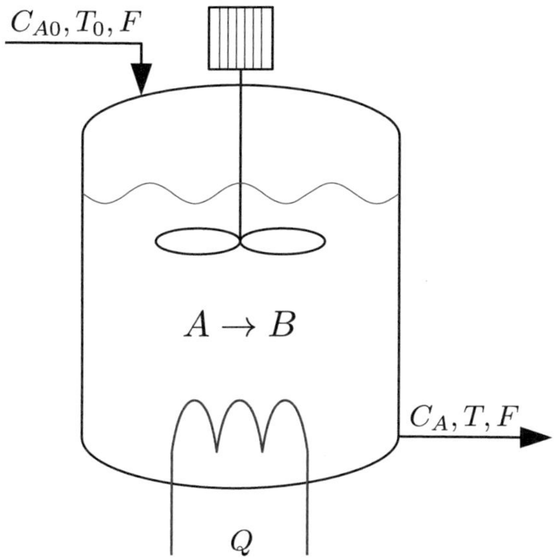

Fig. 1.3 Diagram of a CSTR where a second-order reaction occurs that produces a desired product B from a reactant A

heat rate Q. Applying first principles and standard modeling assumptions, e.g., constant fluid density and heat capacity, which are denoted by ρ_R and C_p, respectively and Arrhenius rate dependence of the reaction rate on temperature, the following system of ordinary differential equations (ODEs) may be obtained that describes the evolution of the CSTR reactant concentration and temperature:

$$\frac{dC_A}{d\bar{t}} = \frac{F}{V_R}(C_{A0} - C_A(\bar{t})) - k_0 e^{-E/RT(\bar{t})}(C_A(\bar{t}))^2 \tag{1.16a}$$

$$\frac{dT}{d\bar{t}} = \frac{F}{V_R}(T_0 - T(\bar{t})) - \frac{(-\Delta H)k_0}{\rho_R C_p} e^{-E/RT(\bar{t})}(C_A(\bar{t}))^2 + \frac{Q}{\rho_R C_p V_R} \tag{1.16b}$$

where $\bar{t}$ is the time, C_A denotes the concentration of A in the reactor, T denotes the temperature of the reactor contents, V_R is the volume of the liquid hold-up in the reactor, k_0 is the rate constant, E is the reaction activation energy, ΔH is the enthalpy of reaction, and R is gas constant.

The ODEs of Eq. 1.16 may be written in dimensionless form by defining $t = \bar{t}F/V_R$, $x_1 = C_A/C_{ref}$, $x_2 = RT/E$, $u = C_{A0}/C_{ref}$, $A_1 = C_{ref}V_R k_0/F$, $A_2 = \Delta H R C_{ref} A_1/(E C_p \rho)$, $x_{20} = RT_0/E + RQ/(C_p \rho E F)$ where C_{ref} is a reference concentration. The resulting dynamic equations in dimensionless form is:

Table 1.2 Process parameters of the CSTR

$k_0 C_{ref}^2$	8.46×10^6	A_1	1.69×10^6
x_{20}	0.050	A_2	1.41×10^4

$$\frac{dx_1}{dt} = -x_1 - A_1 e^{-1/x_2} x_1^2 + u \tag{1.17a}$$

$$\frac{dx_2}{dt} = -x_2 + A_2 e^{-1/x_2} x_1^2 + x_{20} \tag{1.17b}$$

where x_1 is the dimensionless reactant concentration, x_2 is the dimensionless temperature, and A_1, A_2 and x_{20} are constant parameters. The values of the parameters are given in Table 1.2. The input, which is the dimensionless reactant concentration in the reactor inlet, is bounded: $u \in [u_{\min}, u_{max}] = [0.5, 7.5]$. For this example, the production rate of the desired product is assumed to reflect the operating profit of the reactor, which is given by the following function:

$$l_e(x, u) = k_0 C_{ref}^2 e^{-1/x_2} x_1^2. \tag{1.18}$$

Determining the optimal input profile that maximizes the production rate subject to the constraint on the admissible input values is trivial: feed in the maximum amount of material for all time, i.e., set u to u_{max} for all time. A more interesting problem that may lead to a non-trivial solution is to determine the optimal input profile that maximizes the production rate subject to a constraint on the time-averaged amount of material that may fed to the reactor. In the latter problem, a more economical viewpoint is adopted and the problem seeks to determine the optimal method to distribute the material to the reactor. Therefore, the CSTR is assumed to be subject to an input average constraint (dynamic constraint) given by:

$$\frac{1}{t_f} \int_0^{t_f} u(t)\, dt = u_{avg} \tag{1.19}$$

where t_f is the length of operation. For this process (Eq. 1.17), performance metric (Eq. 1.18), and average constraint (Eq. 1.19), forced periodic operation induced by bang-bang type actuation has been shown to improve the average production owing to the second-order dependence of the reaction rate on reactant concentration, e.g., [29–32].

An analysis may be completed to rigorously show that the economic performance, i.e., the average production rate of the product, may be improved by using a time-varying operating strategy (in particular, periodic operation) compared to operating the reactor at steady-state. To show this rigorously, we require some more technical concepts, e.g., the Hamiltonian function, adjoint variables, Pontryagin's maximum principal [53]. Nevertheless, these concepts are not needed later in the book. An auxiliary state is defined for the average constraint:

$$x_3(t) := \frac{1}{t_f} \int_0^t \left(u(t) - u_{avg} \right) \, dt \tag{1.20}$$

which has dynamics:

$$\frac{dx_3(t)}{dt} = \frac{1}{t_f}(u(t) - u_{avg}). \tag{1.21}$$

The non-isothermal CSTR with the stage cost (Eq. 1.18) is a member of a special class of nonlinear systems:

$$\dot{x} = \bar{f}(x) + Bu \tag{1.22}$$

where $\dot{x}$ denotes the time derivative of x, $B \in \mathbb{R}^{n \times m}$ is a constant matrix and $\bar{f} : \mathbb{R}^n \to \mathbb{R}^n$ is a differentiable vector function. Additionally, the stage cost only depends on the states:

$$l_e(x, u) = \bar{l}_e(x) \tag{1.23}$$

where $\bar{l}_e : \mathbb{R}^n \to \mathbb{R}$ is a differentiable function. The Hamiltonian function of the system of Eq. 1.22 and cost of Eq. 1.23 is

$$H(x, u, \lambda) = \bar{l}_e(x) + \lambda^T \bar{f}(x) + \lambda^T Bu \tag{1.24}$$

where λ is the adjoint variable vector that satisfies

$$\dot{\lambda}(t) = -H_x(x(t), u(t), \lambda(t)) \tag{1.25}$$

where H_x denotes the partial derivative of H with respect to x. From Pontryagin's maximum principle [53], a necessary condition can be derived for the optimal control, i.e., the control that maximizes the Hamiltonian:

$$u_i^*(t) = \begin{cases} u_{i,\max}, & \text{if } b_i^T \lambda(t) > 0 \\ u_{i,\min}, & \text{if } b_i^T \lambda(t) < 0 \end{cases} \tag{1.26}$$

where b_i is the i-th column of B. For this class of systems and stage costs, if some time-varying operating policy is the optimal operating strategy, then the operating policy is a bang-bang input policy of Eq. 1.26.

Although the analysis above significantly reduces the space of possible optimal input trajectories, it still yields an infinite space of input trajectories. Thus, consider the following periodic bang-bang input trajectory over one period:

$$u(t) = \begin{cases} u_{\max} & \text{if } t < \tau/2 \\ u_{\min} & \text{else} \end{cases} \tag{1.27}$$

where τ is the period and $t \in [0, \tau)$. The input trajectory of Eq. 1.27 satisfies the average constraint of Eq. 1.19 over each period (in this regard, the length of operation,

t_f, is assumed to be a multiple of τ). For the system of Eq. 1.17 with the input trajectory of Eq. 1.27, there exists a periodic state trajectory for some $\tau > 0$, i.e., it has the property $x(t) = x(t + \tau)$ for all t.

In this example, u_{avg} is taken to be 4.0. The CSTR has an optimal steady-state $x_s^T = [1.182 \ \ 0.073]$ which corresponds to the steady-state input that satisfies the average input constraint ($u_s = u_{avg}$) with a production rate of 14.03. Indeed, the periodic solution of the system of Eq. 1.17 with the input of Eq. 1.27 achieves better economic performance compared to the economic performance at steady-state for some τ. Moreover, the economic performance depends on the period which is shown in Fig. 1.4. Over the range of periods considered (0.5 to 2.4), the period $\tau = 1.20$ yields the best performance (Fig. 1.4). The periodic solution with the input period of $\tau = 1.20$ has an average cost of $\bar{J}_e = 15.20$ which is 8.30 percent better than the performance at the optimal steady-state. Periods greater than 1.96 achieve worse performance compared to that at steady-state. The state, input, and $B^T\lambda = b_1^T\lambda = \lambda_1 + \lambda_3/\tau$ trajectories are given in Fig. 1.5 over one period. From Fig. 1.5, the input trajectory satisfies the necessary condition of Eq. 1.26. From these results, time-varying operation is better than steady-state operation from an economical point of view for this example. If the average constraint of Eq. 1.19 was not imposed, the optimal operating strategy would be steady-state operation at the steady-state corresponding to the input $u_{\max}$. The average constraint plays a crucial role for this particular example.

As pointed out, the above analysis only considers economic performance. If the periodic solution depicted in Fig. 1.5 is indeed optimal or some other bang-bang policy is the best operating strategy, feedback control is needed to force the system state from an initial state to the optimal time-varying solution. Moreover, the control problem becomes more complex when one considers disturbances, plant-

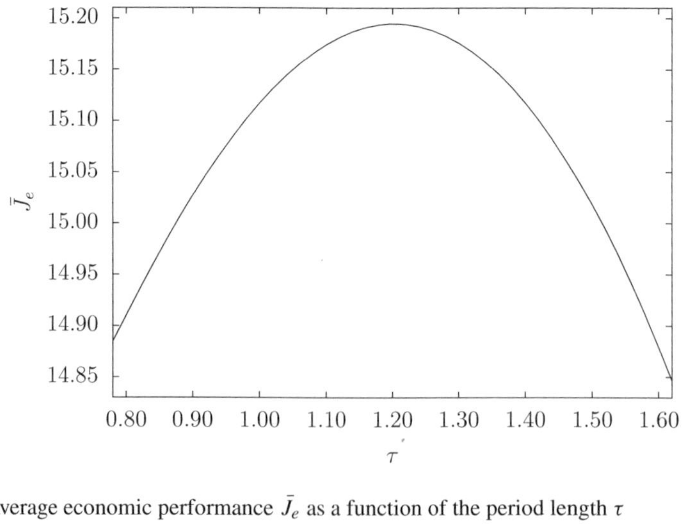

Fig. 1.4 Average economic performance $\bar{J}_e$ as a function of the period length τ

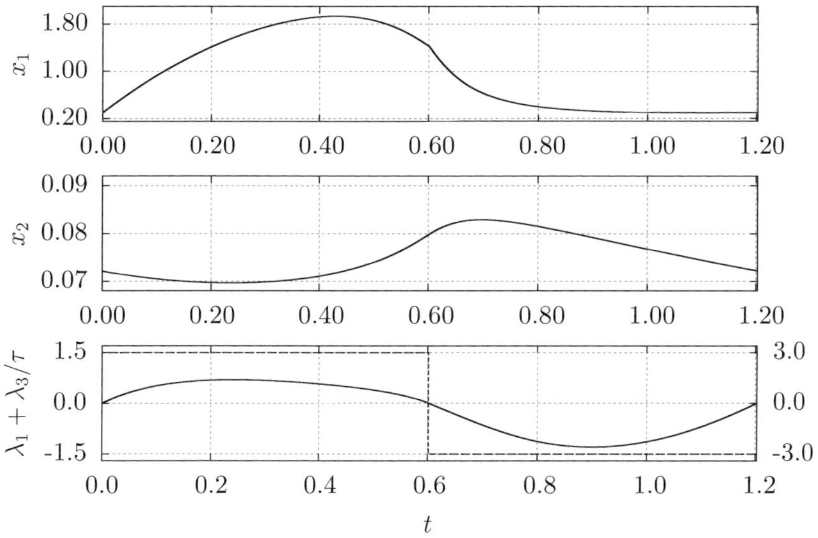

Fig. 1.5 State, input, and $\lambda_1 + \lambda_3/\tau$ trajectories of the CSTR under the bang-bang input policy with period $\tau = 1.20$

model mismatch and other forms of uncertainty, implementability of the computed input trajectory, i.e., bang-bang control may not be implementable in practice, and time-varying economic objectives and constraints. The example further motivates the inquiry and theoretical developments in the context of EMPC systems that dictate time-varying operating policies.

1.4 Objectives and Organization of the Book

This book considers theoretical analysis of closed-loop stability and performance under EMPC, issues related to computational efficiency of EMPC, and chemical process applications controlled by EMPC. Specifically, the objectives of this book are summarized as follows:

1. To develop economic model predictive control methods that address infinite-time and finite-time closed-loop economic performance and time-varying economic considerations.
2. To develop two-layer dynamic economic process optimization and feedback control frameworks that incorporate EMPC with other control strategies allowing for computational efficiency.
3. To develop rigorous output feedback-based EMPC schemes with guaranteed closed-loop stability properties.
4. To address real-time computation of EMPC.

The book is organized as follows. In Chap. 2, a formal definition of the notation is provided. Some definitions and preliminary results on stability and stabilization of nonlinear systems and on tracking MPC are given. The chapter closes with a brief review of nonlinear constrained optimization and solution strategies for dynamic optimization.

In Chap. 3, a brief overview of EMPC methods is provided. In particular, the role of constraints imposed in the optimization problem of EMPC for feasibility, closed-loop stability, and closed-loop performance is explained. Three main types of constraints are considered including terminal equality constraints, terminal region constraints, and constraints designed via Lyapunov-based techniques. EMPC is applied to a benchmark chemical process example to illustrate the effectiveness of time-varying operation to improve closed-loop economic performance compared to steady-state operation and to an open-loop periodic operating policy.

In Chap. 4, a complete discussion of Lyapunov-based EMPC (LEMPC), which was first presented in [27], is given that includes closed-loop stability and robustness properties. LEMPC designs that address closed-loop performance and time-varying economic stage cost function are also addressed in this chapter. The methods are applied to two chemical process examples.

In Chap. 5, output feedback-based EMPC schemes are presented. To provide EMPC with an estimate of the system state from a measurement of the output, a high-gain observer and moving horizon estimation (MHE) are both considered for state estimation. Conditions under which closed-loop stability under the two resulting state estimation-based EMPC schemes are derived. The state estimation-based EMPC schemes are applied to a chemical process example.

In Chap. 6, several two-layer approaches to dynamic economic optimization and control are developed and discussed. The upper layer, utilizing an EMPC, is used to compute economically optimal policies and potentially, also, control actions that are applied to the closed-loop system. The economically optimal policies are sent down to a lower layer MPC scheme which may be a tracking MPC or an EMPC. The lower layer MPC scheme forces the closed-loop state to closely follow the economically optimal policy computed in the upper layer EMPC. The methodologies are applied to several chemical process examples to demonstrate their effectiveness.

In Chap. 7, issues relating to computational efficiency and real-time implementation of EMPC are addressed. First, a composite control structure featuring EMPC is developed for nonlinear two-time-scale systems. The resulting control strategy addresses computational efficiency because it is a distributed control strategy and it has certain numerical advantages explained further in the chapter. Next, an alternative to solving for the control actions for all available inputs in a single optimization problem is discussed. Specifically, several (smaller) optimization problems may formulated and solved to compute the control actions that are applied to the closed-loop system. Owing to the fact that the optimization problems are solved amongst several distributed processors, the resulting strategy is a distributed EMPC (DEMPC) implementation. In this chapter, an application study of DEMPC is presented. Two DEMPC approaches are considered and evaluated with respect to a centralized EMPC implementation. Finally, to address guaranteed closed-loop stability in the presence of computational delay, an implementation strategy is developed which features a triggered evaluation of the LEMPC optimization problem to compute an input trajectory over a finite-time prediction horizon in advance. Closed-loop stability under the real-time LEMPC strategy is analyzed and specific stability conditions are derived.

References

1. Seborg DE, Edgar TF, Mellichamp DA, Doyle FJ (2010) Process Dynamics and Control, 3rd edn. Wiley, New York, NY
2. Marlin TE, Hrymak AN (1996) Real-time operations optimization of continuous processes. In: Proceedings of the fifth international conference on chemical process control, Tahoe City, CA, pp 156–164
3. Darby ML, Nikolaou M, Jones J, Nicholson D (2011) RTO: an overview and assessment of current practice. J Process Control 21:874–884
4. Mayne D, Michalska H (1990) Receding horizon control of nonlinear systems. IEEE Trans Autom Control 35:814–824
5. Henson MA (1998) Nonlinear model predictive control: current status and future directions. Comput Chem Eng 23:187–202
6. Morari M, Lee JH (1999) Model predictive control: past, present and future. Comput Chem Eng 23:667–682
7. Mayne DQ, Rawlings JB, Rao CV, Scokaert POM (2000) Constrained model predictive control: stability and optimality. Automatica 36:789–814
8. Rawlings JB (2000) Tutorial overview of model predictive control. IEEE Control Syst Mag 20:38–52
9. Qin SJ, Badgwell TA (2003) A survey of industrial model predictive control technology. Control Eng Pract 11:733–764
10. Allgöwer F, Zheng A (eds) (2000) Nonlinear model predictive control, Progress in systems and control theorey, vol. 26. Birkäuser Basel, Washington D.C
11. Camacho EF, Alba CB (2013) Model predictive control, 2nd edn. Springer
12. Magni L, Raimondo DM, Allgöwer F (2009) Nonlinear model prediction control: towards new challenging applications, vol 384., Lecture Notes in Control and Information Sciences. Springer, Berlin Heidelberg
13. Rawlings JB, Mayne DQ (2009) Model predictive control: theory and design. Nob Hill Publishing, Madison, WI
14. Christofides PD, Liu J, Muñoz de la Peña D (2011) Networked and distributed predictive control: methods and nonlinear process network applications. Advances in Industrial Control Series. Springer, London, England (2011)
15. Grüne L, Pannek J (2011) Nonlinear model predictive control: theory and algorithms. Communications and control engineering. Springer, London, England
16. Backx T, Bosgra O, Marquardt W (2000) Integration of model predictive control and optimization of processes: enabling technology for market driven process operation. In: Proceedings of the IFAC symposium on advanced control of chemical processes, Pisa, Italy, pp 249–260
17. Kadam JV, Marquardt W (2007) Integration of economical optimization and control for intentionally transient process operation. In: Findeisen R, Allgöwer F, Biegler LT (eds) Assessment and future directions of nonlinear model predictive control, vol 358., Lecture Notes in Control and Information SciencesSpringer, Berlin Heidelberg, pp 419–434
18. Siirola JJ, Edgar TF (2012) Process energy systems: control, economic, and sustainability objectives. Comput Chem Eng 47:134–144
19. Davis J, Edgar T, Porter J, Bernaden J, Sarli M (2012) Smart manufacturing, manufacturing intelligence and demand-dynamic performance. Comput Chem Eng 47:145–156
20. de Gouvêa MT, Odloak D (1998) One-layer real time optimization of LPG production in the FCC unit: procedure, advantages and disadvantages. Comput Chem Eng 22:S191–S198

21. Helbig A, Abel O, Marquardt W (2000) Structural concepts for optimization based control of transient processes. In: Allgöwer F, Zheng A (eds) Nonlinear model predictive control, Progress in Systems and Control Theory, vol. 26. Birkhäuser Basel, pp 295–311
22. Zanin AC, de Gouvêa MT, Odloak D (2002) Integrating real-time optimization into the model predictive controller of the FCC system. Control Eng Pract. 10:819–831
23. Engell S (2007) Feedback control for optimal process operation. J Process Control 17:203–219
24. Rawlings JB, Amrit R (2009) Optimizing process economic performance using model predictive control. In: Magni L, Raimondo DM, Allgöwer F (eds) Nonlinear model predictive control, vol 384., Lecture Notes in Control and Information Sciences. Springer, Berlin Heidelberg, pp 119–138
25. Huang R, Harinath E, Biegler LT (2011) Lyapunov stability of economically oriented NMPC for cyclic processes. J Process Control 21:501–509
26. Angeli D, Amrit R, Rawlings JB (2012) On average performance and stability of economic model predictive control. IEEE Trans Autom Control 57:1615–1626
27. Heidarinejad M, Liu J, Christofides PD (2012) Economic model predictive control of nonlinear process systems using Lyapunov techniques. AIChE J 58:855–870
28. Douglas JM (1967) Periodic reactor operation. Ind Eng Chem Process Design Dev 6:43–48
29. Lee CK, Bailey JE (1980) Modification of consecutive-competitive reaction selectivity by periodic operation. Ind Eng Chem Process Design Dev 19:160–166
30. Sinčić D, Bailey JE (1980) Analytical optimization and sensitivity analysis of forced periodic chemical processes. Chem Eng Sci 35:1153–1161
31. Watanabe N, Onogi K, Matsubara M (1981) Periodic control of continuous stirred tank reactors-I: the Pi criterion and its applications to isothermal cases. Chem Eng Sci 36:809–818
32. Watanabe N, Kurimoto H, Matsubara M, Onogi K (1982) Periodic control of continuous stirred tank reactors-II: cases of a nonisothermal single reactor. Chem Eng Sci 37:745–752
33. Barshad Y, Gulari E (1985) A dynamic study of CO oxidation on supported platinum. AIChE J 31:649–658
34. Silveston PL (1987) Periodic operation of chemical reactors—a review of the experimental literature. Sādhanā 10:217–246
35. Shu X, Rigopoulos K, Çinar A (1989) Vibrational control of an exothermic CSTR: productivity improvement by multiple input oscillations. IEEE Trans Autom Control 34:193–196
36. Sterman LE, Ydstie BE (1990) The steady-state process with periodic perturbations. Chem Eng Sci 45:721–736
37. Sterman LE, Ydstie BE (1991) Periodic forcing of the CSTR: an application of the generalized π-criterion. AIChE J 37:986–996
38. Özgülşen F, Adomaitis RA, Çinar A (1992) A numerical method for determining optimal parameter values in forced periodic operation. Chem Eng Sci 47:605–613
39. Özgülşen F, Kendra SJ, Çinar A (1993) Nonlinear predictive control of periodically forced chemical reactors. AIChE J 39:589–598
40. Özgülşen F, Çinar A (1994) Forced periodic operation of tubular reactors. Chem Eng Sci 49:3409–3419
41. Silveston PL, Hudgins RR, Renken A (1995) Periodic operation of catalytic reactors—introduction and overview. Catal Today 25:91–112
42. Budman H, Kzyonsek M, Silveston P (1996) Control of a nonadiabatic packed bed reactor under periodic flow reversal. Can J Chem Eng 74:751–759
43. Lee JH, Natarajan S, Lee KS (2001) A model-based predictive control approach to repetitive control of continuous processes with periodic operations. J Process Control 11:195–207
44. Natarajan S, Lee JH (2000) Repetitive model predictive control applied to a simulated moving bed chromatography system. Comput Chem Eng 24:1127–1133
45. Budman H, Silveston PL (2008) Control of periodically operated reactors. Chem Eng Sci 63:4942–4954
46. Mancusi E, Altimari P, Russo L, Crescitelli S (2011) Multiplicities of temperature wave trains in periodically forced networks of catalytic reactors for reversible exothermic reactions. Chem Eng J 171:655–668

47. Silveston PL, Hudgins RR (eds) (2013) Periodic operation of reactors. Elsevier, Oxford, England
48. Bailey JE, Horn FJM (1971) Comparison between two sufficient conditions for improvement of an optimal steady-state process by periodic operation. J Optim Theory Appl 7:378–384
49. Bittanti S, Fronza G, Guardabassi G (1973) Periodic control: a frequency domain approach. IEEE Trans Autom Control 18:33–38
50. Bailey JE (1973) Periodic operation of chemical reactors: a review. Chem Eng Commun 1:111–124
51. Guardabassi G, Locatelli A, Rinaldi S (1974) Status of periodic optimization of dynamical systems. J Optim Theory Appl 14:1–20
52. Alfani F, Carberry JJ (1970) An exploratory kinetic study of ethylene oxidation over an unmoderated supported silver catalyst. La Chimica e L'Industria 52:1192–1196
53. Pontryagin LS, Boltyanskii VG, Gamkrelidze RV, Mishchenko EF (1961) Mathematical theory of optimal processes. Fizmatgiz, Moscow

Chapter 2
Background on Nonlinear Systems, Control, and Optimization

This chapter provides a brief review of several concepts that are used throughout this book. The first section presents the notation. In the second section, stability of nonlinear systems is discussed followed by a brief overview of stabilization (control) of nonlinear systems. For a more detailed and complete overview of stability and control of nonlinear systems, the reader is referred to, for example, the classical textbooks [1, 2]. In the last section, a review of nonlinear and dynamic optimization concepts is presented.

2.1 Notation

The set of real numbers is denoted by $\mathbb{R}$, while the set of integers is denoted by $\mathbb{I}$. The symbol $\mathbb{R}_{\geq 0}$ ($\mathbb{I}_{\geq 0}$) is used to denote positive reals (integers), and $\mathbb{R}^n$ is an n-dimensional real (Euclidean) space. The variable t where $t \in \mathbb{R}$ will typically be reserved for time and thus, the notation $x(t) \in \mathbb{R}^n$ represents a time-dependent vector. The symbol $|\cdot|$ denotes the Euclidean norm of a vector, i.e., $|x| = \sqrt{x^T x}$ where $x \in \mathbb{R}^n$ and x^T denotes the transpose of x, and $|\cdot|_Q^2$ denotes the square of a weighted Euclidean norm of a vector, i.e., $|x|_Q^2 = x^T Q x$ where Q is a weighting positive definite matrix. A square diagonal matrix with diagonal elements equal to the elements of a vector v and off-diagonal elements equal to zero is written as $\text{diag}(v)$. An infinite sequence is denoted by $\{t_k\}_{k\geq 0}$, while a finite sequence is written as $\{t_i\}_{i=0}^N$ which describes the sequence: $t_0, t_1, \ldots, t_{N-1}, t_N$.

With regard to functions, a function, $V : \mathbb{R}^n \to \mathbb{R}_{\geq 0}$, is said to be positive definite with respect to $\bar{x} \in \mathbb{R}^n$ if $V(x) > 0$ for all $x \in \mathbb{R}^n$ except for $\bar{x}$ when $V(\bar{x}) = 0$. When a function is positive definite with respect to the origin ($\bar{x} = 0$), the function may be referred to as positive definite, and the distinction that it is positive definite with respect to the origin is omitted. A function, $V : \mathbb{R}^n \to \mathbb{R}_{\leq 0}$, is negative definite (with respect to the origin) if $-V$ is positive definite. A continuous function $\alpha : [0, a) \to \mathbb{R}_{\geq 0}$ is said to be of class $\mathcal{K}$ if it is strictly increasing and

M. Ellis et al., *Economic Model Predictive Control*,
Advances in Industrial Control, DOI 10.1007/978-3-319-41108-8_2

$\alpha(0) = 0$, and it is of class $\mathcal{K}_\infty$ if it is of class $\mathcal{K}$, $a = \infty$, and $\alpha(r) \to \infty$ as $r \to \infty$, i.e., it is radially unbounded. A function $\beta : [0, a) \times \mathbb{R}_{\geq 0} \to \mathbb{R}_{\geq 0}$ is said to be of class-$\mathcal{KL}$ if, for each fixed t, the mapping $\beta(s, t)$ is of class-$\mathcal{K}$ with respect to s and for each fixed s, the mapping $\beta(s, t)$ is non-increasing with respect to t and $\beta(s, t) \to 0$ as $t \to \infty$. The family of piecewise constant, right-continuous functions with period Δ is denoted as $S(\Delta)$. With a slight abuse of notation, we will say $u(\cdot) \in S(\Delta)$ (or simply, $u \in S(\Delta)$) when the vector-valued function $u : [0, N\Delta) \to \mathbb{R}^m$, $u : t \mapsto u(t)$, may be described by

$$u(t) = \bar{u}_i, \text{ for } t \in [i\Delta, (i+1)\Delta)$$

for $i = 0, 1, \ldots, N-1$ where $\Delta > 0$ is the period and $\bar{u}_i \in \mathbb{R}^m$; the appropriate domain of the function u will be implied by the context. The floor and ceiling functions, denoted as $\lfloor a \rfloor$ and $\lceil a \rceil$ for a scalar $a \in \mathbb{R}$, respectively, are the largest integer not greater than a and the smallest integer not less than a, respectively.

The set Ω_r is a level set, also referred to as a level surface or sub-level set in other contexts, of a scalar-valued positive definite function: $\Omega_r := \{x \in \mathbb{R}^n : V(x) \leq r\}$ where $r > 0$. A ball of radius $R > 0$ is given by $B_R := \{x \in \mathbb{R}^n : |x| \leq R\}$. The notation $B \setminus A$ denotes the relative complement of the set A in B, i.e., $B \setminus A = \{x \in B : x \notin A\}$. Finally, for algorithms, the notation $j \leftarrow j + 1$ is used to denote that at the next time step or at the next iteration, the index j is incremented by one.

2.2 Stability of Nonlinear Systems

First, unforced nonlinear systems are considered to present some definitions and stability properties. Specifically, consider the following class of time-invariant nonlinear systems, which is described by the following system of first-order nonlinear ordinary differential equations (ODEs):

$$\dot{x} = f(x) \tag{2.1}$$

where $x \in D \subset \mathbb{R}^n$, $f : D \to \mathbb{R}^n$ is a locally Lipschitz map from a domain $D \subset \mathbb{R}^n$ to $\mathbb{R}^n$. The vector x describes the current state of the system. Thus, x is referred to as the *state* vector, and the space $\mathbb{R}^n$ is referred to as *state-space*. The initial condition of system of Eq. 2.1 is given by $x_0 \in D$, i.e., $x(t_0) = x_0$ where $t_0 \in \mathbb{R}$ is the initial time.

The solution of Eq. 2.1 starting from x_0 at time t_0, is denoted as $x(t, x_0, t_0)$ for $t \in [t_0, t_1]$ with $x(t_0, x_0, t_0) = x_0$ and where $t_1 > t_0$ is the maximal time that the solution exists. The initial time may be taken to be zero with no loss of generality. The solution of Eq. 2.1 is also referred to as the state trajectory, and with slight abuse of notation, the notation of the solution of Eq. 2.1 at time $t \geq t_0$ may be abbreviated to $x(t)$. Two of most fundamental properties of the system of Eq. 2.1 are the existence and uniqueness of a solution to the system of Eq. 2.1 for a given initial condition. If it can be shown that every solution lies in some compact set $\mathbb{X} \subset D$ for all $t \geq t_0$, then a unique solution is guaranteed for all $t \geq t_0$, e.g., [2].

Owing to the fact that the vector field, f, of Eq. 2.1 is nonlinear, the system may possess multiple isolated equilibrium points. Without loss of generality, the origin $x = 0$ is taken to be an equilibrium point of the system of Eq. 2.1, i.e., $f(0) = 0$. If the origin is not the equilibrium point of interest, deviation variables may be introduced such that the origin of the shifted coordinate system is the equilibrium point. For example, consider the system $\dot{x} = f(x)$ with an equilibrium $x_s \neq 0$ ($f(x_s) = 0$). Defining a shifted state $z := x - x_s$, the system may be rewritten in the following coordinates:

$$\dot{z} = f(z + x_s) =: g(z) \tag{2.2}$$

where the equilibrium point of the shifted system is $z = 0$ and $g(0) = 0$.

Within the context of the system of Eq. 2.1, stability of solutions is considered. In particular, the stability of the solution $x \equiv 0$ is considered using Lyapunov stability concepts. The origin of Eq. 2.1 is

- *stable* if, for each $\varepsilon > 0$, there is $\delta(\varepsilon) > 0$ such that

$$|x(0)| < \delta \Rightarrow |x(t)| < \varepsilon, \forall\, t \geq 0 \tag{2.3}$$

- *unstable* if it is not stable
- *locally asymptotically stable* if it is stable and δ may be chosen such that

$$|x(0)| < \delta \Rightarrow \lim_{t \to \infty} |x(t)| = 0 \tag{2.4}$$

- *globally asymptotically stable* if it is stable and $|x(t)| \to 0$ as $t \to 0$ for all $x(0) \in \mathbb{R}^n$
- *locally exponentially stable* if there exist positive real constants δ, c, and λ such that all solutions of Eq. 2.1 with $|x(0)| \leq \delta$ satisfy the inequality:

$$|x(t)| \leq c|x(0)|e^{-\lambda t} \quad \forall\, t \geq 0 \tag{2.5}$$

- *globally exponentially stable* if there exist positive real constants c, and λ such that all solutions of Eq. 2.1 satisfy the inequality:

$$|x(t)| \leq c|x(0)|e^{-\lambda t} \quad \forall\, t \geq 0 \tag{2.6}$$

  for all $x(0) \in \mathbb{R}^n$.

Since the system of Eq. 2.1 is time-invariant, the stability properties above are *uniform*; that is, they do not depend on the initial time. The stability definitions may be written in equivalent forms using so-called comparison functions.

The stability definitions are restated using comparison functions.

Lemma 2.1 ([2, Lemma 4.5]) *The equilibrium point $x = 0$ of Eq. 2.1 is*

- *stable if and only if there exist $\alpha \in \mathcal{K}$ and a positive constant c, such that*

$$|x(t)| \leq \alpha(|x(0)|) \tag{2.7}$$

  for all $t \geq 0$ and $|x(0)| < c$.
- *locally asymptotically stable if and only if there exist $\beta \in \mathcal{KL}$ and a positive constant c such that*

$$|x(t)| \leq \beta(|x(0)|, t) \tag{2.8}$$

  for all $t \geq 0$ and $|x(0)| < c$.
- *globally asymptotically stable if and only if there exist $\beta \in \mathcal{KL}$ such that*

$$|x(t)| \leq \beta(|x(0)|, t) \tag{2.9}$$

  for all $t \geq 0$ and $x(0) \in \mathbb{R}^n$.

When the origin is asymptotically stable, the state-space set of initial conditions where the solution to Eq. 2.1 will asymptotically converge to the origin is of interest. This gives rise to the notion of the *domain of attraction*, which is the set $D = \{x_0 \in \mathbb{R}^n : \lim_{t\to\infty} x(t, t_0, x_0) = 0\}$.

A weaker notion of stability than asymptotic and exponential stability of the origin is boundedness of the solution. Specifically, the solutions of Eq. 2.1 are

- *bounded* if there exists a positive constant c and for every $a \in (0, c)$, there is $\beta(a) > 0$ such that

$$|x(0)| \leq a \Rightarrow |x(t)| \leq \beta, \forall\, t \geq 0 \tag{2.10}$$

- *ultimately bounded* with ultimate bound b if there exist positive constants b and c and for every $a \in (0, c)$ there is $T(a, b) \geq 0$ such that

$$|x(0)| \leq a \Rightarrow |x(t)| \leq b,\ \forall\, t \geq t_0 + T \tag{2.11}$$

For practical systems, global stability properties are often not relevant owing to system constraints. Therefore, we extend the stability concepts to the case where the state of Eq. 2.1 is constrained to be in the set $\tilde{\mathbb{X}} \subset \mathbb{R}^n$. We need the following definition to state the stability properties of the constrained system of Eq. 2.1.

Definition 2.1 A set M is said to be *positively invariant set* with respect to the system of Eq. 2.1 if

$$x(0) \in M \Rightarrow x(t) \in M, \quad \forall\, t \geq 0$$

We will also use the term *forward invariant set* to refer to a positively invariant set. Consider a set $\mathbb{X} \subseteq \tilde{\mathbb{X}}$ to be an positively invariant set for the system of Eq. 2.1 that contains the origin in its interior. Then, the origin is, e.g., [3]:

- *stable* in $\mathbb{X}$ if, for each $\varepsilon > 0$, there is $\delta(\varepsilon) > 0$ such that $B_\delta \subseteq \mathbb{X}$ and

$$|x(0)| < \delta \Rightarrow |x(t)| < \varepsilon, \forall\, t \geq 0 \tag{2.12}$$

- *locally attractive* in $\mathbb{X}$ if there exists a $\eta > 0$ such that $x \in B_\eta \subseteq \mathbb{X}$ implies $|x(t)| \to 0$ as $t \to \infty$
- *attractive* in $\mathbb{X}$ if $|x(t)| \to 0$ as $t \to \infty$ for all $x(0) \in \mathbb{X}$
- *locally asymptotically stable* in $\mathbb{X}$ if it is stable and locally attractive
- *asymptotically stable* in $\mathbb{X}$ if it is stable and attractive
- *locally exponentially stable* in $\mathbb{X}$ if there exist $\eta > 0$, $c > 0$, and $\gamma > 0$ such that

$$|x(t)| \leq c|x(0)|e^{-\lambda t} \quad \forall\, t \geq 0 \tag{2.13}$$

  for all $x(0) \in B_\eta \subseteq \mathbb{X}$
- *exponentially stable* with a region of attraction $\mathbb{X}$ if there exist $c > 0$, and $\gamma > 0$ such that

$$|x(t)| \leq c|x(0)|e^{-\lambda t} \quad \forall\, t \geq 0 \tag{2.14}$$

  for all $x(0) \in \mathbb{X}$.

2.2.1 Lyapunov's Direct Method

For nonlinear systems, stability of the equilibrium points may be characterized in the sense of Lyapunov's direct method. Lyapunov's direct second method uses a scalar-valued positive definite function whose time-derivative is negative (semi-)definite along the state trajectory.

Theorem 2.1 (Lyapunov Stability Theorem, c.f. [2, Theorem 4.1]) *Let $x = 0$ be an equilibrium point for Eq. 2.1 and $D \subset \mathbb{R}^n$ be a domain containing the origin ($x = 0$). Let $V : D \to \mathbb{R}$ be a continuously differentiable positive definite function such that*

$$\dot{V}(x) \leq 0 \tag{2.15}$$

for all $x \in D$. Then, $x = 0$ is stable. If

$$\dot{V}(x) < 0 \tag{2.16}$$

for all $x \in D \setminus \{0\}$, then $x = 0$ is asymptotically stable.

A continuously differentiable positive definition function V as in Theorem 2.1 is called a *Lyapunov function*. The time-derivative of V along the state trajectory x is given by:

$$\dot{V}(x) = \frac{\partial V(x)}{\partial x}\dot{x} = \frac{\partial V(x)}{\partial x} f(x). \tag{2.17}$$

Theorem 2.1 is a sufficient condition for stability and asymptotic stability of the origin. Various converse Lyapunov theorems show that the conditions of Theorem 2.1 are also necessary (under a few additional mild conditions), see, for example, [2, 4–8].

Lyapunov's direct method has an intuitive interpretation by regarding the Lyapunov function as an abstract notion of the total energy of a given system. Specifically, consider any x on the level or Lyapunov surface $V(x) = c$, which is the boundary of the set $\Omega_c = \{x \in \mathbb{R}^n : V(x) \leq c\} \subset D$. When $\dot{V}(x) < 0$ for all $x \in D$, the state trajectory evolves from the boundary of Ω_c to the interior of Ω_c. Over time, the level surface that state trajectory evolves along shrinks to the origin owing to the fact that $\dot{V}(x) < 0$ for all $x \in D$. In other words, the energy of the system decays with time when $\dot{V} < 0$. If, instead, $\dot{V}(x) \leq 0$ for all $x \in D$, this implies that the state trajectory evolves inside the set $\Omega_c \subset D$ without coming out, and the energy over time may only stay the same or decrease (it cannot increase). This in turn means that a trajectory starting from the boundary of Ω_c will stay in the set Ω_c for all time without coming out. In this case, the conclusion that may be made is the origin is stable since the trajectory is contained inside any ball, B_ε, by requiring that the initial state x_0 to lie inside a Lyapunov surface contained in that ball.

2.2.2 LaSalle's Invariance Principle

LaSalle's invariance principle allows for making stronger conclusions about the behavior of solution of Eq. 2.1 when $\dot{V}(x) \leq 0$ for all $x \in D$.

LaSalle's invariance principle states that any state starting in any compact forward invariant subset of D will converge to the largest invariant set where $\dot{V}(x) = 0$.

Theorem 2.2 (LaSalle, c.f. [2, Theorem 4.4]) *Let $\Omega \subset D$ be a compact set that is positively invariant with respect to Eq. 2.1. Let $V : D \to \mathbb{R}$ be a continuously differentiable function such that $\dot{V}(x) \leq 0$ in Ω. Let $E := \{x \in \Omega : \dot{V}(x) = 0\}$ and M be the largest invariant set in E. Then every solution in Ω approaches M as $t \to \infty$.*

A consequence of LaSalle's invariance principle, one may show asymptotic stability of the origin when $M = \{0\}$, i.e., when M is the set containing the point $x = 0$. This result is stated in the following corollary.

Corollary 2.1 (c.f. [2, Corollary 4.1]) *Let $V : D \to \mathbb{R}$ be a continuously differentiable positive definite function on a domain D containing the origin $x = 0$, which is an equilibrium point of Eq. 2.1, such that $\dot{V}(x) \leq 0$ for all $x \in D$. Let $S = \{x \in D : \dot{V}(x) = 0\}$ and suppose that no solution can stay identically in S, other than the trivial solution $x \equiv 0$. Then, the origin is asymptotically stable.*

2.3 Stabilization of Nonlinear Systems

Consider, now, the class of forced nonlinear systems described by the following system of nonlinear ordinary differential equations:

$$\dot{x} = f(x, u, w) \tag{2.18}$$

where $x \in D \subseteq \mathbb{R}^n$ is the state, $u \in \mathbb{U} \subset \mathbb{R}^m$ is the manipulated (control) input, and $w \in \mathbb{R}^l$ is a disturbance. The set of admissible input values $\mathbb{U}$ is assumed to be compact, and the disturbance vector is bounded in the set $\mathbb{W} := \{w \in \mathbb{R}^l : |w| \leq \theta\}$ where $\theta > 0$ bounds the norm of the disturbance vector. Throughout this book, the disturbances vector as in Eq. 2.18 taken to be unknown and un-modeled forcing of the system. Disturbance models, e.g., integrating disturbance models, may readily be incorporated into the model of Eq. 2.18 through augmenting the state vector. However, a complete and thorough discussion of disturbance modeling is beyond the scope of this book and is not considered further. The vector function f is assumed to be locally Lipschitz on $D \times \mathbb{U} \times \mathbb{W}$. Without loss of generality, the origin of the unforced nominal system is assumed to be the equilibrium point of the system of Eq. 2.18, i.e., $f(0, 0, 0) = 0$.

Regarding existence and uniqueness of solutions of the system of Eq. 2.18, first it is important to point out that the input and disturbance trajectories are often not continuous functions of time. In the deterministic framework that we consider, the input and disturbance trajectories require a degree of continuity, and the disturbance may not rigorously be treated as noise. A standing assumption throughout the book is that the disturbance trajectory poses enough continuity to ensure existence of the solution of Eq. 2.18 almost everywhere. In practice such assumption poses little restrictions. For a more complete discussion of conditions that guarantee existence and uniqueness of a solution the interested reader is referred to [9].

2.3.1 *Control Lyapunov Functions*

The concept of control Lyapunov functions is described, which is utilized in many Lyapunov-based control design techniques. For simplicity of presentation, the case of a system with a single input is presented. Nonetheless, this concept extends to systems with multiple inputs. Thus, consider the following single-input system of the form:

$$\dot{x} = f(x, u) \tag{2.19}$$

where $x \in \mathbb{R}^n$, $u \in \mathbb{R}$, and $f(0, 0) = 0$. The control objective considered is to design a feedback control law $h : D \to \mathbb{U}$ that renders the origin of the closed-loop systems given by:

$$\dot{x} = f(x, h(x)) \tag{2.20}$$

globally asymptotically stable.

One potential approach may be to pick a function $V : \mathbb{R}^n \to \mathbb{R}_{\geq 0}$ as a Lyapunov function candidate, and find a control law that guarantees that the time-derivative of the Lyapunov function candidate along the solutions of the closed-loop system of Eq. 2.20 satisfy:

$$\frac{\partial V(x)}{\partial x} f(x, h(x)) \leq -W(x) \tag{2.21}$$

for all $x \in \mathbb{R}^n$ where $W : D \to \mathbb{R}$ is a positive definite function. It may be possible to find a stabilizing control law but Eq. 2.21 may fail to be satisfied for all $x \in \mathbb{R}^n$ because of a poor choice of functions V and W. Therefore, picking a control law that satisfies Eq. 2.21 is a difficult task in general. A system for which a good choice of the functions V and W exist is said to possess a control Lyapunov function.

Definition 2.2 A control Lyapunov function (CLF) for the system of Eq. 2.19 is a smooth positive definite radially unbounded function $V : \mathbb{R}^n \to \mathbb{R}_{\geq 0}$ that satisfies:

$$\inf_{u \in \mathbb{R}} \left\{ \frac{\partial V(x)}{\partial x} f(x, u) \right\} < 0, \ \forall \, x \neq 0. \tag{2.22}$$

Equation 2.22 is necessary and sufficient for the existence of a control law satisfying Eq. 2.21 [10]. Also, it may be shown that the existence of a CLF is equivalent to global asymptotic stabilizability.

For control-affine systems of the form:

$$\dot{x} = f(x) + g(x)u, \tag{2.23}$$

where $f : \mathbb{R}^n \to \mathbb{R}^n$, $g : \mathbb{R}^n \to \mathbb{R}^n$, and $f(0) = 0$. Using the Lie derivative notation:

$$L_f V(x) := \frac{\partial V(x)}{\partial x} f(x),$$
$$L_g V(x) := \frac{\partial V(x)}{\partial x} g(x),$$

the CLF condition of Eq. 2.21 is given by:

$$L_f V(x) + L_g V(x) u \leq -W(x) \tag{2.24}$$

for all $x \in \mathbb{R}^n$. Note that Eq. 2.24 may be satisfied only if:

$$L_g V(x) = 0 \Rightarrow L_f V(x) < 0, \ \forall \, x \neq 0 \tag{2.25}$$

If V is a CLF for the system of Eq. 2.23, then one choice of stabilizing control law is given by Sontag's formula [11]:

$$h(x) = \begin{cases} -\dfrac{L_f V(x) + \sqrt{(L_f V(x))^2 + (L_g V(x))^4}}{(L_g V(x))^2} L_g V(x), & L_g V(x) \neq 0 \\ 0, & L_g V(x) = 0 \end{cases} \tag{2.26}$$

In this case, the positive definite function, W is given by:

$$W(x) = \sqrt{(L_f V(x))^2 + (L_g V(x))^4} > 0,\ x \neq 0 \tag{2.27}$$

While the construction of CLFs is difficult for the general class of nonlinear systems of Eq. 2.19, systematic methods exist for several important classes of nonlinear systems that allow for the construction of CLFs.

2.3.2 Stabilization of Nonlinear Sampled-Data Systems

In the subsequent chapters, EMPC methods are considered. The Lyapunov-based EMPC methods that are presented take advantage of an explicit stabilizing feedback controller. The explicit controller satisfies the following assumption.

Assumption 2.1 There exists a feedback controller $h(x) \in \mathbb{U}$ with $h(0) = 0$ that renders the origin of the closed-loop system of Eq. 2.18 with $u = h(x)$ and $w \equiv 0$ asymptotically stable for all $x \in D_0$ where D_0 is an open neighborhood of the origin.

There are several methods to design an explicit feedback control law, $h : D \to \mathbb{U}$, that renders the origin of Eq. 2.18 asymptotically stable. Specifically, methodologies for (explicit) feedback control design for nonlinear systems include employing linear feedback control techniques, Lyapunov-based control techniques, and geometric control methods, e.g., [12–16].

Applying converse theorems [2, 4], Assumption 2.1 implies that there exists a continuously differentiable Lyapunov function, $V : D \to \mathbb{R}^n$, for the closed-loop system of Eq. 2.18 with $u = h(x) \in \mathbb{U}$ and $w \equiv 0$ such that the following inequalities hold:

$$\alpha_1(|x|) \leq V(x) \leq \alpha_2(|x|), \tag{2.28a}$$

$$\frac{\partial V(x)}{\partial x} f(x, h(x), 0) \leq -\alpha_3(|x|), \tag{2.28b}$$

$$\left|\frac{\partial V(x)}{\partial x}\right| \leq \alpha_4(|x|) \tag{2.28c}$$

for all $x \in D$ where D is an open neighborhood of the origin and α_i, $i = 1, 2, 3, 4$ are functions of class $\mathcal{K}$. A level set of the Lyapunov function Ω_ρ, which defines a subset of D (ideally the largest subset contained in D), is taken to be the stability region of the closed-loop system under the controller $h(x)$. Standard techniques exist for designing a stabilizing control law for various classes of continuous-time nonlinear systems (see, for instance, [1, 2, 13, 15–17] as well as the references contained therein).

While there are no general methods for constructing Lyapunov functions for broad classes of nonlinear systems with constraints, there exists some general methods for constructing Lyapunov functions for certain classes of systems, e.g., Zubov's method [18] and the sum of squares decomposition [19]. Within the context of chemical process control, quadratic Lyapunov functions have been widely used and have been demonstrated to be effective for estimating the region of attraction of a given equilibrium point of a system (see, for example, the numerous examples in [16] as well as the examples of the subsequent chapters of this book).

The explicit controller poses a degree of robustness to disturbances/uncertainty in the sense that when $w \not\equiv 0$, the controller will force the closed-loop state to a small neighborhood of the origin if the bound on the disturbance, θ, is sufficiently small. Moreover, owing to the fact that digital computers are often used in the implementation of controllers, we must also consider the closed-loop stability properties of the controller $h(x)$ applied in a sample-and-hold fashion. When the feedback controller $h(x)$ is applied in a sample-and-hold fashion, the resulting closed-loop system is a nonlinear sampled-data system given by:

$$\dot{x}(t) = f(x(t), h(x(t_k)), w(t)) \tag{2.29}$$

for $t \in [t_k, t_{k+1})$, $t_k = k\Delta$, $k = 0, 1, \ldots$, and $\Delta > 0$ is the sampling period. Regarding the disturbance in Eq. 2.29, in many applications, it is sufficient to take w to be constant over the sampling periods. This is essentially what is done when considering a discrete-time model for a sampled-data system.

Applying standard results on sampled-data systems, e.g., [20–24], it can be shown that when the bound on the disturbances and the sampling period are both sufficiently small the origin is practically stable for all initial conditions in Ω_ρ. More specifically, the state trajectory of Eq. 2.29 starting in Ω_ρ will remain bounded in Ω_ρ and converge to a small compact set containing the origin where it will be maintained thereafter when the bound on the disturbance and the sampling period are sufficiently small. It is important to emphasize that asymptotic stability of the origin of Eq. 2.29 is typically not achieved unless additional conditions hold.

To achieve asymptotic stability of the origin of sampled-data system of Eq. 2.29, a stronger assumption is required. The following assumption and result are stated generally in the sense that no restrictions are placed on the state and input.

Assumption 2.2 There exists a locally Lipschitz feedback controller $u = h(x)$ with $h(0) = 0$ such that the vector field of the closed-loop system $f(x, h(x), 0)$ is continuously differentiable on $\mathbb{R}^n$. Furthermore, the origin of the nominal closed-loop system of Eq. 2.18 ($w \equiv 0$) under the controller $h(x)$ implemented continuously is locally exponentially stable and globally asymptotically stable.

The following theorem characterizes the type of stability achieved when the controller $h(x)$ is applied in a sample-and-hold fashion with a sufficiently small hold period. The result below extends to a more general setting where asynchronous sampling is considered; see, [25] for this more general version of the following result.

Theorem 2.3 *If Assumption 2.2 holds, then given $R > 0$, there exist $\Delta^* > 0$ and $M, \sigma > 0$ such that for $\Delta \in (0, \Delta^*)$ the nominal closed-loop sampled-data system of Eq. 2.29 with arbitrary initial condition $x(0) = x_0 \in B_R$ satisfies the estimate:*

$$|x(t)| \leq M \exp(-\sigma t)|x_0| \tag{2.30}$$

for all $t \geq 0$.

Proof By virtue of Proposition 4.4 of [26], there exists a C^1 positive definite and radially unbounded function $V : \mathbb{R}^n \to \mathbb{R}_{\geq 0}$, constants $\mu,\ \varepsilon > 0$ and a symmetric, positive definite matrix $P \in \mathbb{R}^{n\times n}$ for the nominal closed-loop system of Eq. 2.18 under the controller $h(x)$ implemented continuously such that

$$\frac{\partial V(x)}{\partial x} f(x, h(x), 0) \leq -\mu|x|^2, \quad \text{for all } x \in \mathbb{R}^n, \tag{2.31}$$

$$V(x) = x^T Px, \quad \text{for all } x \in \mathbb{R}^n \text{ with } |x| \leq \varepsilon. \tag{2.32}$$

Let $R > 0$ and define $\hat{\rho} := \max\{V(x) \ :\ x \in B_R\}$. By virtue of Eq. 2.32 and the compactness of $\Omega_{\hat{\rho}}$, there exist constants $c_1,\ c_2 > 0$ and $c_4 > 0$ such that:

$$c_1|x|^2 \leq V(x) \leq c_2|x|^2, \tag{2.33}$$

$$\left|\frac{\partial V(x)}{\partial x}\right| \leq c_4|x| \tag{2.34}$$

for all $x \in \Omega_{\hat{\rho}}$. Since f and h are locally Lipschitz mappings with $f(0, 0, 0) = 0$ and $h(0) = 0$, there exist constants $L,\ M > 0$ such that:

$$|f(x, h(z), 0) - f(x, h(x), 0)| \leq L|x - z|, \tag{2.35}$$

$$|f(x, h(z), 0)| \leq M|x| + M|z| \tag{2.36}$$

for all $x,\ z \in \Omega_{\hat{\rho}}$. Let $\Delta^* > 0$ be sufficiently small so that the following inequality holds:

$$c_4 L \frac{2M\Delta^* \exp(M\Delta^*)}{1 - 2M\Delta^* \exp(M\Delta^*)} < \mu \tag{2.37}$$

In order to prove of estimate of Eq. 2.30, it suffices to show that for every initial condition $x(0) \in \Omega_{\hat{\rho}}$ and for every integer $k \geq 0$ it holds that:

$$\frac{\partial V(x(t))}{\partial x} f(x(t), h(x(t_k)), 0) \leq -\frac{q}{2}|x(t)|^2, \tag{2.38}$$

for all $t \in [t_k, t_{k+1})$ where

$$q := \mu - c_4 L \frac{2M\Delta^* \exp(M\Delta^*)}{1 - 2M\Delta^* \exp(M\Delta^*)} > 0. \tag{2.39}$$

Using Eqs. 2.33 and 2.38, local exponential stability can be established. The proof of Eq. 2.38 is given below for $k = 0$ and $t \in [0, t_1)$. For every other interval, the proof is similar.

If $x(0) = 0$, then Eq. 2.38 trivially holds (since $x(t) = 0$ for $t \in [0, t_1)$). Therefore, consider the case when $x(0) \neq 0$. The proof is made by contradiction. Suppose that there exists $t \in [0, t_1)$ with

$$\frac{\partial V(x(t))}{\partial x} f(x(t), h(x(0)), 0) > -\frac{q}{2}|x(t)|^2.$$

The case that $x(t)$ is not defined for some $t \in [0, t_1)$ is also covered by this assumption. Define

$$a := \inf \left\{ t \in [0, t_1) : \frac{\partial V(x(t))}{\partial x} f(x(t), h(x(0)), 0) > -\frac{q}{2}|x(t)|^2 \right\}.$$

A standard continuity argument in conjunction with the fact that

$$\frac{\partial V(x(0))}{\partial x} f(x(0), h(x(0)), 0) \leq -\mu|x(0)|^2 < -\frac{q}{2}|x(0)|^2$$

shows that $a \in (0, t_1)$ and that

$$\frac{\partial V(x(t))}{\partial x} f(x(t), h(x(0)), 0) \leq -\frac{q}{2}|x(t)|^2$$

for all $t \in [0, a]$ with $(\partial V(x(a))/\partial x) f(x(a), h(x(0)), 0) = -\frac{q}{2}|x(a)|^2$. Moreover, for all $t \in [0, a]$ the inequality of Eq. 2.38 implies that $V(x(t)) \leq V(x(0)) \leq \hat{\rho}$. Therefore, $x(t) \in \Omega_{\hat{\rho}}$ for all $t \in [0, a]$. Using inequalities Eqs. 2.31, 2.34, 2.35, we obtain:

$$\frac{\partial V(x(t))}{\partial x} f(x(t), h(x(0)), 0) \leq -\mu|x(t)|^2 + c_4 L|x(t)||x(t) - x(0)| \tag{2.40}$$

for all $t \in [0, a]$. Using Eq. 2.36 and since $a \leq t_1 \leq \Delta^*$, a bound on the difference between $x(t)$ and $x(0)$ is obtained:

$$|x(t) - x(0)| \leq 2M\Delta^*|x(0)| + M \int_0^t |x(\tau) - x(0)|\, d\tau \tag{2.41}$$

for all $t \in [0, a]$. Applying the Gronwall-Bellman lemma to Eq. 2.41, we obtain:

$$|x(t) - x(0)| \leq 2M\Delta^* \exp(M\Delta^*)|x(0)| \tag{2.42}$$

for all $t \in [0, a]$. Using Eq. 2.42, the triangle inequality and the fact that

$$2M\Delta^* \exp(M\Delta^*) < 1$$

which is implied by Eq. 2.37, we get for all $t \in [0, a]$:

$$|x(t) - x(0)| \leq \frac{2M\Delta^* \exp(M\Delta^*)}{1 - 2M\Delta^* \exp(M\Delta^*)}|x(t)|. \tag{2.43}$$

Thus, using Eqs. 2.40, 2.43 and the fact that

$$q := \mu - c_4 L \frac{2M\Delta^* \exp(M\Delta^*)}{1 - 2M\Delta^* \exp(M\Delta^*)} > 0$$

we get for all $t \in [0, a]$:

$$\frac{\partial V(x(t))}{\partial x} f(x(t), h(x(0)), 0) \leq -q|x(t)|^2. \tag{2.44}$$

Consequently, we must have:

$$\frac{\partial V(x(a))}{\partial x} f(x(a), h(x(0)), 0) \leq -q|x(a)|^2 \leq -\frac{q}{2}|x(a)|^2. \tag{2.45}$$

Since $(\partial V(x(a))/\partial x) f(x(a), h(x(0)), 0) = -\frac{q}{2}|x(a)|^2$, we get $x(a) = 0$. However, this contradicts Eq. 2.42 (since Eq. 2.42 in conjunction with the fact that

$$2M\Delta^* \exp(M\Delta^*) < 1$$

implies that $|x(a) - x(0)| < |x(0)|$), which completes the proof.

Explicit feedback controllers that may be designed to satisfy Assumption 2.2 include, for example, feedback linearizing controller and some Lyapunov-based controllers [2, 15]. Owing to the input constraints, it may not be possible to design a controller $h(x)$ that achieves global asymptotic stability of the origin. In this case, we must modify the assumption which is considered in the following corollary.

Corollary 2.2 *Suppose there exists a locally Lipschitz feedback controller* $u = h(x)$ *with* $h(0) = 0$ *for the system of Eq. 2.18 that renders the origin of the nominal closed-loop system under continuous implementation of the controller* $h(x)$ *locally exponentially stable. More specifically, there exist constants* $\rho > 0$, $c_i > 0$, $i = 1, 2, 3, 4$ *and a continuously differentiable Lyapunov function* $V : \mathbb{R}^n \to \mathbb{R}_+$ *such that the following inequalities hold:*

$$c_1 |x|^2 \leq V(x) \leq c_2 |x|^2, \tag{2.46a}$$

$$\frac{\partial V(x)}{\partial x} f(x, h(x), 0) \leq -c_3 |x|^2, \tag{2.46b}$$

$$\left| \frac{\partial V(x)}{\partial x} \right| \leq c_4 |x|, \tag{2.46c}$$

for all $x \in \Omega_\rho$. *There exists* $\Delta^* > 0$ *and* $M, \sigma > 0$ *such that for all* $\Delta \in (0, \Delta^*)$ *the estimate of Eq. 2.30 holds for the nominal closed-loop sampled-data system of Eq. 2.29 with arbitrary initial condition* $x(0) \in \Omega_\rho$.

Proof The proof follows along the same lines of Theorem 2.3 and shows that V is a Lyapunov function for the closed-loop sampled-data system and takes advantage of the compactness of the set Ω_ρ to establish an exponentially decaying estimate for the state trajectory of the closed-loop sample-data system for any initial condition $x(0) \in \Omega_\rho$.

Remark 2.1 Sufficient conditions such that there exists a function V satisfying the inequalities of Eq. 2.46 are when $x = 0$ is a locally exponentially stable (LES) equilibrium point for the closed-loop system $\dot{x} = f(x, h(x), 0)$ and the mapping $f(x, h(x), 0)$ is continuously differentiable on $\mathbb{R}^n$. Indeed, by Lemma 8.1 in [2] the region of attraction A of $x = 0$ is an open, connected, invariant set. Let $r > 0$ be such that the set $S = \{x \in \mathbb{R}^n : |x| \leq r\}$ is contained in the region of attraction A. Then LES and compactness of S imply that an exponential bound holds for the solutions of the closed-loop system $\dot{x} = f(x, h(x), 0)$ with initial conditions $x(0) \in S$. It follows from Theorem 4.14 in [2] that there exists a Lyapunov function V for the closed-loop system $\dot{x} = f(x, h(x), 0)$ that satisfies inequalities of Eq. 2.46 for certain constants $c_1, c_2, c_3, c_4 > 0$ and for all $x \in \text{int}(S)$ ($\text{int}(S)$ denotes the interior of S). Let $R < r$ be an arbitrary positive number and define $V(x) = V(\text{Proj}(x))$ for all $x \in \mathbb{R}^n$, where $\text{Proj}(x)$ denotes the projection on the closed ball of radius R centered at $x = 0$. Then all inequalities of Eq. 2.46 hold with arbitrary $\rho < c_1 R^2$.

2.3.3 *Tracking Model Predictive Control*

Designing an explicit feedback control such as one that satisfies Assumption 2.1 to stabilize the origin of the system of Eq. 2.18 has many advantages such as it may be shown to possess robustness to disturbances and sample-and-hold or discrete-time implementation. However, the most significant drawback of such an approach to controller design is that performance considerations and system constraints are not explicitly handled in a general framework. For example, consider that the system of Eq. 2.18 is subject to the following constraint:

$$(x(t), u(t)) \in \mathbb{Z} \tag{2.47}$$

for all $t \geq 0$ where $\mathbb{Z}$ is assumed to be compact, which accounts, for example, state, input, and other process constraints. One such mathematical framework that allows for one to explicitly account for these considerations is optimization. This is the framework employed in model predictive control.

Tracking model predictive control (MPC), also referred to as receding horizon control, is an on-line optimization-based control technique that optimizes a performance index or cost function over a prediction horizon by taking advantage of a dynamic nominal process model, i.e., Eq. 2.18 with $w \equiv 0$, while accounting for system/process constraints, e.g., [27–32]. The main objective of tracking MPC is to steer the system to and maintain operation thereafter at the economically optimal steady-state or the economically optimal trajectory computed in an upper-layer optimization problem (real-time optimization). To manage the trade-off between the speed of response of the closed-loop system and the amount of control energy required to generate the response, MPC is typically formulated with a quadratic objective function which penalizes the deviations of the state and inputs from their corresponding optimal steady-state or reference values over the prediction horizon. Within this book, the term *tracking MPC* will refer to both regulation MPC or MPC that forces a system to steady-state and tracking MPC or MPC that forces a system track a reference trajectory.

The tracking MPC problem is given by the following dynamic optimization problem:

$$\min_{u \in S(\Delta)} \int_{t_k}^{t_{k+N}} l_T(\tilde{x}(\tau), u(\tau))\, d\tau \tag{2.48a}$$

$$\text{s.t.} \quad \dot{\tilde{x}}(t) = f(\tilde{x}(t), u(t), 0) \tag{2.48b}$$

$$\tilde{x}(t_k) = x(t_k) \tag{2.48c}$$

$$(x(t), u(t)) \in \mathbb{Z}, \ \forall\, t \in [t_k,\ t_{k+N}) \tag{2.48d}$$

where

$$l_T(x, u) = |x|_{Q_c}^2 + |u|_{R_c}^2 \tag{2.49}$$

and Q_c is a positive semidefinite matrix and R_c is a positive definite matrix that manage the trade-off between the speed of response and the cost of control action. Given that the cost function is positive definite with respect to the origin, which is the steady-state of the system of Eq. 2.18, the global minimum of the cost function occurs at the optimal steady-state. The stage cost function of Eq. 2.49 may be readily extended to be positive definite with respect to a reference trajectory. The state trajectory $\tilde{x}$ is the predicted evolution of the state using the nominal dynamic model ($w \equiv 0$) of Eq. 2.18 under the piecewise constant input profile computed by the MPC. The initial condition on the dynamic model are given in Eq. 2.48c which are obtained at each sampling period through a measurement. The constraints of Eq. 2.48d are the system/process constraints, e.g., input and state constraints.

MPC is the resulting control law when the problem of Eq. 2.48 computes the control action applied to the system in a receding horizon fashion. Specifically, at the sampling time t_k, the problem of Eq. 2.48 is initialized with a state feedback measurement and the problem is solved. The optimal input trajectory, i.e., the optimal solution, is denoted by $u^*(t|t_k)$ and defined for $t \in [t_k, t_{k+N})$. A brief overview of methods for solving such dynamic optimization problems of the form of Eq. 2.48 is given in Sect. 2.4.4. The (constant) input trajectory $u^*(t|t_k)$ defined for $t \in [t_k, t_{k+1})$, which may be denoted by $u^*(t_k|t_k)$, is send to the control actuators to be implemented on the system for $t \in [t_k, t_{k+1})$. At t_{k+1}, the problem is re-initialized with an updated measurement and the problem of Eq. 2.48 is re-solved by shifting the horizon one sampling period into the future. Thus, the resulting input trajectory under MPC is given by:

$$u(t) = u^*(t_k|t_k), \ \forall\, t \in [t_k, t_{k+1}). \tag{2.50}$$

When the prediction horizon N is finite, it is well-known that the MPC scheme of Eq. 2.48 may not be stabilizing, e.g., [29]. To handle guaranteed stabilization of the closed-loop system when N is finite, various constraints and variations to the cost function may be made to guarantee stability such as using a sufficiently long prediction horizon, incorporating terminal constraints and/or a terminal cost in the optimization problem, or the use of contractive constraints (see, for example, [29], and the references therein).

2.3.4 Tracking Lyapunov-Based MPC

To address stability of the closed-loop system with tracking model predictive control (MPC) and recursive feasibility, one tracking MPC technique unites the stability and robustness properties of the Lyapunov-based controller, i.e., a control law that satisfies Assumption 2.1, with the optimal control properties of model predictive control (MPC) [21, 33–35]. The resulting tracking MPC is called Lyapunov-based MPC (LMPC) and is characterized by the following optimization problem:

$$\min_{u \in S(\Delta)} \int_{t_k}^{t_{k+N}} l_T(\tilde{x}(\tau), u(\tau))\, d\tau \tag{2.51a}$$

$$\text{s.t. } \dot{\tilde{x}}(t) = f(\tilde{x}(t), u(t), 0) \tag{2.51b}$$

$$\tilde{x}(t_k) = x(t_k) \tag{2.51c}$$

$$u(t) \in \mathbb{U}, \ \forall\, t \in [t_k, t_{k+N}) \tag{2.51d}$$

$$\frac{\partial V(x(t_k))}{\partial x} f(x(t_k), u(t_k), 0) \le \frac{\partial V(x(t_k))}{\partial x} f(x(t_k), h(x(t_k)), 0) \tag{2.51e}$$

where $\tilde{x}$ is the predicted state trajectory over the prediction horizon with the computed input trajectory by the LMPC, and $N > 0$ is the number of sampling periods in the finite prediction horizon. The constraint of Eq. 2.51d is the input constraint, while the constraint of Eq. 2.51e is a contractive constraint for guaranteed stability that is explained further below.

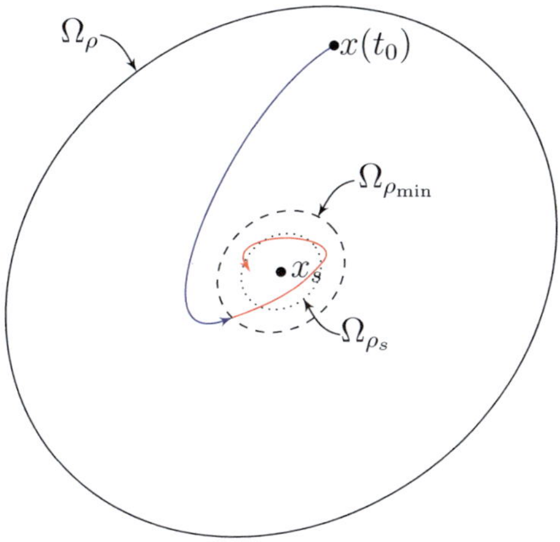

Fig. 2.1 A state-space illustration of a closed-loop state trajectory under LMPC

Specifically, the constraint of Eq. 2.51e ensures that the LMPC computes a control action for the first sampling period that decreases the Lyapunov function by at least the rate achieved by the Lyapunov-based controller at t_k. The Lyapunov-based constraint of Eq. 2.51e is a contractive constraint and ensures that the Lyapunov function decays until the closed-loop state converges to a small neighborhood of steady-state. Moreover, from the Lyapunov-based constraint, the LMPC inherits the closed-loop stability and robustness properties and the stability region Ω_ρ of the Lyapunov-based controller in the sense that for any initial condition $x(0) \in \Omega_\rho$, the closed-loop system state is guaranteed to converge to a small neighborhood of the origin and the optimization problem of Eq. 2.51 is guaranteed to be feasible.

Figure 2.1 gives an illustration of the closed-loop state trajectory under LMPC. The state trajectory starts in $\Omega_\rho \setminus \Omega_{\rho_s}$ whereby in this region the Lyapunov function is guaranteed to decay with time. Once the state trajectory converges to Ω_{ρ_s}, the Lyapunov function is no longer guaranteed to decay owing to the sampling-and-hold implementation of LMPC and the effect of persistent disturbances. However, the state will be maintained in a small forward invariant set $\Omega_{\rho_{\min}} \supset \Omega_{\rho_s}$ when the the sampling period and the bound on the disturbance are sufficiently small.

2.4 Brief Review of Nonlinear and Dynamic Optimization

Although this book does not directly deal with developing nonlinear and dynamic optimization techniques, a brief review of nonlinear optimization (also commonly referred to as nonlinear programming) and dynamic optimization/optimal control concepts is provided in this section. The presentation is meant to demonstrate to the reader how one may approach obtaining a solution to dynamic optimization problems which is required to understand the concepts presented in the subsequent chapters. This section includes definitions, optimality conditions, nonlinear optimization solution techniques, and practical dynamic optimization strategies. For a comprehensive and detailed presentation on optimization methods, the reader is referred to one of the many textbooks on the topic, e.g., [36–40]. For more details relating to dynamic optimization or optimal control, see, for instance, [37, 41–43].

2.4.1 Notation

For a vector $x \in \mathbb{R}^n$, $x \geq 0$ means component-wise inequality, i.e., $x_i \geq 0$, $i = 1, \ldots, n$. The transpose of a vector or matrix is denoted $(\cdot)^T$, e.g., the transpose of $x \in \mathbb{R}^n$ is denoted x^T. The *gradient* (n-dimensional vector) of a differentiable scalar-valued function $f : \mathbb{R}^n \to \mathbb{R}$ evaluated at $x \in \mathbb{R}^n$ is denoted as

$$\nabla f(x) := \left[\frac{\partial f(x)}{\partial x_1} \; \frac{\partial f(x)}{\partial x_2} \; \cdots \; \frac{\partial f(x)}{\partial x_n} \right]^T .$$

When a scalar-valued differentiable function has multiple arguments, for example, $f : \mathbb{R}^{n_x} \times \mathbb{R}^{n_y} \to \mathbb{R}$, $f : (x, y) \mapsto f(x, y)$, the notation $\nabla_x f(x, y)$ may be used to denote the gradient of f with respect to x. For a vector-valued differentiable function $g : \mathbb{R}^n \to \mathbb{R}^m$, the *gradient matrix* is an $n \times m$ matrix whose ith column is the gradient vector $\nabla g_i(x)$ ($i = 1, \ldots, m$):

$$\nabla g(x) := \left[\nabla g_1(x) \; \cdots \; \nabla g_m(x) \right],$$

while the *Jacobian* of g is

$$\frac{\partial g(x)}{\partial x} := \begin{bmatrix} \frac{\partial g_1(x)}{\partial x_1} & \cdots & \frac{\partial g_1(x)}{\partial x_n} \\ \vdots & \ddots & \vdots \\ \frac{\partial g_m(x)}{\partial x_1} & \cdots & \frac{\partial g_m(x)}{\partial x_n} \end{bmatrix} .$$

With the definitions above, the gradient matrix is the transpose of the Jacobian:

$$\nabla g(x)^T = \frac{\partial g(x)}{\partial x} .$$

The *Hessian matrix* of a scalar-valued differentiable function $f : \mathbb{R}^n \to \mathbb{R}$ is denoted as

$$\nabla_{xx} f(x) = \begin{bmatrix} \frac{\partial^2 f(x)}{\partial x_1^2} & \cdots & \frac{\partial^2 f(x)}{\partial x_1 \partial x_n} \\ \vdots & \ddots & \vdots \\ \frac{\partial^2 f(x)}{\partial x_n \partial x_1} & \cdots & \frac{\partial^2 f(x)}{\partial x_n^2} \end{bmatrix}.$$

2.4.2 Definitions and Optimality Conditions

Consider the following nonlinear constrained optimization problem:

$$\begin{aligned} \min_{x \in \mathbb{R}^n} \quad & f(x) \\ \text{s.t.} \quad & g(x) \leq 0 \\ & h(x) = 0 \end{aligned} \tag{2.52}$$

where $x \in \mathbb{R}^n$ is the decision variable or the unknown variable to be determined that minimizes the objective function ($f : \mathbb{R}^n \to \mathbb{R}$) while satisfying the inequality constraints ($g : \mathbb{R}^n \to \mathbb{R}^{n_g}$) and the equality constraints ($h : \mathbb{R}^n \to \mathbb{R}^{n_h}$). The objective function is also referred to as the cost function or cost functional in the context of dynamic optimization problems. Here and elsewhere in the book, the usage of the notation "min" in Eq. 2.52 is more aligned with that typically found in the engineering literature, that is, it refers to the greatest lower bound or infimum of $f(x)$ over $\mathbb{X}$. Nevertheless, in the application studies contained in this book, the optimization problems are formulated in a manner that guarantee that they may be numerically solved in the sense that $f^* = \inf_{x \in \mathbb{X}} f(x)$ where $\mathbb{X} = \{x \in \mathbb{R}^n : g(x) \leq 0,\ h(x) = 0\}$ is non-empty, f^* is finite, and there exists a vector $x^* \in \mathbb{X}$ where the minimum is attained. The issue of the existence of a minimizing vector will not be treated in depth.

The functions f, g, and h are assumed to be continuously differentiable. A vector $x \in \mathbb{R}^n$ is said to be a *feasible* point if $g(x) \leq 0$ and $h(x) = 0$. The set of all feasible points or the *feasible set* to the problem of Eq. 2.52 is the set $\mathbb{X} \subseteq \mathbb{R}^n$. For the problem of Eq. 2.52 to be meaningful, the feasible set must be non-empty. Otherwise, the problem of Eq. 2.52 is said to be *infeasible*. Feasibility of the optimization problems formulated in this book will be carefully examined which is a crucial property for control purposes. A vector $x^* \in \mathbb{X}$ is said to be a *local minimum* if there exists $\varepsilon > 0$ such that $f(x^*) \leq f(x)$ for all $x \in \mathbb{X}$ with $|x - x^*| < \varepsilon$. A vector $x^* \in \mathbb{X}$ is said to be a *global minimum* if $f(x^*) \leq f(x)$ for all $x \in \mathbb{X}$. A local or global minimum is called a *strict* minimum if the inequalities are strict for all $x \neq x^*$. For a given local or global minimum $x^* \in \mathbb{X}$, $f(x^*)$ is called the local or global optimal objective function value or optimal value.

In subsequent chapters, non-convex dynamic nonlinear optimization problems will be considered. It is sufficient, for purposes of this book, to understand a non-convex optimization problem as one possibly having multiple local minima. For non-convex problems, most general nonlinear optimization solvers are capable of computing a local solution to the problem. Generally, no guarantee can be made that the computed local solution is or is not a global solution without further analysis. To ensure that a global solution is returned, one needs to employ more advanced techniques that are typically more computationally expensive, (see, for example, [44, 45] on global optimization techniques). Owing to this consideration, local minima will be of interest in this book. Also, in the subsequent chapters, the term *optimal solution* may be used to refer to a local minimum of an optimization problem, and the explicit distinction that the minimum is a local minimum may be omitted.

While minimization problems are treated here, maximization problems, e.g., $\max_{x \in \mathbb{X}} f(x)$ may readily be converted into a minimization problem by minimizing the negative of the objective function, e.g., $\min_{x \in \mathbb{X}} -f(x)$. The optimal solution of each optimization problem are the same, and the optimal objective function value of the maximization problem is equal to the negative of the optimal value of the minimization problem. Thus, there is no loss of generality by considering only minimization problems.

To present general necessary and sufficient optimality conditions for optimality, some regularity conditions or constraint qualifications must be satisfied. First, the active set is defined. For a feasible vector $x \in \mathbb{X}$, the index set of active inequality constraints is defined as $\mathscr{A}(x) := \{j \in \{1, \ldots, n_g\} : g_j(x) = 0\}$. For all $j \notin \mathscr{A}(x) \setminus \{1, \ldots, n_g\}$, the jth inequality constraint is said to be *inactive* at x, i.e., $g_j(x) < 0$. Since the equality constraints are always active, the *active set* includes all the active inequality constraints and all equality constraints. The linear independence constraint qualification (LICQ) holds at $x \in \mathbb{X}$ if the gradients of all active constraints are linearly independent at x, that is, the vectors $\nabla g_j(x)$, $j \in \mathscr{A}(x)$ and $\nabla h_i(x)$, $i = 1, \ldots, n_h$ are linearly independent.

The Lagrangian function of the problem of Eq. 2.52 is given by

$$\mathscr{L}(x, \lambda, \nu) = f(x) + \lambda^T g(x) + \nu^T h(x) \tag{2.53}$$

where $\lambda \in \mathbb{R}^{n_g}$ and $\nu \in \mathbb{R}^{n_h}$ are the Lagrange multipliers. Necessary and sufficient optimality conditions have been derived for the problem of Eq. 2.52. These conditions are not only fundamental to the theory of optimization, but also, allow for the development of computational algorithms that are capable of computing solutions to the optimization problem of Eq. 2.52. The Karush-Kuhn-Tucker (KKT) optimality conditions [46, 47] are first-order necessary conditions for a (local) solution to the problem of Eq. 2.52.

Theorem 2.4 (KKT Conditions, e.g., [36, Proposition 3.3.1]) *Let $x^* \in \mathbb{R}^n$ be a local minimum of the problem of Eq. 2.52 and the LICQ holds at x^*. Then there exist unique $\lambda^* \in \mathbb{R}^{n_g}$ and $\nu^* \in \mathbb{R}^{n_h}$ such that:*

$$\nabla_x \mathscr{L}(x^*, \lambda^*, \nu^*) = 0 \tag{2.54a}$$

$$g(x^*) \leq 0 \tag{2.54b}$$

$$h(x^*) = 0 \tag{2.54c}$$

$$\lambda^* \geq 0 \tag{2.54d}$$

$$\lambda_i^* g_i(x^*) = 0, \; i = 1, \; \ldots, \; n_h \tag{2.54e}$$

If in addition f, h, and g are twice continuously differentiable, then

$$y^T \nabla_{xx} \mathscr{L}(x^*, \lambda^*, \nu^*) y \geq 0, \tag{2.55}$$

for all $y \in \mathbb{R}^n$ such that

$$\begin{aligned} \nabla h_i(x^*)^T y = 0, \forall\, i = 1, \ldots, n_h, \\ \nabla g_j(x^*)^T y = 0, \forall\, j \in \mathscr{A}(x^*). \end{aligned} \tag{2.56}$$

As pointed out in Theorem 2.4, LICQ at a local minimum x^* guarantees existence of Lagrange multipliers. It can be shown that the multipliers are unique as well if the KKT conditions are satisfied and the LICQ holds at x^*. Any triple (x^*, λ^*, ν^*) satisfying the KKT conditions is said to be a *KKT point*. The KKT conditions mean: the gradient of the Lagrangian with respect to the decision variable must vanish at the KKT point, the primal problem, i.e., Eq. 2.52, must be feasible at a KKT point, the dual problem (not discussed here) must be feasible at the KKT point, and *complementarity* or *complementary slackness*, i.e., the condition of Eq. 2.54e, must hold at the KKT point. Since the KKT conditions are necessary conditions, not all KKT points are local minimums. Often second-order necessary conditions are included with the KKT conditions, which is the condition of Eq. 2.55. The second-order necessary conditions mean that it is necessary for a local minimum that the Hessian of the Lagrangian must be positive semidefinite in all feasible directions. If the Hessian of the Lagrangian is shown to be positive definite for a KKT point and *strict complementarity* or *strict complementary slackness* holds, i.e., $\lambda^* > 0$ if $g_i(x^*) = 0$ and $\lambda_i^* = 0$ if $g_i(x^*) < 0$, it can be concluded that the KKT point is a local minimum. This is stated in the following second order sufficient optimality conditions.

Theorem 2.5 (Second Order Optimality Conditions, e.g., [36, Proposition 3.3.2]) *Let the triple (x^*, λ^*, ν^*) be a KKT point that also satisfies:*

$$y^T \nabla_{xx} \mathscr{L}(x^*, \lambda^*, \nu^*) y > 0 \tag{2.57}$$

for all $y \in \mathbb{R}^n$ such that

$$\nabla h_i(x^*)^T y = 0, \;\; \forall\, i = 1, \ldots, n_h, \tag{2.58}$$

$$\nabla g_j(x^*)^T y = 0, \;\; \forall\, j \in \mathscr{A}(x^*), \tag{2.59}$$

$$\lambda_j > 0, \quad j \in \mathscr{A}(x^*). \tag{2.60}$$

Then x^* is a strict local minimum of Eq. 2.52.

Numerous variants of the optimality conditions and constraint qualifications given above exist (see, for example, [36]).

2.4.3 Nonlinear Optimization Solution Strategies

The KKT conditions form a set of nonlinear equations and many computation methods for solving for a local minimum of Eq. 2.52 seek to find a solution to the KKT conditions. However, the inequality constraints and the complementary slackness condition, which poses a non-differentiability in the equations, must be handled carefully, and one cannot simply solve the KKT conditions directly in general. Owing to the fact that these conditions form a set of nonlinear equations, the most widely adapted method employed to solve the KKT conditions is Newton's method. More precisely, variants of Newton's method are typically used. From a high level perspective, most nonlinear optimization solvers utilize the user-supplied input information shown in Fig. 2.2 to compute a solution to an optimization problem. The input information includes the functions and their corresponding derivatives. The Hessian of the Lagrangian may also be supplied to the solver. However, in some algorithms such as quasi-Newton methods, the Hessian is approximated in the algorithm. In this section, a basic review of Newton's method is given along with the basic concepts of two widely employed solution techniques for solving nonlinear optimization problems. The two methods include sequential quadratic programming (SQP) and interior point (IP) methods.

2.4.3.1 Newton's Method

The core of most nonlinear optimization solution strategies relies on some variant of Newton's method to solve a set of nonlinear algebraic equations. The standard Newton method is presented to facilitate the discussion of SQP and IP methods for

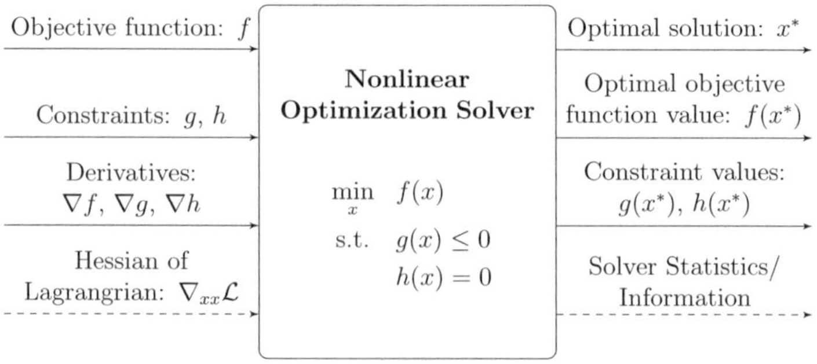

Fig. 2.2 Typical inputs and outputs of a nonlinear optimization solver

solving nonlinear optimization problems. However, it is important to emphasize that key modifications are made to ensure computational efficiency, robustness, etc. To review Newton's method, consider the following nonlinear algebraic equation:

$$F(y) = 0 \tag{2.61}$$

where $F : \mathbb{R}^{n_y} \to \mathbb{R}^{n_y}$ is a vector-valued differentiable function.

Newton's method is an iterative algorithm that is initialized with a starting guess $y_0 \in \mathbb{R}^{n_y}$. At each iteration, the following system of linear equations, which is a linearized version of Eq. 2.61 around the iterate $y_k \in \mathbb{R}^{n_y}$, is solved:

$$\frac{\partial F(y_k)}{\partial y} d_k = -F(y_k) \tag{2.62}$$

where k denotes the iteration number and $d_k \in \mathbb{R}^{n_y}$ is the unknown variable. At the next iteration, i.e., the $(k+1)$th iteration, the iterate is updated as follows:

$$y_{k+1} = y_k + d_k. \tag{2.63}$$

From the update formula of Eq. 2.63, the variable d_k can be interpreted as a descent direction of Newton's method. One of the most advantageous properties of Newton's method is that it has a locally quadratic convergence rate meaning for an initial guess y_0 that is sufficiently close to the solution of Eq. 2.61, Newton's method will converge at a quadratic rate.

To demonstrate applying Newton's method to solving an optimization problem, consider the following unconstrained problem:

$$\min_{x \in \mathbb{R}^n} f(x) \tag{2.64}$$

where f is a twice differentiable scalar-valued function. The necessary conditions for optimality at a point $x^* \in \mathbb{R}^n$ are

$$\nabla f(x^*) = 0, \tag{2.65a}$$

$$\nabla^2 f(x^*) \geq 0. \tag{2.65b}$$

Applying Newton's method to solve the nonlinear equations, we have the following update:

$$x_{k+1} = x_k - \nabla_{xx} f(x_k)^{-1} \nabla f(x_k). \tag{2.66}$$

Owing to the fact that the Hessian, i.e., $\nabla_{xx} f$ in the unconstrained case or $\nabla_{xx}\mathscr{L}$ in the constrained case, may be expensive to compute, quasi-Newton methods have been designed to approximate the Hessian; see, for example, [40]. Let $H_k > 0$ be the approximation of the Hessian at iteration k. Then, the update takes form of:

$$x_{k+1} = x_k - H_k^{-1} \nabla f(x_k). \tag{2.67}$$

The fundamental requirement for the convergence of Newton's method algorithm is that the initial guess supplied to the algorithm be sufficiently close to the solution. Globalization strategies are used to allow for convergence to a solution from initial guesses that are not close to the solution. Numerous numerical nonlinear optimization solvers have been developed that are equipped with various globalization strategies. Globalization strategies are briefly discussed here (see, [36, 37, 40], for the details), and they include algorithm monitoring strategies to decide if a computed iterate update is acceptable and modification strategies to modify the iterate updates.

In the first category, merit functions and filter methods are used as a measure of the progress of the algorithm. This adds logic to the algorithm to decide if the step/update is acceptable (loosely speaking, defining the step as d_k in Eq. 2.63). *Merit functions* are scalar-valued functions that are typically chosen to have the same local minimum as the nonlinear optimization problem. At each iteration, the step is accepted if the update yields a decrease in the merit function. Otherwise, the step is rejected. However, merit functions may lead to rejecting pure Newton steps near the optimum and thus, slow down the convergence of the algorithm. As an alternative, *filter methods* treat making the objective function as small as possible and reducing the constraint violations as equal goals. In a filter method, a *filter* keeps track of previous iterates with the best objective function value and amount of the constraint violation. A step is accepted if the update yields a better objective function value or smaller constraint violation. If the update yields a step that is such that one of the previous iterates have a better objective function value and constraint violation, the step is rejected.

In the second category, strategies are used to modify the step or update of the iterates. *Line search* methods take potentially shortened steps if necessary. In other words, line search methods add a dampening factor α in the update formula. The update formula for the iterates is given by:

$$y_{k+1} = y_k + \alpha d_k \tag{2.68}$$

where $\alpha \in (0, 1]$ is selected by the line search method. On the other hand, *trust region* methods recognize that Newton's method utilizes a linearization of the nonlinear function. The resulting linearization is only valid in a neighborhood of the current iterate. Thus, it restricts selection of the step d_k in a small region of the current iterate y_k.

2.4.3.2 Sequential Quadratic Programming

One of the two main solution strategies of nonlinear constrained optimization is to consider successive linearization of the KKT conditions (Eq. 2.54). It turns out that the linearized KKT conditions are the KKT conditions for the following quadratic program (QP) (for a complete derivation one may refer to [37]):

$$\begin{aligned}
\min_{d_x} \quad & \nabla f(x_k) d_x + \frac{1}{2} d_x^T H_k d_x \\
\text{s.t.} \quad & g(x_k) + \nabla g(x_k)^T d_x \le 0 \\
& h(x_k) + \nabla h(x_k)^T d_x = 0
\end{aligned} \tag{2.69}$$

where H_k is either the exact or an approximation of the Hessian of the Lagrangian evaluated at iteration k, i.e., $\nabla_{xx}\mathscr{L}(x_k, \lambda_k, \nu_k)$. If H_k is positive semidefinite, the problem of Eq. 2.69, which is a quadratic program (QP), is convex and efficient methods exist that can readily solve the quadratic program to global optimality. This approach to solving a nonlinear optimization problem is referred to as sequential quadratic programming. Many primal-dual methods used to solve each QP work to find a KKT point of the KKT conditions of Eq. 2.69. The KKT conditions are given by:

$$\begin{aligned}
H_k d_x + \nabla f(x_k) + \nabla g(x_k) d_\lambda + \nabla h(x_k) d_\nu &= 0 \\
g(x_k) + \nabla g(x_k)^T d_x &\le 0 \\
h(x_k) + \nabla h(x_k)^T d_x &= 0 \\
d_\lambda &\ge 0 \\
\left[g(x_k) + \nabla g(x_k)^T d_x \right]_i d_{\lambda,i} &= 0, \, i = 1, \ldots, n_g
\end{aligned} \tag{2.70}$$

With d_x, d_ν, and d_λ, the primal and dual iterates, i.e., x_k, λ_k, and ν_k, are updated. Using this type of the solution strategy, the active set is automatically discovered once the algorithm converges. Under strict complementarity, the solutions of the QP subproblems converge to a local solution to the nonlinear optimization problem once the iterates x_k are in the neighborhood of x^*, e.g. [37].

2.4.3.3 Interior Point Methods

The second widely used class of numerical nonlinear optimization solvers is based on interior point methods. In a number of applications in this book, an open-source interior-point solver, called Ipopt [48], is applied to solve the MPC (EMPC) problems. One interpretation of interior point methods is that the inequality constraints are replaced by the barrier function. The barrier function has the property that the function becomes very large in value as one of the constraint values goes to zero, i.e., if the barrier function is denoted as B, then $B(x) \to \infty$ if $g_i(x) \to 0$ for some $i \in \{1, \ldots, n_g\}$. One widely used barrier function is the logarithmic function:

$$B(x) = -\sum_{i=1}^{n_g} \ln(-g_i(x)). \tag{2.71}$$

The resulting optimization problem is obtained:

$$\begin{aligned} \min_{x} \quad & f(x) - \tau \sum_{i=1}^{n_h} \log(-g_i(x)) \\ \text{s.t.} \quad & g(x) = 0 \end{aligned} \tag{2.72}$$

where $\tau > 0$ is a parameter. In particular, to solve the original nonlinear optimization problem for a local solution, a sequence of modified problems of the form of Eq. 2.72 are solved for a given parameter $\tau > 0$. Under certain conditions, it may be shown that the solution of the original nonlinear optimization problem is the same as the one of the modified problem when the parameter τ approaches zero, e.g., [37, 40].

Another interpretation of interior point methods is that they replace the non-smooth complementary slackness condition of the KKT conditions (Eq. 2.54e) by a smooth approximation. Specifically, the smooth approximation of the KKT conditions is given by:

$$\nabla f(x) + \nabla g(x)\lambda + \nabla h(x)\nu = 0 \tag{2.73a}$$

$$h(x) = 0 \tag{2.73b}$$

$$\lambda_i g_i(x) + \tau = 0, \quad i = 1, \ldots, n_g \tag{2.73c}$$

where $\tau > 0$ is a smoothing parameter. From the last condition, $\lambda_i = -\tau / g_i(x)$, $i = 1, \ldots, n_g$, and thus, the modified KKT conditions of Eq. 2.73 are the KKT conditions of the modified nonlinear optimization problem of Eq. 2.72.

2.4.4 Dynamic Optimization

The specific class of optimization problems that will be of interest in this book is dynamic optimization problems. In particular, the EMPC methods presented in subsequent chapters require the repeated solution of a dynamic optimization problem to compute control actions to apply to a dynamic system. Dynamic optimization or optimal control problems are optimization problems that have a dynamic model embedded in the problem. At this point, the literature on dynamic optimization is vast and impossible to summarize in this brief overview. For general references on theoretical and applied optimal control, the interested reader is referred to one of the many texts on the subject, for example, [41, 49–52]. Important early results that helped shape the foundations of optimal control include optimal control based on the Hamilton-Jacobi-Bellman equation and dynamic programming [53], Pontryagin's maximum principle [54], and the linear quadratic regulator [55].

In this section, direct methods are considered, which are the most commonly employed solution technique for dynamic optimization problem in practical applications. Direct methods first discretize a continuous-time dynamic model and then, use a nonlinear optimization solver to solve the resulting nonlinear optimization problem. Besides direct methods other methods exist including dynamic programming [53] and solving the Hamilton-Jacobi-Bellman partial differential equations, and indirect methods, which optimize first and then, discretize. The latter methods include setting up the so-called Euler-Lagrange differential equations and applying Pontyragin's Maximum Principle [54], which is a necessary condition, to solve for the optimal control. For a comprehensive review on dynamic optimization with a particular focus on the application of solutions method to moving/receding horizon problems within the context of chemical processes, the reader is referred to the review [42].

Consider the following nonlinear dynamic system:

$$\dot{x} = f(x, u), \quad x(t_0) = x_0 \tag{2.74}$$

where $x \in \mathbb{X} \subseteq \mathbb{R}^n$ is the state vector, $u \in \mathbb{U} \subset \mathbb{R}^m$ is the input vector, and $f : \mathbb{X} \times \mathbb{U} \to \mathbb{X}$. For simplicity, the sets $\mathbb{X}$ and $\mathbb{U}$ are assumed to be compact sets and the vector field f is assumed to satisfy enough smoothness assumptions so that a unique solution to Eq. 2.74 exists over the interval $[t_0, t_f]$ with an initial condition x_0 and piecewise continuous input function $u : [t_0, t_f] \to \mathbb{U}$. The smoothness properties will also be needed to solve the resulting optimization problems below. The solution is denoted as $x(\cdot, x_0, u(\cdot))$, i.e., $x(t, x_0, u(\cdot))$ denotes the solution at $t \in [t_0, t_f]$ and $x(t_0, x_0, u(\cdot)) = x_0$.

Since the purpose of this section is to highlight the various computational approaches to solving dynamic optimization problems, a simple dynamic optimization problem is considered. Nevertheless, more complex problems may be considered while utilizing the presented techniques, e.g., problems with algebraic constraints like path or end-point constraints. Specifically, consider a dynamic optimization problems in Mayer form:

$$\begin{aligned} \min_{x(\cdot),u(\cdot)} \quad & \phi(x(t_f)) \\ \text{s.t.} \quad & \dot{x}(t) = f(x(t), u(t)), \ x(t_0) = x_0 \\ & x(t) \in \mathbb{X}, \ u(t) \in \mathbb{U}, \forall\, t \in [t_0, t_f] \end{aligned} \tag{2.75}$$

where $x_0 \in \mathbb{X}$ denotes the initial condition, which also could be a decision variable in the optimization problem. For simplicity, the initial condition will be assumed to be fixed in the remainder of this chapter. The set constraints are assumed to take the form of inequality constraints, i.e., of the form $h_x(x) \le 0$ where $\mathbb{X} = \{x \in \mathbb{R}^n : h_x(x) \le 0\}$ and similarly, for the input constraint with $h_u(u) \le 0$. The existence of minimizing trajectories for the problem of Eq. 2.75 is assumed; for conditions that guarantee existence of a solution, the interested reader is referred to [56, 57].

It is important to point out that if one seeks the solution to a dynamic optimization problem that optimizes an objective function of the form, i.e., Bolza form:

$$J(x_0, u(\cdot)) = \int_{t_0}^{t_f} l(x(t), u(t))\, dt + V_f(x(t_f)), \tag{2.76}$$

one could readily convert this type problem into Mayer form by defining the state vector as:

$$\bar{x} := \begin{bmatrix} x \\ \bar{\phi} \end{bmatrix} \tag{2.77}$$

with dynamics:

$$\dot{\bar{x}} = \begin{bmatrix} f(x, u) \\ l(x, u) \end{bmatrix}. \tag{2.78}$$

Then, the Mayer term is given by $\phi(\bar{x}(t_f)) := \bar{\phi}(t_f) + V(x(t_f))$.

The major difference between the optimization problem of Eq. 2.52 and the optimization problem of Eq. 2.75 is the presence of the dynamic model embedded in the optimization problem. To avoid an infinite dimensional optimization problem, the control function $u(\cdot)$ must be parameterized by a finite set of parameters, which is referred to as control vector parameterization. The most widely used control vector parameterization is zeroth-order hold, i.e., the input trajectory is assumed to take the form of:

$$u(t) = \bar{u}_i \tag{2.79}$$

for $t \in [\tau_i, \tau_{i+1})$ where $\bar{u}_i \in \mathbb{U}$, $\tau_i := i\Delta + t_0$ for $i = 0, \ldots, N-1$, $\tau_N = t_f$, and $\Delta > 0$ is the hold period. In what follows, the family of (possibly vector-valued) functions that take the form of Eq. 2.79 is generally denoted in this book by $S(\Delta)$ where $\Delta > 0$ is the hold period. For the remainder of this chapter, zeroth order hold control vector parameterization is assumed for simplicity.

Given that the dynamic model embedded in the problem of Eq. 2.75 may be nonlinear, an analytic solution is often difficult to obtain for a given initial condition and input trajectory. Therefore, some numerical method that obtains the solution of the dynamic model is required to solve the optimization problem. The choice of numerical solution techniques used to solve the optimization problem substantially influences the computational efficiency and therefore, is an important implementation consideration. From a nonlinear optimization point of view, dynamic optimization problems typically have a high degree of sparsity and therefore, using sparsity-exploiting nonlinear optimization solvers may also be an important implementation constraint. On the other hand, from a numerical integration standpoint, the most computationally expensive part tends to lie in solving the dynamic model. In particular, computing sensitivity information of the dynamic model tends to be the most computationally expensive step. Therefore, selecting the numerical solver is critical for the success of the solution technique. Below three solution techniques are briefly described.

2.4.4.1 Single Shooting Approach

With a given input trajectory and initial condition, the dynamic model of Eq. 2.74 may be solved forward in time using an ODE solver, i.e., numerical integration method,

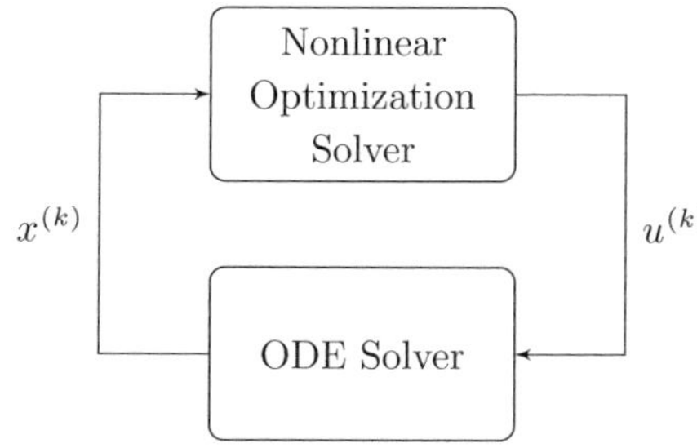

Fig. 2.3 In a single shooting approach, the input trajectory is computed by the nonlinear solver at each iteration. The input trajectory is passed to an ODE solver to compute the corresponding state trajectory, and the corresponding sensitivity information

to obtain the solution over the time interval $[t_0, t_f]$. In this respect, the solution to the dynamic model is a function of the input trajectory and of the initial condition. After the solution to the dynamic model is obtained from the ODE solver, the input trajectory may be updated using a nonlinear optimization solver. These concepts are used in the design of a solution strategy to the optimization problem of Eq. 2.75, which is the single shooting approach.

A block diagram of the methodology is given in Fig. 2.3. At each iteration, the model is first solved over the interval to obtain $x^{(k)}(t, x_0, u^{(k)}(\cdot))$ for $t \in [t_0, t_f]$ where the notation $x^{(k)}$ and $u^{(k)}$ denote the state and input trajectory at the kth iteration of the nonlinear solver. With $x^{(k)}$ and $u^{(k)}$, the objective value, the state and input constraint values, and the sensitivity information (first and second-order derivatives) of the problem of Eq. 2.75 are computed and the nonlinear optimization solver computes the updated input trajectory for the next iteration. The algorithm repeats until the solver converges to a local optimal input trajectory.

For the sake of simplicity, the input trajectory is assumed to be the only decision variable of the dynamic optimization problem and piecewise constant input trajectory is assumed. The resulting formulation of the optimization is given by:

$$\begin{aligned} \min_{\bar{u}_0,\ldots,\bar{u}_{N-1}} \quad & \phi(x(t_f, x_0, u(\cdot))) \\ \text{s.t.} \quad & u(t) = \bar{u}_i, \ \forall\, t \in [\tau_i, \tau_{i+1}), \ i \in \mathbb{I}_{0:N-1} \\ & h_u(\bar{u}_i) = 0, \ \forall\, i \in \mathbb{I}_{0:N-1} \\ & h_x(x(\tau_j, x_0, u(\cdot))) \le 0, \ \forall\, j \in \mathbb{I}_{0:N_x} \end{aligned} \tag{2.80}$$

where $\tau_j \in [t_0, t_f]$ for $j = 0, \ldots, N_x$ denotes the time grid that the state constraints are imposed with $\tau_0 = t_0$ and $\tau_{N_x} = t_f$. In many cases, taking the time grids used for the control parameterizations and for imposing the state constraints to be equal yields acceptable results.

As a first-pass implementation, one may employ a finite-difference method to approximate sensitivity information of the problem of Eq. 2.80. However, this tends to be inefficient, result in large numerical error, and yield unreliable performance, e.g., [37]. Instead, one may obtain exact first-order sensitivity information from direct sensitivity, adjoint sensitivity, or automatic (algorithmic) differentiation, e.g., [37, 41]. While methods exist that are capable of computing the exact Hessian of the Lagrangian of the problem of Eq. 2.80, efficient methods for obtaining an approximation of the Hessian have been developed such as the Broyden-Fletcher-Goldfarb-Shanno (BFGS) algorithm. Moreover, the BFGS method tends to yield good computational performance; see, for example, [36, 40].

The key advantages of the single shooting approach to solving a dynamic optimization problem relative to the other two approaches described below are that the method tends to be the easiest to implement and the dynamic model is satisfied (up to numerical precision) at each iteration of the solution method. However, the single shooting method tend to be substantially slower than the multiple shooting and collocation approaches described below. Also, the input trajectory being computed by the nonlinear optimization solver at each iteration of the solution method is an open-loop one, i.e., the input trajectory is first specified by the optimization solver and then, the dynamic equations are solved forward in time with the given open-loop input trajectory. Therefore, solving dynamic optimization problems with a single shooting method when the dynamic model is open-loop unstable may result in numerical problems like unbounded solutions being computed or convergence failure.

2.4.4.2 Multiple Shooting Approach

The multiple shooting method [58] serves as an alternative to a single shooting method. Instead of solving for the solution of the dynamic model over the entire time interval $[t_0, t_f]$, the time interval may be divided into subintervals and the dynamic model may be initialized and solved within each of these subintervals. For simplicity of the presentation, the time horizon $[t_0, t_f]$ is divided into N intervals of constant size $\Delta > 0$, referred to as nodes. While the node intervals are taken to be equally spaced and equal to the hold period of the controls, neither assumption is required for implementation. Let $\tau_i := i\Delta + t_0$ for $i = 0, \ldots, N$. At the beginning of the ith subinterval, the dynamic model is initialized with an initial condition denoted by $s_i \in \mathbb{X}$, which is determined by the optimization solver, and the solution, $x(t, s_i, \bar{u}_i)$ defined for $t \in [\tau_i, \tau_{i+1}]$ where $\bar{u}_i$ denotes the constant input applied over the ith subinterval, is computed by employing a numerical integration method. To ensure that the dynamic model is satisfied over the entire time interval, a constraint is imposed in the optimization problem to ensure that the initial condition specified for the ith subinterval is equal to $x(\tau_i, s_{i-1}, \bar{u}_{i-1})$. In other words, the solutions of each subinterval are pieced together by imposing a constraint in the optimization problem to obtain the solution over the entire interval $[t_0, t_f]$.

The resulting optimization problem for the multiple shooting approach is given by:

$$\min_{\substack{s_0,\ldots,s_N, \\ \bar{u}_0,\ldots,\bar{u}_{N-1}}} \quad \phi(s_N) \tag{2.81a}$$

$$\text{s.t.} \quad s_0 - x_0 = 0 \tag{2.81b}$$

$$x(\tau_{i+1}, s_i, \bar{u}_i) - s_{i+1} = 0, \ \forall\, i \in \mathbb{I}_{0:N-1} \tag{2.81c}$$

$$h_u(\bar{u}_i) = 0, \ \forall\, i \in \mathbb{I}_{0:N-1} \tag{2.81d}$$

$$h_x(s_i) \leq 0, \ \forall\, i \in \mathbb{I}_{1:N} \tag{2.81e}$$

where the constraint of Eq. 2.81b ensures that the initial condition for the first subinterval is equal to x_0 and the constraint of Eq. 2.81c ensures that the solution value at τ_{i+1} is equal to the initial condition specified for the $(i + 1)$th subinterval. The constraints of Eqs. 2.81d–2.81e are the input and state constraints, respectively. The state constraint of Eq. 2.81e may readily be extended so that it is imposed over a different time grid like that of the problem of Eq. 2.80.

The multiple shooting method has advantages over a single shooting method in that, loosely speaking, the open-loop instabilities and nonlinearities are distributed amongst the nodes of the time grid. Since solving the dynamic model and the corresponding sensitivity information of the dynamic equations is typically the most computationally expensive task when solving dynamic optimization problems, the multiple shooting method offers a clear way to parallelize the one of the most computationally expensive calculation. Additionally, the problem presents a high-degree of sparsity which may be exploited. Therefore, although the problem of Eq. 2.81 clearly has more decision variables than the problem of Eq. 2.80, the problem of Eq. 2.81 tends to be more computationally efficient than a single shooting approach owing to the aforementioned reasons. The main disadvantage of the method are that the iterates of the method do not necessarily satisfy the dynamic model in the sense each iterate may not satisfy the constraint of Eq. 2.81c.

2.4.4.3 Orthogonal Collocation Approach

The third technique to solve the optimization problem of Eq. 2.75 is to employ orthogonal collocation to obtain the solution to the dynamic model of Eq. 2.74 [59, 60]. Orthogonal collocation approximates the solution of a system of the form of Eq. 2.74 with an interpolating polynomial, e.g., Lagrange polynomials. The coefficients of the polynomial are adjusted such that the interpolating polynomial satisfies the dynamic equation at the collocation points, which are points along the time horizon chosen on the basis of a quadrature rule.

Similar to the multiple shooting approach, the time interval $[t_0, t_f]$ is divided into N subintervals of length $\Delta > 0$. Again, for simplicity of presentation, the subintervals are assumed to be of equal length and constant, i.e., may not be adjusted by the optimization solver. The method extends to the more general case when the subintervals are not of equal length and may be a decision variable of the optimization problem; please see, for example, [37] and the references therein. Let $\tau_i = t_0 + i\Delta$, $i = 0, 1, \ldots, N$, and the interval $[\tau_i, \tau_{i+1}]$ is the ith subinterval. Within the interval $[\tau_i, \tau_{i+1}]$, n_c collocation points are chosen. Let $p_i(t, c_i)$ be an interpolating polynomial defined for the ith subinterval, i.e., $t \in [\tau_i, \tau_{i+1}]$, that is parameterized by a coefficient vector c_i and τ_i^j denotes the jth collocation point in the ith subinterval where $j = 1, 2, \ldots, n_c$. The solution of Eq. 2.74, obtained through collocation, is computed by solving the following equations:

$$p_i(\tau_i, c_i) = s_i \tag{2.82a}$$

$$\dot{p}_i(\tau_i^j, c_i) = f(p_i(\tau_i^j, c_i), \bar{u}_i), \ \forall j \in \mathbb{I}_{1:n_c}. \tag{2.82b}$$

The optimization problem solved through the collocation approach is given by:

$$\min_{\substack{s_0, \ldots, s_N \\ \bar{u}_0, \ldots, \bar{u}_{N-1} \\ c_0, \ldots, c_{N-1}}} \quad \phi(s_N) \tag{2.83a}$$

$$\text{s.t.} \quad s_0 - x_0 = 0 \tag{2.83b}$$

$$p_i(\tau_i, c_i) - s_i = 0, \ \forall\, i \in \mathbb{I}_{0:(N-1)} \tag{2.83c}$$

$$\dot{p}_i(\tau_{i,j}, c_i) - f(p_i(\tau_{i,j}, c_i), \bar{u}_i) = 0, \ \forall j \in \mathbb{I}_{1:n_c}, \ \forall\, i \in \mathbb{I}_{0:(N-1)} \tag{2.83d}$$

$$p_i(\tau_{i+1}, c_i) - s_{i+1} = 0, \ \forall\, i \in \mathbb{I}_{0:N-1} \tag{2.83e}$$

$$h_u(\bar{u}_i) = 0, \ \forall\, i \in \mathbb{I}_{0:N-1} \tag{2.83f}$$

$$h_x(s_i) \leq 0, \ \forall\, i \in \mathbb{I}_{1:N} \tag{2.83g}$$

where the decision variables of the optimization problem are states values at the nodes, s_i for $i = 0, \ldots, N$, the input trajectory parameterization vectors, $\bar{u}_k$ for $k = 0, 1, \ldots, N - 1$, and the coefficients of the interpolating polynomial, c_i for $i = 0, 1, \ldots, (N - 1)$.

Owing to the similarities in the structures of the problems of Eqs. 2.81 and 2.83, the orthogonal collocation approach has similar advantages and disadvantages to the ones of the multiple shooting approach. Since the solution to the dynamic model is approximated as a polynomial, analytic computation of the sensitivity information is perhaps easier to obtain than the shooting method approaches because the latter may use a general ODE solver. Within the context of solving the problem of Eq. 2.83, the use of a sparsity exploiting optimization solver is critical because the resulting optimization problem is large-scale with a high degree of sparsity.

References

1. Isidori A (1995) Nonlinear control systems: an introduction, 3rd edn. Springer, New York, NY
2. Khalil HK (2002) Nonlinear systems, 3rd edn. Prentice Hall, Upper Saddle River, NJ
3. Rawlings JB, Mayne DQ (2009) Model predictive control: theory and design. Nob Hill Publishing, Madison, WI
4. Massera JL (1956) Contributions to stability theory. Ann Math 64:182–206
5. Krasovskii NN (1963) Stability of motion. Stanford University Press, Stanford
6. Hahn W (1967) Stability of motion. Springer, New York
7. Lin Y, Sontag E, Wang Y (1996) A smooth converse Lyapunov theorem for robust stability. SIAM J Control Optim 34:124–160
8. Teel AR, Praly L (1999) Results on converse Lyapunov functions from class-KL estimates. In: Proceedings of the 38th IEEE conference on decision and control, vol. 3. Phoenix, AZ, pp 2545–2550
9. Hale J (1980) Ordinary differential equations, 2nd edn. Krieger Publishing Company
10. Artstein Z (1983) Stabilization with relaxed controls. Nonlinear Anal: Theory Meth Appl 7:1163–1173
11. Sontag ED (1989) A 'universal' construction of Artstein's theorem on nonlinear stabilization. Syst Control Lett 13:117–123
12. Kravaris C, Kantor JC (1990) Geometric methods for nonlinear process control. 1. Background. Ind Eng Chem Res 29:2295–2310
13. Kravaris C, Kantor JC (1990) Geometric methods for nonlinear process control. 2. Controller synthesis. Ind Eng Chem Res 29:2310–2323
14. Sepulchre R, Janković M, Kokotović PV (1997) Constructive nonlinear control. Communications and control engineering. Springer, London, England
15. Kokotović P, Arcak M (2001) Constructive nonlinear control: a historical perspective. Automatica 37:637–662
16. Christofides PD, El-Farra NH (2005) Control of nonlinear and hybrid process systems: designs for uncertainty, constraints and time-delays. Springer, Berlin, Germany
17. Isidori A (1989) Nonlinear control systems, 2nd edn. Springer, New York
18. Dubljević S, Kazantzis N (2002) A new Lyapunov design approach for nonlinear systems based on Zubov's method. Automatica 38:1999–2007
19. Papachristodoulou A, Prajna S (2002) On the construction of Lyapunov functions using the sum of squares decomposition. In: Proceedings of the 41st IEEE conference on decision and control. Las Vegas, NV, pp 3482–3487
20. Teel AR, Nešić D, Kokotovic PV (1998) A note on input-to-state stability of sampled-data nonlinear systems. In: Proceedings of the 37th IEEE conference on decision and control, vol. 3, pp 2473–2478
21. Muñoz de la Peña D, Christofides PD (2008) Lyapunov-based model predictive control of nonlinear systems subject to data losses. IEEE Trans Autom Control 53:2076–2089
22. Laila DS, Nešić D, Astolfi A (2006) Sampled-data control of nonlinear systems. In: Loría A, Lamnabhi-Lagarrigue F, Panteley E (eds) Advanced Topics in Control Systems Theory, vol 328., Lecture Notes in Control and Information Science. Springer, London, pp 91–137
23. Nešić D, Teel AR (2001) Sampled-data control of nonlinear systems: an overview of recent results. In: Moheimani SOR (ed) Perspectives in robust control, vol 268., Lecture Notes in Control and Information Sciences. Springer, London, pp 221–239
24. Nešić D, Teel AR, Carnevale D (2009) Explicit computation of the sampling period in emulation of controllers for nonlinear sampled-data systems. IEEE Trans Autom Control 54:619–624
25. Ellis M, Karafyllis I, Christofides PD (2014) Stabilization of nonlinear sampled-data systems and economic model predictive control application. In: Proceedings of the American Control conference. Portland, OR, pp 5594–5601
26. Karafyllis I, Grüne L (2011) Feedback stabilization methods for the numerical solution of ordinary differential equations. Discrete Continuous Dyn Syst 16:283–317

27. García CE, Prett DM, Morari M (1989) Model predictive control: theory and practice-A survey. Automatica 25:335–348
28. Morari M, Lee JH (1999) Model predictive control: past, present and future. Comput Chem Eng 23:667–682
29. Mayne DQ, Rawlings JB, Rao CV, Scokaert POM (2000) Constrained model predictive control: stability and optimality. Automatica 36:789–814
30. Rawlings JB (2000) Tutorial overview of model predictive control. IEEE Control Syst Mag 20:38–52
31. Qin SJ, Badgwell TA (2003) A survey of industrial model predictive control technology. Control Eng Pract 11:733–764
32. Camacho EF, Alba CB (2013) Model predictive control, 2nd edn. Springer
33. Mhaskar P, El-Farra NH, Christofides PD (2005) Predictive control of switched nonlinear systems with scheduled mode transitions. IEEE Trans Autom Control 50:1670–1680
34. Mhaskar P, El-Farra NH, Christofides PD (2006) Stabilization of nonlinear systems with state and control constraints using Lyapunov-based predictive control. Syst Control Lett 55:650–659
35. Christofides PD, Liu J, Muñoz de la Peña D (2011) Networked and distributed predictive control: methods and nonlinear process network applications. Advances in Industrial Control Series. Springer, London, England
36. Bertsekas DP (1999) Nonlinear programming, 2nd edn. Athena Scientific, Belmont, MA
37. Biegler LT (2010) Nonlinear programming: concepts, algorithms, and applications to chemical processes. SIAM, Philadelphia, PA
38. Boyd S, Vandenberghe L (2004) Convex optimization. Cambridge University Press, Cambridge, UK
39. Luenberger DG (2003) Linear and nonlinear programming, 2nd edn. Kluwer Academic Publishers, Boston, MA
40. Nocedal J, Wright S (2006) Numerical optimization, 2nd edn. Springer, New York, NY
41. Betts JT (2010) Practical methods for optimal control and estimation using nonlinear programming, 2nd edn. Society for Industrial and Applied Mathematics, Philadelphia, PA
42. Binder T, Blank L, Bock HG, Bulirsch R, Dahmen W, Diehl M, Kronseder T, Marquardt W, Schlöder JP, von Stryk O (2001) Introduction to model based optimization of chemical processes on moving horizons. In: Grötschel M, Krumke SO, Rambau J (eds) Online optimization of large scale systems. Springer, Berlin Heidelberg, pp 295–339
43. Diehl M, Ferreau HJ, Haverbeke N (2009) Efficient numerical methods for nonlinear MPC and moving horizon estimation. In: Magni L, Raimondo DM, Allgöwer F (eds) Nonlinear model predictive control, vol 384., Lecture Notes in Control and Information Sciences. Springer, Berlin Heidelberg, pp 391–417
44. Horst R, Tuy H (1996) Global optimization: deterministic approaches. Springer
45. Floudas CA (2000) Deterministic global optimization: theory, methods and applications. Nonconvex Optimization and its Application, vol 37. Kluwer Academic Publishers (2000)
46. Karush W (1939) Minima of functions of several variables with inequalities as side conditions. Master's thesis, University of Chicago
47. Kuhn HW, Tucker AW (1951) Nonlinear programming. In: Proceedings of the Second Berkeley symposium on mathematical statistics and probability. University of California Press, Berkeley, CA , pp 481–492
48. Wächter A, Biegler LT (2006) On the implementation of an interior-point filter line-search algorithm for large-scale nonlinear programming. Math Program 106:25–57
49. Bertsekas DP (2005) Dynamic programming and optimal control, vol 1, 3rd edn. Athena Scientific, Belmont, MA
50. Bryson AE, Ho YC (1975) Applied optimal control: optimization. Estimation and control. Taylor & Francis Group, New York
51. Kirk DE (2012) Optimal control theory: an introduction. Dover Publications, Mineola, NY
52. Lewis FL, Vrabie DL, Syrmos VL (2012) Optimal control, 3rd edn. Wiley, Hoboken, New Jersey
53. Bellman RE (1957) Dynamic programming. Princeton University Press, Princeton, N.J

54. Pontryagin LS, Boltyanskii VG, Gamkrelidze RV, Mishchenko EF (1961) Mathematical theory of optimal processes. Fizmatgiz, Moscow
55. Kalman RE (1960) Contributions to the theory of optimal control. Boletin de la Sociedad Matematica Mexicana 5:102–119
56. Lee EB, Markus L (1967) Foundations of optimal control theory. SIAM series in applied mathematics. Wiley, New York, NY
57. Vinter R (2010) Optimal control. Modern Birkhäuser Classics. Birkhäuser, Boston, MA
58. Bock HG, Plitt KJ (1984) A multiple shooting algorithm for direct solution of optimal control problems. In: Proceedings of the 9th IFAC world congress. Budapest, Hungary, pp 242–247
59. Biegler LT (2007) An overview of simultaneous strategies for dynamic optimization. Chem Eng Process 46:1043–1053
60. Biegler LT (1984) Solution of dynamic optimization problems by successive quadratic programming and orthogonal collocation. Comput Chem Eng 8:243–247

Chapter 3
Brief Overview of EMPC Methods and Some Preliminary Results

This chapter contains a brief background on EMPC methods. The background on EMPC methods is meant to provide context to the EMPC design methodologies of the subsequent chapters. However, it is not meant to be comprehensive and rigorous. For a more comprehensive and rigorous review, please refer to the reviews [1, 2] as well as [3], which primarily focuses on the role of constraints in various EMPC formulations.

3.1 Background on EMPC Methods

A brief overview of EMPC methods is provided in this section.

3.1.1 Class of Nonlinear Systems

The class of systems considered is described by the system of nonlinear ordinary differential equations (ODEs):

$$\dot{x} = f(x, u, w) \tag{3.1}$$

where $x \in \mathbb{X} \subset \mathbb{R}^n$ denotes the state vector, $u \in \mathbb{U} \subset \mathbb{R}^m$ denotes the manipulated (control) input vector, and $w \in \mathbb{W} \subset \mathbb{R}^l$ denotes the disturbance vector. The set of admissible input values $\mathbb{U}$ is assumed to be compact, and the disturbance vector is restricted to take values in the set $\mathbb{W} := \{w \in \mathbb{R}^l : |w| \leq \theta\}$ where $\theta > 0$ bounds the norm of the disturbance vector. The vector function $f : \mathbb{X} \times \mathbb{U} \times \mathbb{W} \to \mathbb{X}$ is locally Lipschitz on $\mathbb{X} \times \mathbb{U} \times \mathbb{W}$. A state measurement is synchronously sampled at sampling instances denoted by the sequence $\{t_k\}_{k \geq 0}$ where $t_k := k\Delta$, $k \in \mathbb{I}_{\geq 0}$, and $\Delta > 0$ is the sampling period (the initial time is taken to be zero). The assumption of state feedback is standard owing to the fact that the separation principle does not

M. Ellis et al., *Economic Model Predictive Control*,
Advances in Industrial Control, DOI 10.1007/978-3-319-41108-8_3

generally hold for nonlinear systems. Nevertheless, some rigorous output feedback implementations of EMPC exist, e.g., [4–6] and the design of such EMPC systems is also discussed in Chap. 5.

The system of Eq. 3.1 is equipped with a function $l_e : \mathbb{X} \times \mathbb{U} \rightarrow \mathbb{R}$ that is continuous over its domain, which reflects the process/system economics. The function $l_e(\cdot)$ is used as a stage cost in a model predictive control (MPC) framework and is referred to as the economic stage cost.

Additionally, the system of Eq. 3.1 may be subject to constraints other than the input and state constraints. Collecting all the constraints including the input, state, and additional constraints, the constraints may be written generally as:

$$(x, u) \in \mathbb{Z} \tag{3.2}$$

where $\mathbb{Z}$ is assumed to be compact. Economic considerations may motivate the need for dynamic constraints such as the average constraints of Eqs. 1.15 and 1.19 of the two examples of Chap. 1. These constraints may be readily incorporated into the EMPC by augmenting the state dynamic equations.

An optimal (minimizing) steady-state pair (x_s^*, u_s^*) with respect to the economic stage cost is assumed to exist and to be unique. Specifically, the economically optimal steady-state and steady-state input pair is:

$$(x_s^*, u_s^*) = \underset{(x_s, u_s) \in \mathbb{Z}}{\arg\min} \, \{l_e(x_s, u_s) : f(x_s, u_s, 0) = 0\} \, . \tag{3.3}$$

If the minimizing pair is not unique, let (x_s^*, u_s^*) denote one of the minimizing steady-state pairs.

As pointed out in Sect. 2.3.3, MPC schemes are typically implemented by solving an optimization problem at discrete-time steps. Zeroth-order hold is usually used for control vector parameterization. Under these conditions, the resulting closed-loop system under MPC is a nonlinear sampled-data system given by:

$$\dot{x}(t) = f(x(t), \kappa(x(t_k)), w(t)) \tag{3.4}$$

for $t \in [t_k, t_{k+1})$, $t_k = k\Delta$, $k = 0, 1, \ldots$, $\Delta > 0$ is the sampling period, and $\kappa(\cdot)$ is the implicit control law resulting from EMPC. Discrete-time versions of Eq. 3.4 are often considered for designing EMPC methods whereby w is taken to be constant over a sampling period. The discrete-time version of Eq. 3.4 is written with slight abuse of notation as:

$$x(k + 1) = f_d(x(k), u(k), w(k)) \tag{3.5}$$

where k denotes the kth time step, i.e., it represents t_k in real-time, and f_d is assumed to be continuous over its domain. In practice, approximate discrete-time models are often used because the exact discrete-time model of Eq. 3.4 may be difficult to obtain. Also, the nonlinear continuous-time model must be solved using a numerical ODE solver. Nevertheless, this usually works well in practice, and the numerical error associated with the approximate discrete-time model may be made small with the precision of modern computers. The interested reader is referred to, for example, [7] for conditions under which stability is guaranteed when using an approximate discrete-time model in controller design for sampled-data systems.

3.1.2 EMPC Methods

Economic model predictive control is an MPC method that uses the economic stage cost in its formulation. The EMPC problem, with a finite-time prediction horizon, may be broadly characterized by the following optimal control problem (OCP):

$$\min_{u \in S(\Delta)} \int_{t_k}^{t_{k+N}} l_e(\tilde{x}(t), u(t))\, dt + V_f(\tilde{x}(t_{k+N})) \tag{3.6a}$$

$$\text{s.t.} \quad \dot{\tilde{x}}(t) = f(\tilde{x}(t), u(t), 0) \tag{3.6b}$$

$$\tilde{x}(t_k) = x(t_k) \tag{3.6c}$$

$$(\tilde{x}(t), u(t)) \in \mathbb{Z}, \ \forall\, t \in [t_k,\ t_{k+N}) \tag{3.6d}$$

$$\tilde{x}(t_{k+N}) \in \mathbb{X}_f \tag{3.6e}$$

where the decision variable of the optimization problem is the piecewise constant input trajectory over the prediction horizon, i.e., the time interval $[t_k, t_{k+N})$ (zeroth-order hold is assumed), and $\tilde{x}$ denotes the predicted state trajectory over the prediction horizon. With slight abuse of notation, the discrete-time version of the EMPC problem is:

$$\min_{\mathbf{u}} \sum_{j=k}^{N-1} l_e(\tilde{x}(j), u(j)) + V_f(\tilde{x}(j+N)) \tag{3.7a}$$

$$\text{s.t.} \quad \tilde{x}(j+1) = f_d(\tilde{x}(j), u(j), 0) \tag{3.7b}$$

$$\tilde{x}(k) = x(t_k) \tag{3.7c}$$

$$(\tilde{x}(j), u(j)) \in \mathbb{Z}, \ j = k, k+1, \ldots, k+N-1 \tag{3.7d}$$

$$\tilde{x}(k+N) \in \mathbb{X}_f \tag{3.7e}$$

where $\mathbf{u} = \{u(k), u(k+1), \ldots, u(k+N-1)\}$ is the decision variable, $\tilde{x}$ denotes the predicted state sequence, and N is the number of sampling times in the prediction horizon.

The cost functional of Eq. 3.6a (Eq. 3.7a) consists of the economic stage cost with a terminal cost/penalty $V_f : \mathbb{X}_f \to \mathbb{R}$. The nominal dynamic model of Eq. 3.6b (Eq. 3.7b) is used to predict the future evolution of the system and is initialized with a state measurement of Eq. 3.6c (Eq. 3.7c). When available, disturbance estimates or predictions may be incorporated in the model to provide a better prediction of the system evolution. The constraint of Eq. 3.6d (Eq. 3.7d) represents the process/system constraints. Finally, the constraint of Eq. 3.6e (Eq. 3.7e) is a terminal constraint, which enforces that the predicted state at the end of horizon be contained in a terminal set, $\mathbb{X}_f$. The assumptions on the terminal cost and set for EMPC methods that employ these are presented below.

Regarding solving the optimal control problem of EMPC, higher-order control parameterizations may also be considered. As previously mentioned, the model of Eq. 3.6b is solved and the cost functional is evaluated using a numerical integrator. Also, the constraint of Eq. 3.6d is typically imposed at several points along the prediction horizon, e.g., at the sampling times of the horizon. For formulating a solvable optimization problem with modern nonlinear optimization solvers, the functions f (or f_d) and l_e usually need to satisfy additional smoothness assumptions. With these implementation details and additional assumptions, the resulting optimization problem is finite dimensional and a local minima may be computed.

EMPC is typically implemented with a receding horizon implementation to better approximate the infinite-horizon solution and to ensure robustness of the control solution to disturbances and open-loop instabilities. At a sampling time t_k, the EMPC receives a state measurement, which is used to initialize the model of Eq. 3.6b. The OCP of Eq. 3.6 is solved on-line for a (local) optimal piecewise input trajectory, denoted by $u^*(t|t_k)$ for $t \in [t_k, t_{k+N})$ or $u^*(j|k)$ for $j = k, k+1, \ldots, k+N-1$ in discrete-time. The control action computed for the first sampling period of the prediction horizon, $u^*(t_k|t_k)$ ($u^*(k|k)$), is sent to the control actuators to be implemented over the sampling period from t_k to t_{k+1}. At the next sampling time, the OCP of Eq. 3.6 is re-solved after receiving a new measurement and by shifting the prediction horizon into the future by one sampling period.

EMPC, which consists of the on-line solution of the OCP of Eq. 3.6 along with a receding horizon implementation, results in an implicit state feedback law $u(t) = \kappa(x(t_k))$ for $t \in [t_k, t_{k+1})$ (or similarly $u(k) = \kappa(x(k))$ in discrete-time). From a theoretical perspective, three fundamental issues are considered and addressed with respect to EMPC. The first consideration is the feasibility of the optimization problem (both initial and recursive feasibility are considered). Second, if Eq. 3.6 is recursively feasible, it is important to consider the stability properties of the closed-loop system under EMPC. In general, one may not expect that EMPC will force the state to a desired steady-state. The last theoretical consideration is closed-loop economic performance under EMPC.

Within the context of EMPC, closed-loop performance typically means the average closed-loop economic performance. Over a finite-time operating interval of length t_f, the average performance is defined by the following index:

$$\bar{J}_e := \frac{1}{t_f} \int_0^{t_f} l_e(x(t), u(t))\, dt \tag{3.8}$$

where x and u are the closed-loop state and input trajectories, respectively, and over an infinite-time operating interval, the infinite-time (asymptotic) average economic performance is given by:

$$\bar{J}_{e,\infty} := \limsup_{t_f \to \infty} \frac{1}{t_f} \int_0^{t_f} l_e(x(t), u(t))\, dt \; . \tag{3.9}$$

It is straightforward to extend these definitions to discrete-time systems.

For several reasons explored throughout this book such as model nonlinearities, non-convexity of the stage cost, and average constraints, it may be optimal with respect to the economic stage cost to operate the system in a complex or time-varying fashion. That is, the optimal operating strategy may not be steady-state operation. While the instantaneous stage cost under EMPC at any time may be better or worse than the stage cost at the economically optimal steady-state and steady-state input pair, the average economic performance under the time-varying operating policy dictated by EMPC over the length of operation may be better than that achieved by operation at the economically optimal steady-state.

The notion of optimal steady-state operation is made precise. To maintain consistency with the literature, the definitions of optimal steady-state operation and suboptimal off steady-state operation is presented here for discrete-time systems. As in [8], for a bounded signal $v : \mathbb{I}_{\geq 0} \to \mathbb{R}^{n_v}$, the set of asymptotic averages is given by:

$$\mathrm{Av}[v] := \left\{ \bar{v} \in \mathbb{R}^{n_v} : \exists t_n \to \infty : \lim_{n \to \infty} \frac{\sum_{k=0}^{t_n - 1} v(k)}{t_n} = \bar{v} \right\} . \tag{3.10}$$

Definition 3.1 ([8]) The system of Eq. 3.5 with $w \equiv 0$ is optimally operated at steady-state with respect to the stage cost $l_e(x, u)$, if for any solution satisfying $(x(k), u(k)) \in \mathbb{Z}$ for all $k \in \mathbb{I}_{\geq 0}$, the following holds:

$$\mathrm{Av}[l_e(x, u)] \subseteq [l_e(x_s^*, u_s^*), \infty) \tag{3.11}$$

If, in addition, $\mathrm{Av}[l_e(x, u)] \subseteq (l_e(x_s^*, u_s^*), \infty)$ or $\liminf_{k \to \infty} |x(k) - x_s^*| = 0$, then the system of Eq. 3.5 is sub-optimally operated off steady-state.

In some of the examples considered in this book, the system being considered is not optimally operated at steady-state.

Regarding closed-loop stability under EMPC, one may easily construct examples of systems and stage costs of the form described above where the resulting closed-loop system is unstable without additional assumptions and conditions. Theoretical investigations on EMPC that do not incorporate additional stability and terminal constraints exist including the work of [9–11]. These EMPC methods require that the resulting EMPCs have a sufficiently long horizon as well as certain controllability assumptions and turnpike conditions be satisfied to guarantee closed-loop stability and performance properties.

Moreover, even though EMPC optimizes the process/system economics, it does so over a finite-time prediction horizon. Over long periods of operation, no conclusion, in general, may be made on closed-loop performance under EMPC (without additional constraints or conditions). For provable results on feasibility, closed-loop stability, and closed-loop performance under EMPC, typically, additional stability and/or performance constraints are added to the formulation of EMPC. These formulations are discussed in the subsequent sections.

To address closed-loop stability, one may consider employing an infinite-horizon in the EMPC. This may be a more appropriate prediction horizon because many chemical processes are continuously operated over long periods of time (practically infinite time). At least intuitively, the resulting control law will provide some form of closed-loop stability assuming the existence of a solution to the infinite-horizon EMPC as well as the ability to solve for a solution on-line. However, it is usually difficult to solve an OCP with an infinite-horizon. To overcome this problem, two approaches include: (1) approximating the infinite-horizon with a sufficiently long finite-time horizon and (2) dividing the infinite-horizon into a finite-time horizon and estimating the infinite-horizon tail through an auxiliary control law or with modeling-based techniques, e.g., [12–20]. Although some of these EMPC schemes may be computationally tractable, the use of constraints typically enables shorter prediction horizons, which may reduce the on-line computation relative to those that require sufficiently long horizons. Thus, only EMPC systems formulated with constraints to provide guaranteed closed-loop properties are considered in the remainder of this chapter.

3.1.2.1 EMPC with an Equality Terminal Constraint

Much of the recent theoretical work on EMPC investigates the extension of stabilizing elements used in tracking MPC to EMPC such as adding a terminal constraint and/or terminal cost (see, for instance, [21] for more details on the use of terminal constraints and/or a terminal cost within the context of tracking MPC). Numerous EMPC formulations and theoretical developments which include a terminal constraint and/or terminal cost have been proposed and studied, e.g., [1, 8, 13, 14, 22–39]. There are two main types of EMPC with terminal constraints: (1) EMPC with an equality terminal constraint, and (2) EMPC with a terminal region constraint. In this subsection, the former type of EMPC is considered which is an EMPC that is described by the following optimization problem:

$$\min_{u \in S(\Delta)} \int_{t_k}^{t_{k+N}} l_e(\tilde{x}(t), u(t)) \, dt \tag{3.12a}$$

$$\text{s.t. } \dot{\tilde{x}}(t) = f(\tilde{x}(t), u(t), 0) \tag{3.12b}$$

$$\tilde{x}(t_k) = x(t_k) \tag{3.12c}$$

$$(\tilde{x}(t), u(t)) \in \mathbb{Z}, \ \forall \, t \in [t_k, \, t_{k+N}) \tag{3.12d}$$

$$\tilde{x}(t_N) = x_s^* \tag{3.12e}$$

where the constraint of Eq. 3.12e forces that the predicted state trajectory to converge to the optimal steady-state at the end of the finite-time horizon. In this formulation,

$\mathbb{X}_f = \{x_s^*\}$ and $V_f \equiv 0$. The formulation of the EMPC in discrete-time is analogous to that of Eq. 3.7 with $V_f \equiv 0$ and $\mathbb{X}_f = \{x_s^*\}$. For EMPC with an equality terminal constraint, the terminal cost is often omitted as it is not required for stability and performance guarantees.

Feasibility. EMPC with a terminal equality constraint is (initially) feasible for any initial state in $\mathbb{X}_N \in \mathbb{R}^n$ which denotes the feasible region of EMPC of Eq. 3.12. The feasible region is also the domain of attraction of the closed-loop system, and it depends on the prediction horizon. An explicit characterization of $\mathbb{X}_N$ is difficult in general. Recursive feasibility, i.e., feasibility at each subsequent sampling time, of EMPC with an equality terminal constraint is guaranteed for the nominally operated system ($w \equiv 0$) for any initial state $x(0) \in \mathbb{X}_N$. This follows from the fact that a feasible solution to the EMPC may be constructed from the solution obtained at the previous sampling time. Namely, $u(t) = u^*(t|t_{k-1})$ for $t \in [t_k, t_{k+N-1})$ and $u(t) = u_s^*$ for $t \in [t_{k+N-1}, t_{k+N})$ is a feasible solution for the EMPC at t_k because it satisfies the constraints and the terminal constraint of Eq. 3.12e. However, recursive feasibility is harder to show, in general, when $w \not\equiv 0$.

Closed-Loop Stability. With respect to closed-loop stability, a weak notion of stability follows from the EMPC with terminal constraint formulation. If the initial state is in the feasible region, the closed-loop state trajectory remains contained in the feasible region under nominal operation. For stronger stability properties, e.g., asymptotic stability of x_s^*, additional assumptions on the closed-loop system must be satisfied. To discuss this issue, nonlinear discrete-time systems are considered. Discrete-time systems are considered here to maintain consistency with the literature on the topic. Nonetheless, some of these conditions and results have been extended to continuous-time systems, e.g., [40]. One condition that leads to stronger stability properties is the notion of dissipativity which has been extended to EMPC. Dissipativity was originally presented in [41] for continuous-time systems and then, extended to discrete-time systems [42]. It is worth pointing out that the notion of the available storage with respect to a supply rate function is a useful tool with which the dissipativity of a system can be assessed [41, 42].

Definition 3.2 ([8]) The system of Eq. 3.5 is *strictly dissipative* with respect to a supply rate $s : \mathbb{X} \times \mathbb{U} \to \mathbb{R}$ if there exist a storage function $\lambda : \mathbb{X} \to \mathbb{R}$, which is assumed to be continuous over its domain, and a positive definite function $\beta : \mathbb{X} \to \mathbb{R}_{\geq 0}$ such that

$$\lambda(f_d(x, u, 0)) - \lambda(x) \leq -\beta(x) + s(x, u) \tag{3.13}$$

for all $(x, u) \in \mathbb{Z}$.

If the system of Eq. 3.5 is strictly dissipative with a supply rate:

$$s(x, u) = l_e(x, u) - l_e(x_s^*, u_s^*) \tag{3.14}$$

then, the optimal steady-state is asymptotically stable for the closed-loop system under EMPC with an equality terminal constraint [8]. Moreover, a Lyapunov function for the closed-loop system was derived using the cost functional of the so-called rotated cost function [8]:

$$L(x, u) := l_e(x, u) + \lambda(x) - \lambda(f_d(x, u, 0)) . \tag{3.15}$$

The idea of using the rotated cost function to construct a Lyapunov function for the closed-loop system was originally proposed in [14]. However, it relied on strong duality of the steady-state optimization problem, which is a stronger assumption than strict dissipativity.

Closed-loop Performance. Utilizing the optimal input trajectory at t_k (or time step k in discrete-time) as a feasible solution to the EMPC at the next sampling period, one may upper bound the difference between the cost functional value at the next sampling time and at the current sampling time under nominal operation. The optimal input trajectory in discrete-time is denoted $u^*(j|k)$ for $j = k,\ k+1,\ \ldots,\ k+N-1$, and the optimal cost functional value at time step k is denoted:

$$L_e^*(x(k)) = \sum_{j=k}^{k+N-1} l_e(x^*(j|k), u^*(j|k)) , \tag{3.16}$$

where $u^*(\cdot|k)$ is the optimal input sequence (trajectory), $x^*(\cdot|k)$ is the corresponding state sequence starting at $x(k)$, and $x(k)$ denotes the closed-loop state at time step k. Using the bound on the difference between the two consecutive cost functional values, the closed-loop average economic performance may be bounded:

$$\frac{1}{T}\sum_{k=0}^{T-1} l_e(x(k), u^*(k|k)) \le l_e(x_s^*, u_s^*) + \frac{L_e^*(x(0)) - L_e^*(x(T))}{T} \tag{3.17}$$

where $x(k)$ is the closed-loop state at time step k, and $T \in \mathbb{i}_{\ge 0}$ is the length of operation. From Eq. 3.17, the effect of the second term of the right-hand side dissipates with longer (but finite) operation. For infinite-time, the average economic performance is bounded by:

$$\limsup_{T\to\infty} \frac{1}{T}\sum_{k=0}^{T-1} l_e(x(k), u^*(k|k)) \le l_e(x_s^*, u_s^*) , \tag{3.18}$$

that is, the asymptotic average performance is no worse than that at the steady-state pair (x_s^*, u_s^*) [8].

The EMPC of Eq. 3.12 requires that the predicted state converge to steady-state at the end of the prediction horizon. This may restrict the feasible region. An interesting extension to the terminal equality constraint employing a generalized (equality) terminal constraint with an appropriately designed terminal cost [31, 34]. The generalized terminal constraint replaces the constraint of Eq. 3.12e with the following constraint:

$$f(\tilde{x}(t_{k+N}), u(t_{k+N-1})) = 0 \tag{3.19}$$

which allows for the predicted terminal state be forced to any admissible steady-state. The resulting EMPC equipped with an appropriately designed terminal cost shares many of the same properties, but clearly, a larger feasible region than the EMPC of Eq. 3.12 [31, 34].

If the dynamic constraints are imposed on the system taking the form of average constraints, [8, 37] provide methodologies for EMPC with an equality terminal constraint to ensure that the average constraint is satisfied asymptotically and over finite-time operating horizons, respectively. Also, the use of a periodic terminal equality constraint has been considered in [8, 39].

3.1.2.2 EMPC with a Terminal Region Constraint

As previously pointed out, EMPC of Eq. 3.12 requires that the initial state be sufficiently close to the steady-state such that it is possible to reach the steady-state in N sampling times. This type of constraint may limit the feasible region [25]. Numerically computing a solution that satisfies such a constraint exactly may also be challenging. Therefore, terminal region constraints may be employed in EMPC.

One such method is a terminal region constraint designed via an auxiliary local control law. The terminal region is designed to be a forward invariant set for the nonlinear system under the local control law. The local control law can, for instance, be designed on the basis of the linearization of the system around the optimal steady-state. The terminal region is denoted as $\mathbb{X}_f$ and the resulting EMPC is given by the following problem:

$$\min_{u \in S(\Delta)} \int_{t_k}^{t_{k+N}} l_e(\tilde{x}(t), u(t))\, dt + V_f(\tilde{x}(t_{k+N})) \tag{3.20a}$$

$$\text{s.t.}\ \dot{\tilde{x}}(t) = f(\tilde{x}(t), u(t), 0) \tag{3.20b}$$

$$\tilde{x}(t_k) = x(t_k) \tag{3.20c}$$

$$(\tilde{x}(t), u(t)) \in \mathbb{Z},\ \forall\, t \in [t_k,\ t_{k+N}) \tag{3.20d}$$

$$\tilde{x}(t_N) \in \mathbb{X}_f \tag{3.20e}$$

where Eq. 3.20e is the terminal region constraint. The discrete-time version of the problem takes the same form as Eq. 3.7. In general, for closed-loop stability and performance, the terminal cost is such that $V_f \not\equiv 0$.

In [25], a procedure to design a local control law, a terminal region constraint, and a terminal cost for EMPC satisfying the assumption below was proposed:

Assumption 3.1 There exist a compact terminal region $\mathbb{X}_f \subset \mathbb{R}^n$, containing the point x_s^* in its interior, and control law $h_L : \mathbb{X}_f \rightarrow \mathbb{U}$, such that for the discrete-time system of Eq. 3.5:

$$V_f(f_d(x, h_L(x), 0)) \leq V_f(x) - l_e(x, h_L(x)) + l_e(x_s^*, u_s^*) \tag{3.21}$$

for all $x \in \mathbb{X}_f$.

Feasibility. For nominal operation, if the EMPC with a terminal region is initially feasible, the EMPC will be recursively feasible. This may be shown by using similar recursive arguments as those used in showing the feasibility of the EMPC with the equality terminal constraint. If $u^*(t|t_{k-1})$ for $t \in [t_{k-1}, t_{k+N-1})$ is the optimal input trajectory at t_{k-1}, then at t_k, a feasible solution is $u(t) = u^*(t|t_{k-1})$ for $t \in [t_k, t_{k+N-1})$ and $u(t) = h_L(\tilde{x}(t_{k+N-1}))$ for $t \in [t_{k+N-1}, t_{k+N})$ where $\tilde{x}(t_{k+N-1})$ is the predicted state at t_{k+N-1}. For recursive feasibility when $w \not\equiv 0$, one EMPC methodology designed with a terminal region constraint was presented in [43].

Closed-Loop Stability. The closed-loop stability properties of EMPC with a terminal constraint designed to satisfy Assumption 3.1 is similar to those of EMPC with an equality terminal constraint. For nominal operation, the closed-loop state trajectory will stay in the feasible region. If the system of Eq. 3.5 is strictly dissipative with supply rate of Eq. 3.14, the steady-state is asymptotically stable under EMPC with a terminal region constraint [25]; see, also, [40] which extends these results to continuous-time systems.

Closed-loop Performance. If the local control law, terminal cost, and terminal region are designed such that Assumption 3.1 is satisfied, the bound on asymptotic average performance of Eq. 3.18 holds [25]. For finite-time, a similar bound to the bound of Eq. 3.17 may be derived for the closed-loop system under EMPC with a terminal cost and a terminal region constraint.

The notion of a generalized terminal constraint has also been extended to terminal region constraints for EMPC in [44].

3.1.2.3 EMPC Designed with Lyapunov-Based Techniques

The feasible region of EMPC with a terminal region constraint, while larger than the feasible region of EMPC with an equality terminal constraint, depends on the prediction horizon length. Moreover, the feasible region of both EMPC formulations is difficult to characterize. As an alternative to overcome these challenges, one may consider designing an explicit nonlinear control law for the nonlinear system of Eq. 3.1 and constructing a Lyapunov function for the resulting closed-loop system consisting of the system of Eq. 3.1 under the explicit control law. With the control law and Lyapunov function, a region constraint may be designed to be imposed within EMPC. Because the control law and Lyapunov function are derived for the nonlinear system of Eq. 3.1, they can be used to provide an estimate of the region of attraction of the nonlinear system. The resulting EMPC is the so-called Lyapunov-based EMPC (LEMPC) [4, 45–50]. LEMPC is discussed in-depth in Chap. 4. However, it is important to point out that LEMPC is a dual-mode control strategy. Under the first mode of operation, the LEMPC may dictate a time-varying operating policy to optimize the economics within the region derived using the nonlinear control law. If steady-state operation is desired, the second mode of operation, defined by a contractive

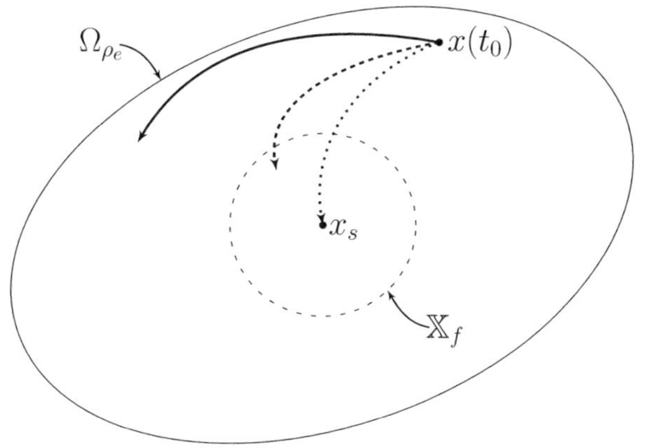

Fig. 3.1 An illustration of possible open-loop predicted trajectories under EMPC formulated with a terminal constraint (*dotted*), under EMPC formulated with a terminal region constraint (*dashed*), and under LEMPC (*solid*)

constraint, is used to ensure that the closed-loop state trajectory converges to a small neighborhood of the steady-state. In contrast to the aforementioned EMPC methods, no dissipativity requirement is needed to accomplish steady-state operation.

3.1.2.4 Comparison of the Open-Loop Predicted State Trajectory

The EMPC formulations of Eqs. 3.12 and 3.20, and the LEMPC (described in detail in the next chapter) may result in different open-loop predicted state trajectories which are illustrated in Fig. 3.1 (Ω_{ρ_e} denotes the region constraint in LEMPC). Nonetheless, if the prediction horizon is sufficiently long, the closed-loop behavior of the system under the various EMPC formulations would (intuitively) be expected to be similar because for a long prediction horizon, the EMPC solution starts to closely approximate the infinite horizon solution and the effect on the closed-loop behavior of the terminal conditions of the open-loop predicted trajectory is less significant than the corresponding effect for shorter prediction horizons.

3.2 Application of EMPC to a Chemical Process Example

To motivate the use of EMPC over conventional control methods that enforce steady-state operation as well as periodic operation, EMPC is applied to the benchmark example of Sect. 1.3.1. The reactor has an asymptotically stable steady-state:

$$x_s^T = [0.998\ 0.424\ 0.032\ 1.002] \tag{3.22}$$

which corresponds to the steady-state input:

$$u_{s,1} = 0.35,\ u_{s,2} = 0.5 \tag{3.23}$$

where throughout this study the coolant temperature is fixed to its nominal value of $u_{s,3} = 1.0$. The control objective considered here is to optimize the time-averaged yield of ethylene oxide by operating the reactor in a time-varying fashion around the stable steady-state. Owing to practical considerations, the average amount of ethylene that may be fed into the reactor over the length of operation is constrained to be equal to that when uniformly distributing the reactant material to the reactor. Mathematically, this constraint is given by the following integral constraint:

$$\frac{1}{t_f} \int_0^{t_f} u_1(t) u_2(t)\, dt = u_{s,1} u_{s,2} = 0.175 \tag{3.24}$$

where $u_{s,1}$ and $u_{s,2}$ are the steady-state inlet volumetric flow rate and ethylene concentration, respectively. Since the average ethylene fed to the reactor is fixed, which fixes the denominator of the yield, the economic stage cost used in the formulation of the EMPC is given by:

$$l_e(x, u) = -x_3 x_4 u_1 \, . \tag{3.25}$$

For the periodic operating policy, a similar periodic operating strategy as that proposed in [51] which varies the inlet feed flow rate and feed concentration in an open-loop periodic fashion as shown in Fig. 3.2. The parameters used for the periodic control strategy are $\tau = 46.8$, $a_1 = 0.073$, $a_2 = 0.500$, $a_3 = 0.514$, and $a_4 = 0.941$, which are similar parameters to the ones used in [51]. It is important to note that the periodic control strategy of Fig. 3.2 with the aforementioned parameters satisfies the integral constraint of Eq. 3.24.

To compare steady-state operation and periodic operating strategy with the operating policy achieved under EMPC, an EMPC is designed for the reactor system with a sampling period of $\Delta = 0.1$. To enforce that the integral constraint of Eq. 3.24 be

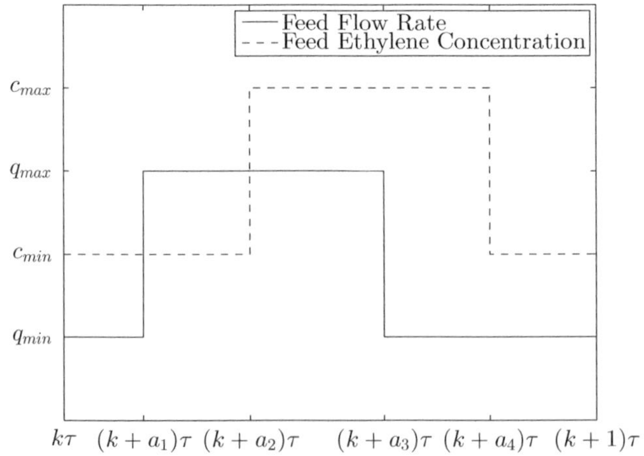

Fig. 3.2 Design of the open-loop periodic operation strategy over one period τ

satisfied over each operating window of length $\tau = 46.8$, the EMPC is formulated with a shrinking horizon, i.e., at $t = 0$, the horizon is set to $N_0 = 468$. At the next sampling time ($t = 0.1$), the horizon is decreased by one ($N_1 = 467$). At subsequent sampling times, the prediction horizon is similarly decreased by one sampling period. At $t = 46.8$, the horizon is reset to $N_{468} = 468$. For simplicity of notation, let j be the number of operating windows of length $\tau = 46.8$ that have elapsed and the EMPC considered in this example is given by the following formulation:

$$\min_{u \in S(\Delta)} \; -\int_{t_k}^{(j+1)\tau} \tilde{x}_3(t)\tilde{x}_4(t)u_1(t)\,dt \tag{3.26a}$$

$$\text{s.t. } \dot{\tilde{x}}(t) = f(\tilde{x}(t), u(t), 0) \tag{3.26b}$$

$$\tilde{x}(t_k) = x(t_k) \tag{3.26c}$$

$$u_1(t) \in [0.0704, 0.7042], \; \forall\, [t_k, (j+1)\tau) \tag{3.26d}$$

$$u_2(t) \in [0.2465, 2.4648], \; \forall\, [t_k, (j+1)\tau) \tag{3.26e}$$

$$\frac{1}{\tau}\int_{t_k}^{(j+1)\tau} u_1(t)u_2(t)\,dt = 0.175 - \frac{1}{\tau}\int_{j\tau}^{t_k} u_1^*(t)u_2^*(t)\,dt \tag{3.26f}$$

where $u_1^*(t)$ and $u_2^*(t)$ denotes the inputs applied to the system over the time $j\tau$ to t_k and $(j+1)\tau$ denotes the end of the operating window.

The catalytic reactor system is initialized at

$$x_0^T = [0.997 \; 1.264 \; 0.209 \; 1.004] \tag{3.27}$$

which corresponds to an initial state on the stable limit cycle that the process with the periodic strategy follows. Simulations are carried out with the periodic operating strategy and the EMPC of Eq. 3.26 over 10 operating windows. The evolution of the reactor for both cases is given in Fig. 3.3 with the open-loop periodic operation and

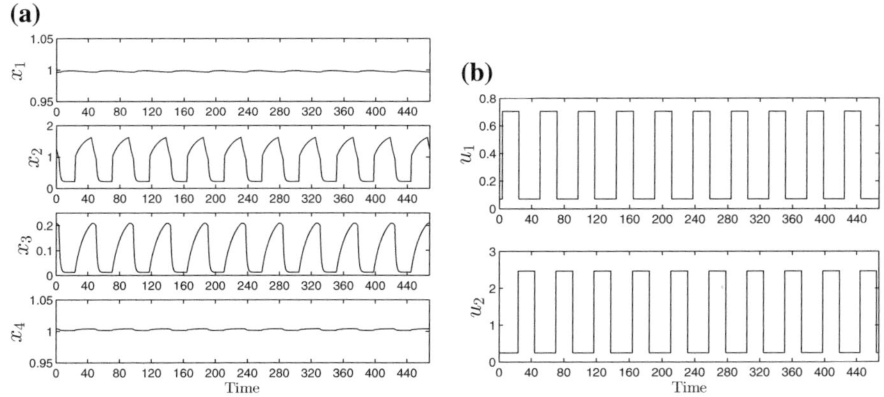

Fig. 3.3 The open-loop reactor **a** state trajectories and **b** input trajectories with the periodic operating strategy resulting from open-loop manipulation of the inputs as shown in Fig. 3.2

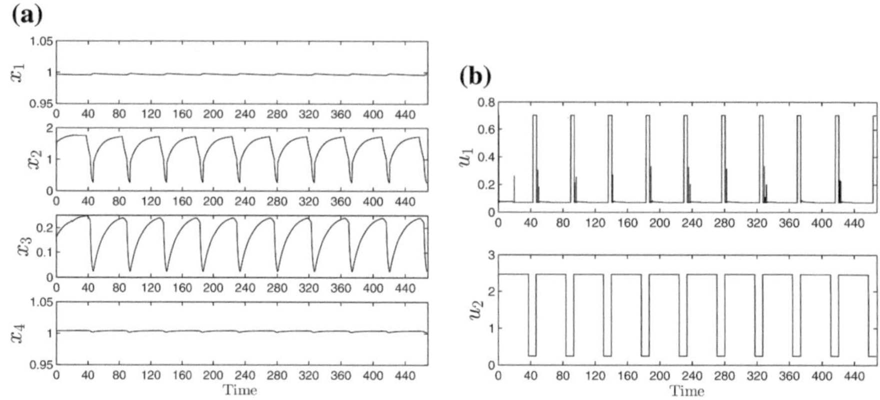

Fig. 3.4 The closed-loop reactor **a** state trajectories and **b** input trajectories with EMPC of Eq. 3.26

Fig. 3.5 State-space evolution in the $x_2 - x_3$ phase plane of the reactor system under the EMPC of Eq. 3.26 and with the periodic operating strategy resulting from open-loop manipulation of the inputs as shown in Fig. 3.2

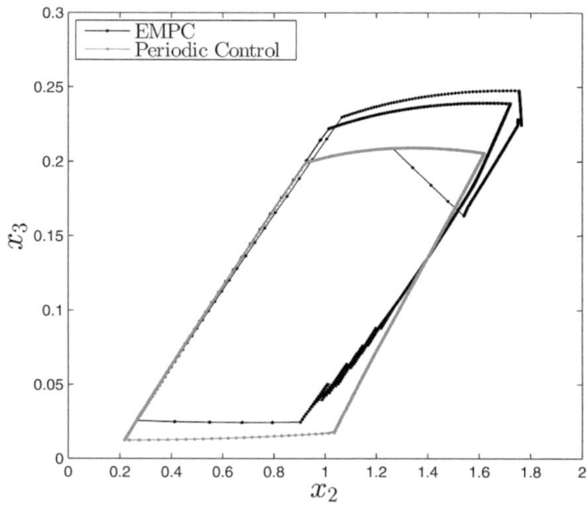

Fig. 3.4 under EMPC. The state-space evolution of the two strategies are shown in the $x_2 - x_3$ phase plane in Fig. 3.5. From these figures, the system with the two operating strategies approaches different periodic trajectories. Under EMPC, the time-averaged yield over the entire time interval of the simulation is 9.97 % compared to 7.93 % with the periodic operation. If the reactor is initialized with the same initial point and the material is instead distributed uniformly over the length of operation, the average yield is 6.63 %. On the other hand, initializing the system at the steady-state and maintain the system at steady-state thereafter achieves a yield of 6.41 %. Therefore, operation under EMPC has a clear performance benefit relative to steady-state operation and the open-loop periodic operating strategy.

References

1. Rawlings JB, Angeli D, Bates CN (2012) Fundamentals of economic model predictive control. In: Proceedings of the 51st IEEE conference on decision and control, Maui, Hawaii, pp 3851–3861
2. Ellis M, Durand H, Christofides PD (2014) A tutorial review of economic model predictive control methods. J Process Control 24:1156–1178
3. Ellis M, Durand H, Christofides PD (2016) Elucidation of the role of constraints in economic model predictive control. Annu Rev Control 41:208–217
4. Heidarinejad M, Liu J, Christofides PD (2012) State-estimation-based economic model predictive control of nonlinear systems. Syst Control Lett 61:926–935
5. Ellis M, Zhang J, Liu J, Christofides PD (2014) Robust moving horizon estimation based output feedback economic model predictive control. Syst Control Lett 68:101–109
6. Zhang J, Liu S, Liu J (2014) Economic model predictive control with triggered evaluations: state and output feedback. J Process Control 24:1197–1206
7. Nešić D, Teel AR, Kokotović PV (1999) Sufficient conditions for stabilization of sampled-data nonlinear systems via discrete-time approximations. Syst Control Lett 38:259–270
8. Angeli D, Amrit R, Rawlings JB (2012) On average performance and stability of economic model predictive control. IEEE Tran Autom Control 57:1615–1626
9. Grüne L (2013) Economic receding horizon control without terminal constraints. Automatica 49:725–734
10. Grüne L, Stieler M (2014) Asymptotic stability and transient optimality of economic MPC without terminal conditions. J Process Control 24:1187–1196
11. Faulwasser T, Korda M, Jones CN, Bonvin D (2014) Turnpike and dissipativity properties in dynamic real-time optimization and economic MPC. In: Proceedings of the 53rd IEEE annual conference on decision and control, Los Angeles, CA, pp 2734–2739
12. Würth L, Rawlings JB, Marquardt W (2007) Economic dynamic real-time optimization and nonlinear model-predictive control on the infinite horizon. In: Proceedings of the 7th IFAC international symposium on advanced control of chemical processes, Istanbul, Turkey, pp 219–224
13. Huang R, Harinath E, Biegler LT (2011) Lyapunov stability of economically oriented NMPC for cyclic processes. J Process Control 21:501–509
14. Diehl M, Amrit R, Rawlings JB (2011) A Lyapunov function for economic optimizing model predictive control. IEEE Trans Autom Control 56:703–707
15. Huang R, Biegler LT, Harinath E (2012) Robust stability of economically oriented infinite horizon NMPC that include cyclic processes. J Process Control 22:51–59
16. Mendoza-Serrano DI, Chmielewski DJ (2012) HVAC control using infinite-horizon economic MPC. In: Proceedings of the 51st IEEE conference on decision and control. Maui, Hawaii, pp 6963–6968
17. Mendoza-Serrano DI, Chmielewski DJ (2013) Demand response for chemical manufacturing using economic MPC. In: Proceedings of the American control conference, Washington, D.C., pp 6655–6660
18. Omell BP, Chmielewski DJ (2013) IGCC power plant dispatch using infinite-horizon economic model predictive control. Ind Eng Chem Res 52:3151–3164
19. Würth L, Wolf IJ, Marquardt W (2013) On the numerical solution of discounted economic NMPC on infinite horizons. In: Proceedings of the 10th IFAC international symposium on dynamics and control of process systems, Bombay, Mumbai, India, pp 209–214
20. Würth L, Marquardt W (2014) Infinite-horizon continuous-time NMPC via time transformation. IEEE Trans Autom Control 59:2543–2548
21. Mayne DQ, Rawlings JB, Rao CV, Scokaert POM (2000) Constrained model predictive control: stability and optimality. Automatica 36:789–814
22. Rawlings JB, Bonné D, Jørgensen JB, Venkat AN, Jørgensen SB (2008) Unreachable setpoints in model predictive control. IEEE Trans Autom Control 53:2209–2215

23. Rawlings JB, Amrit R (2009) Optimizing process economic performance using model predictive control. In: Magni L, Raimondo DM, Allgöwer F (eds) Nonlinear model predictive control, vol 384, Lecture notes in control and information sciences. Springer, Berlin, pp 119–138
24. Ferramosca A, Rawlings JB, Limon D, Camacho EF (2010) Economic MPC for a changing economic criterion. In: Proceedings of the 49th IEEE conference on decision and control, Atlanta, GA, pp 6131–6136
25. Amrit R, Rawlings JB, Angeli D (2011) Economic optimization using model predictive control with a terminal cost. Annu Rev Control 35:178–186
26. Lee J, Angeli D (2011) Cooperative distributed model predictive control for linear plants subject to convex economic objectives. In: Proceedings of the 50th IEEE conference on decision and control and European control conference, Orlando, FL, pp 3434–3439
27. Müller MA, Allgöwer F (2012) Robustness of steady-state optimality in economic model predictive control. In: Proceedings of the 51st IEEE conference on decision and control, Maui, Hawaii, pp 1011–1016
28. Driessen PAA, Hermans RM, van den Bosch PPJ (2012) Distributed economic model predictive control of networks in competitive environments. In: Proceedings of the 51st IEEE conference on decision and control, Maui, HI, pp 266–271
29. Lee J, Angeli D (2012) Distributed cooperative nonlinear economic MPC. In: Proceedings of the 20th international symposium on mathematical theory of networks and systems, Melbourne, Australia
30. Amrit R, Rawlings JB, Biegler LT (2013) Optimizing process economics online using model predictive control. Comput Chem Eng 58:334–343
31. Fagiano L, Teel AR (2013) Generalized terminal state constraint for model predictive control. Automatica 49:2622–2631
32. Gopalakrishnan A, Biegler LT (2013) Economic nonlinear model predictive control for periodic optimal operation of gas pipeline networks. Comput Chem Eng 52:90–99
33. Hovgaard TG, Boyd S, Larsen LFS, Jørgensen JB (2013) Nonconvex model predictive control for commercial refrigeration. Int J Control 86:1349–1366
34. Müller MA, Angeli D, Allgöwer F (2013) Economic model predictive control with self-tuning terminal cost. Eur J Control 19:408–416
35. Müller MA, Angeli D, Allgöwer F (2013) On convergence of averagely constrained economic MPC and necessity of dissipativity for optimal steady-state operation. In: Proceedings of the American control conference, Washington, D.C., pp 3147–3152
36. Baldea M, Touretzky CR (2013) Nonlinear model predictive control of energy-integrated process systems. Syst Control Lett 62:723–731
37. Müller MA, Angeli D, Allgöwer F (2014) Transient average constraints in economic model predictive control. Automatica 50:2943–2950
38. Wolf IJ, Muñoz DA, Marquardt W (2014) Consistent hierarchical economic NMPC for a class of hybrid systems using neighboring-extremal updates. J Process Control 24:389–398
39. Zanon M, Gros S, Diehl M (2013) A Lyapunov function for periodic economic optimizing model predictive control. In: Proceedings of the 52nd IEEE conference on decision and control, Florence, Italy, pp 5107–5112
40. Alessandretti A, Aguiar A, Jones C (2014) An economic model predictive control scheme with terminal penalty for continuous-time systems. In: Proceedings of the 53rd IEEE conference on decision and control, Los Angeles, CA, pp 2728–2733
41. Willems JC (1972) Dissipative dynamical systems part I: general theory. Arch Ration Mech Anal 45:321–351
42. Byrnes CI, Lin W (1994) Losslessness, feedback equivalence, and the global stabilization of discrete-time nonlinear systems. IEEE Trans Autom Control 39:83–98
43. Bayer FA, Müller MA, Allgöwer F (2014) Tube-based robust economic model predictive control. J Process Control 24:1237–1246
44. Müller MA, Angeli D, Allgöwer F (2014) On the performance of economic model predictive control with self-tuning terminal cost. J Process Control 24:1179–1186

45. Heidarinejad M, Liu J, Christofides PD (2012) Economic model predictive control of nonlinear process systems using Lyapunov techniques. AIChE J 58:855–870
46. Chen X, Heidarinejad M, Liu J, Christofides PD (2012) Distributed economic MPC: application to a nonlinear chemical process network. J Process Control 22:689–699
47. Heidarinejad M, Liu J, Christofides PD (2013) Algorithms for improved fixed-time performance of Lyapunov-based economic model predictive control of nonlinear systems. J Process Control 23:404–414
48. Heidarinejad M, Liu J, Christofides PD (2013) Economic model predictive control of switched nonlinear systems. Syst Control Lett 62:77–84
49. Ellis M, Heidarinejad M, Christofides PD (2013) Economic model predictive control of nonlinear singularly perturbed systems. J Process Control 23:743–754
50. Ellis M, Christofides PD (2014) Economic model predictive control with time-varying objective function for nonlinear process systems. AIChE J 60:507–519
51. Özgülşen F, Adomaitis RA, Çinar A (1992) A numerical method for determining optimal parameter values in forced periodic operation. Chem Eng Sci 47:605–613

Chapter 4
Lyapunov-Based EMPC: Closed-Loop Stability, Robustness, and Performance

4.1 Introduction

Within chemical process industries, many chemical processes are safety critical, and maintaining safe and stable operation is the highest priority of a control system. Given that EMPC may operate a process/system in a consistently dynamic fashion to optimize the economics, maintaining the closed-loop state trajectory in a well-defined state-space region, where a degree of robustness to uncertainty is achieved, is one method to achieve safe and stable operation under EMPC. This objective is the main motivating factor in designing Lyapunov-based EMPC (LEMPC). LEMPC is a dual-mode control strategy that allows for time-varying operation while maintaining the closed-loop state in a compact state-space set. If it is desirable to force the closed-loop state to a steady-state at any point over the length of operation, the second mode of operation of the LEMPC may be used and will steer the closed-loop state to a small neighborhood of the steady-state.

In this chapter, several LEMPC designs are developed. The LEMPC designs, which are capable of optimizing closed-loop performance with respect to general economic considerations for nonlinear systems, address recursive feasibility of the optimization problem at each sampling time, closed-loop stability, and closed-loop performance. The fundamental design concept employed in the LEMPC designs is based on uniting receding horizon control with explicit Lyapunov-based nonlinear controller design techniques. These techniques allow for an explicit characterization of the stability region of the closed-loop system. Other considerations including asynchronous and delayed sampling and time-varying economic stage cost functions are also addressed in this chapter.

In all cases considered, sufficient conditions are derived such that the closed-loop nonlinear system under the LEMPC designs possess a specific form of closed-loop stability and robustness to be made precise in what follows. A critical property of the sufficient conditions derived for closed-loop stability is that they do not rely on solving the LEMPC problem to optimality at each sampling time, i.e., suboptimal solutions also stabilize the closed-loop system. In other words, feasibility of the

M. Ellis et al., *Economic Model Predictive Control*,
Advances in Industrial Control, DOI 10.1007/978-3-319-41108-8_4

solution returned by the LEMPC and not optimality implies closed-loop stability under LEMPC. This is a property initially investigated within the context of tracking MPC [1]. Owing to the LEMPC design methodology, a feasible solution to the LEMPC may always be readily computed. Terminal constraint design for LEMPC is also addressed. The terminal constraint imposed in the LEMPC problem allows for guaranteed finite-time and infinite-time closed-loop economic performance improvement over a stabilizing controller. The LEMPC methodologies are applied to chemical process examples to demonstrate, evaluate, and analyze the closed-loop properties of the systems controlled by LEMPC. Also, the closed-loop properties are compared to traditional/conventional approaches to optimization and control, i.e., steady-state optimization and tracking MPC.

4.2 Lyapunov-Based EMPC Design and Implementation

4.2.1 Class of Nonlinear Systems

The class of nonlinear systems considered is described by the following state-space model:

$$\dot{x} = f(x, u, w) \tag{4.1}$$

where $x \in \mathbb{X} \subset \mathbb{R}^n$ denotes the state vector, $u \in \mathbb{U} \subset \mathbb{R}^m$ denotes the control (manipulated) input vector, and $w \in \mathbb{W} \subset \mathbb{R}^l$ denotes the disturbance vector. The control inputs are restricted to a nonempty compact set $\mathbb{U}$. The disturbance is bounded, i.e., $\mathbb{W} := \{w \in \mathbb{R}^l : |w| \leq \theta\}$ where $\theta > 0$ bounds the norm of the disturbance vector. The vector field f is assumed to be a locally Lipschitz vector function on $\mathbb{X} \times \mathbb{U} \times \mathbb{W}$. Without loss of generality, the origin is an equilibrium point of the unforced nominal system, i.e., $f(0, 0, 0) = 0$, and the initial time is taken to be zero, i.e., $t_0 = 0$. State measurements of the system are assumed to be available synchronously at sampling times denoted by the time sequence $\{t_k\}_{k\geq 0}$ where $t_k = k\Delta, k = 0, 1, \ldots$ and $\Delta > 0$ is the sampling period. Robustness of the controlled system under LEMPC with respect to asynchronous and delayed state measurements will be considered in Sect. 4.3.2. To describe the system economics, e.g., operating profit or operating cost, the system of Eq. 4.1 is equipped with a time-invariant cost function $l_e : \mathbb{X} \times \mathbb{U} \to \mathbb{R}$ which is a measure of the instantaneous system economics. The function l_e is referred to as the economic cost function and is continuous on $\mathbb{X} \times \mathbb{U}$.

4.2.2 Stabilizability Assumption

The existence of an explicit controller $h : \mathbb{X} \to \mathbb{U}$, which renders the origin of the nominal closed-loop system asymptotically stable, is assumed. This assumption is a stabilizability assumption for the nonlinear system of Eq. 4.1 and is similar to the assumption that the pair (A, B) is stabilizable in the case of linear systems. Throughout the monograph, the explicit controller may be referred to as the stabilizing controller or the Lyapunov-based controller. When convenient, the notation $h(x)$ may be used when referring to the explicit controller. However, this notation refers to the controller itself, which is a mapping from $\mathbb{X}$ to $\mathbb{U}$. Even though the explicit controller is referred to as the Lyapunov-based controller, it may be designed using any controller design techniques and not just Lyapunov-based techniques.

Applying converse Lyapunov theorems, e.g., [2–7], the stabilizability assumption implies that there exists a continuously differentiable Lyapunov function $V : D \to \mathbb{R}$ for the nominal closed-loop system, i.e., $\dot{x} = f(x, h(x), 0)$, that satisfies the inequalities:

$$\alpha_1(|x|) \leq V(x) \leq \alpha_2(|x|) \tag{4.2a}$$

$$\frac{\partial V(x)}{\partial x} f(x, h(x), 0) \leq -\alpha_3(|x|) \tag{4.2b}$$

$$\left| \frac{\partial V(x)}{\partial x} \right| \leq \alpha_4(|x|) \tag{4.2c}$$

for all $x \in D \subset \mathbb{R}^n$ where $\alpha_i \in \mathcal{K}$ for $i = 1, 2, 3, 4$ and D is an open neighborhood of the origin. The region $\Omega_\rho \subset D$ such that also $\Omega_\rho \subseteq \mathbb{X}$ is called the stability region of the closed-loop system under the Lyapunov-based controller, and is an estimate of the region of attraction of the nonlinear system of Eq. 4.1. Since the stability region depends on the explicit controller, the choice and design of the controller plays a significant role in the estimated region of attraction. The case that the set $\mathbb{X}$ represents explicit hard state constraints is discussed further in Sect. 4.2.5.

4.2.3 LEMPC Formulation

In the LEMPC design, the LEMPC optimizes the economic cost function, which is used as the stage cost in the EMPC. Lyapunov-based MPC techniques, e.g., [8–10], are employed in the EMPC design to take advantage of the stability properties of the Lyapunov-based controller. The LEMPC is equipped with two operation modes. Under the first operation mode, the LEMPC optimizes the economic cost function while maintaining the system state within the stability region Ω_ρ. The LEMPC may dictate a general time-varying operating policy under the first operation mode. Under the second operation mode, the LEMPC optimizes the economic cost function while ensuring that the computed control action for the closed-loop system forces the state along a path that causes the Lyapunov function value to decay. The first and second operation mode of the LEMPC will be referred to as mode 1 and mode 2 operation of the LEMPC, respectively, and are defined by specific Lyapunov-based constraints imposed in the LEMPC optimization problem.

To enforce convergence of the closed-loop state to the steady-state (if desirable), the LEMPC is formulated with a switching time t_s. From the initial time to time t_s, the LEMPC may dictate a time-varying operating policy to optimize the economics while maintaining the closed-loop state in Ω_ρ. After the time t_s, the LEMPC operates exclusively in the second operation mode and calculates the inputs in a way that the state of the closed-loop system is steered to a neighborhood of the steady-state. For the sake of simplicity, the switching time t_s is an integer multiple of the sampling period (Δ) of the LEMPC. This assumption poses little practical restrictions.

LEMPC is an EMPC scheme that uses the Lyapunov-based controller to design two regions of operation where closed-loop stability of the system of Eq. 4.1 under the LEMPC and recursive feasibility of the optimization problem are guaranteed for operation in the presence of bounded disturbances. The formulation of the LEMPC optimization is:

$$\min_{u \in S(\Delta)} \int_{t_k}^{t_{k+N}} l_e(\tilde{x}(\tau), u(\tau))\, d\tau \tag{4.3a}$$

$$\text{s.t. } \dot{\tilde{x}}(t) = f(\tilde{x}(t), u(t), 0) \tag{4.3b}$$

$$\tilde{x}(t_k) = x(t_k) \tag{4.3c}$$

$$u(t) \in \mathbb{U},\ \forall\, t \in [t_k, t_{k+N}) \tag{4.3d}$$

$$V(\tilde{x}(t)) \le \rho_e,\ \forall\, t \in [t_k, t_{k+N}) \quad \text{if } V(x(t_k)) \le \rho_e \text{ and } t_k < t_s \tag{4.3e}$$

$$\frac{\partial V(x(t_k))}{\partial x} f(x(t_k), u(t_k), 0) \le \frac{\partial V(x(t_k))}{\partial x} f(x(t_k), h(x(t_k)), 0) \quad \text{if } V(x(t_k)) > \rho_e \text{ or } t_k \ge t_s \tag{4.3f}$$

where the decision variable of the optimization problem is the piecewise constant input trajectory over the prediction horizon and $N < \infty$ denotes the number of sampling periods in the prediction horizon. The notation $\tilde{x}$ is used to denote the predicted (open-loop) state trajectory. While zeroth-order hold is assumed, higher-order control vector parameterization may also be employed. Moreover, the theoretical analysis may also apply to the case that a higher-order parameterization is used because zeroth-order hold may, for some control parameterizations, be a conservative approximation of the higher-order control vector parameterization method.

The LEMPC dynamic optimization problem of Eq. 4.3 minimizes a cost functional (Eq. 4.3a) consisting of the economic cost function; that is, l_e is used as the stage cost in the LEMPC. The nominal system model is the constraint of Eq. 4.3b and is used to predict the evolution of the system under the computed input trajectory over the prediction horizon. The dynamic model is initialized with a state measurement obtained at the current sampling period (Eq. 4.3c). The constraint of Eq. 4.3d limits the computed control actions to be in the set of the available control actions.

Mode 1 operation of the LEMPC is defined by the constraint of Eq. 4.3e and is active when the current state is inside a predefined subset of the stability region $\Omega_{\rho_e} \subset \Omega_\rho$ and $t_k < t_s$. Since an economic performance benefit may be realized when operating the system of Eq. 4.1 in a consistently dynamic fashion compared to operating the system at the economically optimal steady-state, i.e., a steady-state that minimizes the economic stage cost amongst all of the admissible steady-states, mode 1 is used to allow the LEMPC to enforce a potentially dynamic or transient operating policy. The set Ω_{ρ_e} is designed such that if the current state $x(t_k) \in \Omega_{\rho_e}$ and the predicted state at the next sampling time $\tilde{x}(t_{k+1}) \in \Omega_{\rho_e}$, then the actual (closed-loop) state at the next sampling time, which may be forced away from Ω_{ρ_e} by a bounded disturbance/uncertainty, will be in Ω_{ρ}. The maximum size of Ω_{ρ_e} depends on the bound on the disturbance and sampling period size; please refer to Eq. 4.21 of Sect. 4.3.1.

To maintain boundedness of the closed-loop state within a well-defined state-space set, the second mode is used, which is defined by the constraint of Eq. 4.3f. This constraint forces the computed control action by the LEMPC to decrease the Lyapunov function by at least the decay rate achieved by the Lyapunov-based controller. Owing to the properties of the Lyapunov-based controller implemented in a sample-and-hold fashion with a sufficiently small sampling period, the Lyapunov function value under the LEMPC operating in mode 2 will decrease over the sampling period when the constraint of Eq. 4.3f is active and when the state at t_k is outside a small compact set containing the steady-state (this set is defined as Ω_{ρ_s} in Theorem 4.1). If steady-state operation is desired, i.e., enforcing the closed-loop state to a neighborhood of the steady-state, selecting the switching time to be finite will guarantee that LEMPC forces the state to converge to a small forward invariant set containing the steady-state.

The two tuning parameters of LEMPC, besides the user-defined economic cost function, are the switching time t_s and the set Ω_{ρ_e}. If $t_s = 0$, the LEMPC will always operate in mode 2. This may be desirable if steady-state operation is expected and/or if it is the best operating strategy. If $t_s \to \infty$, the LEMPC may dictate a time-varying operating policy over the entire length of operation. An intermediate choice for the switching time ($t_s \in (0, \infty)$) may be used to balance the trade-off between achieving better economic performance through time-varying operation and excessive control actuator wear required to enforce the time-varying operating policy. The other tuning parameter of LEMPC is ρ_e which does not need to be chosen so that Ω_{ρ_e} is the largest subset of Ω_ρ such that the state at the next sampling time is guaranteed to be in Ω_ρ under mode 1 operation of the LEMPC. A larger set Ω_{ρ_e} may allow for better closed-loop economic performance. On the other hand, a smaller set Ω_{ρ_e} may allow for more robustness to uncertainty.

Remark 4.1 As pointed out in Sect. 2.4.4, it is beyond the scope of this book to thoroughly discuss conditions that guarantee the existence of a solution to general optimal control problems. For mode 2 of the LEMPC, one may require that the state trajectory be constrained to be contained in Ω_ρ for all $t \in [t_k, t_{k+N})$. Then, it suffices to point out that owing to standard continuity arguments, the compactness of Ω_ρ, Ω_{ρ_e}, and $\mathbb{U}$, and the ability to show the existence of a feasible solution to conclude that a solution to Eq. 4.3 exists in either operation mode of the LEMPC. The interested reader is referred to, for example, [11, 12] for a more thorough discussion on the existence of a solution to optimal control problems.

Remark 4.2 While the stage cost function is assumed to be continuous on its domain and the vector function f is assumed to be locally Lipschitz on its domain, stronger smoothness assumptions are typically needed to solve the resulting nonlinear optimization problem with standard methods. Also, it is not possible to numerically impose the constraint of Eq. 4.3e for all $t \in [t_k, t_{k+N})$. One relaxation of the constraint is impose the constraint at each sampling time of the prediction horizon. This methodology has demonstrated to yield acceptable results for several examples contained in this book. Imposing the state constraint in this manner and owing to the zeroth-order hold control parameterization ensures that the optimization problem of Eq. 4.3 is a finite-dimensional one.

4.2.4 Implementation Strategy

The LEMPC is implemented in a receding horizon fashion. At each sampling time, the LEMPC receives a state measurement $x(t_k)$, solves the optimization problem of Eq. 4.3, and sends the control action for the first sampling period of the prediction horizon to be implemented by the control actuators from t_k to t_{k+1}. At the next sampling time, the LEMPC receives a state measurement $x(t_{k+1})$ and solves the optimization problem again by rolling the horizon one sampling period into the future. The optimal input trajectory computed by the LEMPC at a given sampling time t_k is denoted as $u^*(t|t_k)$ and is defined for $t \in [t_k, t_{k+N})$. The control action that is sent at time t_k to the control actuators to be applied over the sampling period from t_k to t_{k+1} is denoted as $u^*(t_k|t_k)$. The receding horizon fashion implementation of the dual-mode LEMPC is stated formally in the following algorithm:

1. At a sampling time t_k, the controller receives the state measurement $x(t_k)$. Go to Step 2.
2. If $t_k < t_s$, go to Step 3. Else, go to Step 3.2.
3. If $x(t_k) \in \Omega_{\rho_e}$, go to Step 3.1. Else, go to Step 3.2.
    3.1. Mode 1 operation of the LEMPC is active, i.e., Eq. 4.3e is imposed in the optimization problem and Eq. 4.3f is inactive. Go to Step 4.
    3.2. Mode 2 operation of the LEMPC is active, i.e., Eq. 4.3f is imposed in the optimization problem and Eq. 4.3e is inactive. Go to Step 4.
4. The LEMPC of Eq. 4.3 is solved to compute an optimal input trajectory $u^*(t|t_k)$ for $t \in [t_k, t_{k+N})$ and sends the control action $u^*(t_k|t_k)$ computed for the first sampling period of the prediction horizon to be applied to the closed-loop system over the sampling period (from t_k to t_{k+1}). Go to Step 5.
5. Go to Step 1 ($k \leftarrow k + 1$).

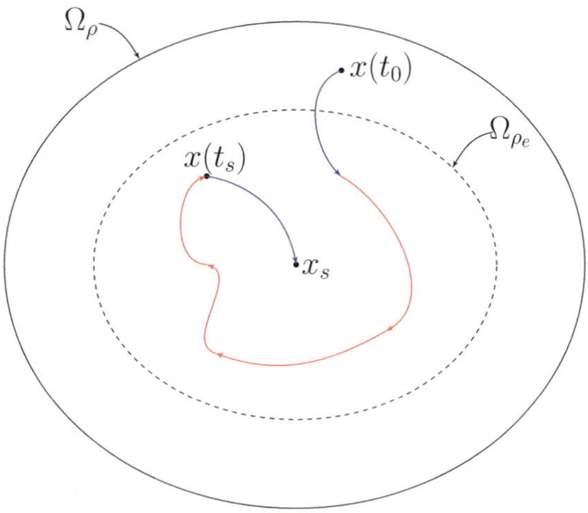

Fig. 4.1 An illustration of the state-space evolution of a system under LEMPC. The *red trajectory* represents the state trajectory under mode 1 operation of the LEMPC, and the *blue trajectory* represents the state trajectory under mode 2 operation

The notation $k \leftarrow k + 1$ used in Step 5 of the algorithm means that k is set to $k + 1$ for the next time through the algorithm loop. In other words, k is set to $k + 1$ before returning to Step 1.

An illustration of the possible evolution of a system under LEMPC is shown in Fig. 4.1. At the initial time, t_0, the state is outside Ω_{ρ_e}. The contractive constraint of Eq. 4.3f is active to steer the state to Ω_{ρ_e}. Once the state is in Ω_{ρ_e}, the LEMPC computes control actions using mode 1 operation, i.e., the constraint of Eq. 4.3e is active. Under this mode of operation, the LEMPC dictates a time-varying operating policy. After t_s, the contractive constraint (Eq. 4.3f) is imposed at all subsequent sampling times to ensure that the closed-loop state trajectory converges to a small neighborhood of the steady-state.

4.2.5 Satisfying State Constraints

While it may not appear that hard state constraints are included in the LEMPC problem of Eq. 4.3, hard constraints may be accounted for through the design of Ω_ρ, which extends the ideas of imposing state constraints from (tracking) Lyapunov-based MPC [8]. Specifically, define the set Φ_u as the set in state-space that includes all the states where $\dot{V} < 0$ under the Lyapunov-based controller $h(x)$. Since the Lyapunov-based controller accounts for the input constraints, the set Φ_u also accounts for the inputs constraints.

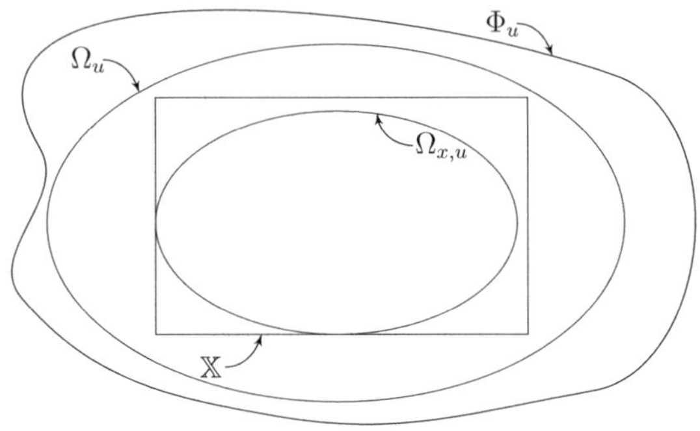

Fig. 4.2 An illustration of the various state-space sets described for enforcing state constraints with LEMPC. The case when $\mathbb{X} \subset \Phi_u$ is depicted in this illustration

First, consider the case where $\Phi_u \subseteq \mathbb{X}$ ($\mathbb{X}$ is the state constraint set). This means that any initial state starting in the region $\mathbb{X} \setminus \Phi_u$ will satisfy the state constraint. However, the time-derivative of the Lyapunov function may be positive and thus, it may not be possible to stabilize the closed-loop system under the Lyapunov-based controller starting from such initial conditions. Therefore, the stability region used in the formulation of the LEMPC for this case is $\Omega_\rho = \Omega_{x,u} = \{x \in \mathbb{R}^n : V(x) \leq \rho_{x,u}\}$ where $\rho_{x,u}$ is chosen such that $\Omega_{x,u} \subseteq \Phi_u$ because there exists a feasible input trajectory that maintains the state in $\Omega_{x,u}$ for all initial conditions in $\Omega_{x,u}$ while satisfying both the state and input constraints. The notation $\rho_{x,u}$ denotes that both state and input constraints are accounted for in the design of the region $\Omega_{x,u}$.

Now, consider the case where $\mathbb{X} \subset \Phi_u$. This case is depicted in Fig. 4.2. For any initial state starting outside $\mathbb{X}$, the state constraint will be violated from the onset. Also, for any initial state in the set $\mathbb{X}$, it is not possible, in general, to guarantee that the set $\mathbb{X}$ is forward invariant because there may exist a stabilizing state trajectory, i.e., a trajectory where $\dot{V} < 0$ along this trajectory, that goes outside of the set $\mathbb{X}$ before it enters back into the set $\mathbb{X}$ to converge to the origin. For the case with hard constraints, define the set Ω_ρ as $\Omega_\rho = \Omega_{x,u} = \{x \in \mathbb{R}^n : V(x) \leq \rho_{x,u}\}$ where $\rho_{x,u}$ is such that $\Omega_{x,u} \subseteq \mathbb{X}$. Since Φ_u cannot be computed in practice, the set $\Omega_u := \{x \in \mathbb{R}^n : V(x) \leq \rho_u\}$ where ρ_u is such that $\dot{V} < 0$ for all states in Ω_u under the Lyapunov-based controller $h(x)$ which accounts for the input constraint only may be used. An illustration of the set definitions is provided in Fig. 4.2.

On the other hand, consider the case that it is desirable to impose the state constraints as soft constraints. This may potentially allow for a larger region (Ω_ρ) at the expense that the state constraints may not be satisfied for all time. If $\Phi_u \subset \mathbb{X}$, Ω_ρ is constructed such that $\Omega_\rho \subset \mathbb{X}$, and the state constraint is always satisfied for any initial conditions in Ω_ρ. From a closed-loop stability stand-point, it is not desirable to increase the size of the set Ω_ρ because there may be states outside of Ω_ρ where $\dot{V} > 0$. Therefore, consider the more interesting case that $\mathbb{X} \subset \Phi_u$. If the state constraints are imposed as soft constraints, then one could potentially take $\Omega_\rho = \Omega_u \subseteq \Phi_u$ where $\mathbb{X} \subset \Omega_\rho$. For any initial state in Ω_ρ, the amount of time that the state constraint (imposed as a soft constraint) will be violated is finite. This statement holds because for any state in $\Omega_\rho \setminus \mathbb{X}$, $\dot{V} < 0$ and therefore, the state trajectory evolves along a path where the Lyapunov function value decays over time. Eventually, the state trajectory will converge to a level set $\Omega_{\rho'}$ in finite-time that is contained in $\mathbb{X}$ (assuming that the contractive constraint of the LEMPC is active when $x(t_k) \in \Omega_\rho \setminus \mathbb{X}$).

The following example illustrates the above methodology to satisfy hard state constraints.

Example 4.1 Consider the scalar system:

$$\dot{x} = x + u \tag{4.4}$$

which has an open-loop unstable steady-state at the origin. If the system was subject to the following constraints on the state and input: $x \in \mathbb{X} = [-2, 2]$ and $u \in \mathbb{U} = [-1, 1]$. For any initial state $x_0 \notin [-1, 1]$, the state will diverge, i.e., some initial states that initially satisfy the state constraint will not continue satisfying the state constraints over the length of operation. Following the approach detailed above, the nonlinear control law $u = -\text{sat}(Kx)$ where $\text{sat}(\cdot)$ is the saturation function and $K > 1$ is a tuning parameter renders the origin of Eq. 4.4 locally exponentially stable while satisfying the input constraints. The quadratic function:

$$V(x) = x^2 \tag{4.5}$$

is a Lyapunov function for the closed-loop system. Moreover, the stability region accounting for the input constraints, i.e., region where $\dot{V} < 0$, is $\Omega_u = \{x \in \mathbb{R} : V(x) \le \rho_u\}$ where ρ_u is chosen such that $\rho_u < 1$. Since $\Omega_u \subset \mathbb{X}$, Ω_ρ may be taken to be Ω_u. If, instead, $\mathbb{X} = [-0.9, 0.9]$, then $\mathbb{X} \subset \Omega_u$ and Ω_ρ may be taken to be $\{x \in \mathbb{R} : V(x) \le 0.81\}$, i.e., $\Omega_\rho = \mathbb{X}$. In either situation, one may verify that for any initial state in Ω_ρ, the closed-loop state trajectory will remain bounded in Ω_ρ and converge exponentially to the origin without violating the state constraints. If one were to apply the LEMPC of Eq. 4.3 to the system of Eq. 4.4 designed based on the nonlinear control law, the Lyapunov function of Eq. 4.5, and the region Ω_ρ, the closed-loop state trajectory for all initial conditions in Ω_ρ is guaranteed to satisfy the state constraint for all times.

4.2.6 Extensions and Variants of LEMPC

A few extensions and variants of the LEMPC formulation of Eq. 4.3 and implementations are discussed below.

- For closed-loop economic performance reasons, terminal conditions, e.g., a terminal constraint or a terminal cost, may be added to the problem of Eq. 4.3 which is discussed in Sect. 4.4. Also, it is possible to design such terminal conditions by taking advantage of the availability of the explicit controller.
- For all $t < t_s$, the LEMPC may dictate a time-varying operating policy. If the LEMPC enforces that the closed-loop state trajectory evolves along a path close

to the boundary of Ω_{ρ_e}, it is possible that the LEMPC continuously switches between mode 1 and mode 2 operation. The successive switching between mode 1 and mode 2 could lead to undesirable closed-loop behavior such as performance deterioration. For these cases, various modifications could be made to LEMPC to avoid this behavior. Owing to the fact that the economic cost is user-defined, one could add penalization terms to the stage cost of the LEMPC that penalize the closeness of the predicted state to the boundary of Ω_{ρ_e}. Then, the LEMPC will ideally maintain the closed-loop state in the interior of Ω_{ρ_e} preventing the state from coming out of Ω_{ρ_e}.

- For closed-loop stability, the constraint $\tilde{x}(t) \in \Omega_{\rho_e}$ (Eq. 4.3e) only needs to hold for all $t \in [t_k, t_{k+1})$ under mode 1 operation, i.e., the predicted state at the next sampling time must be contained in Ω_{ρ_e}. However, the constraint of Eq. 4.3e is included for all $t \in [t_k, t_{k+N})$ for two reasons. First, it ensures that the predicted state trajectory remains in a compact set. Second, it allows the LEMPC to optimize the input trajectory with respect to (ideally) a better prediction of the closed-loop behavior (recall that the closed-loop state must remain in Ω_ρ and under mode 1 operation, it is desirable to maintain the closed-loop state in the interior of Ω_{ρ_e} to avoid continuous switching between mode 1 and mode 2 operation). Nonetheless, for practical implementation, one may consider imposing the constraint at a few time instances along the prediction horizon, which will decrease the number of constraints relative to imposing the constraint at every sampling time instance, for example.
- For mode 2 operation, one may need to enforce a similar constraint as the mode 1 constraint to maintain the predicted state in Ω_ρ, i.e., to enforce $\tilde{x}(t) \in \Omega_\rho$ for $t \in [t_k, t_{k+N})$. This ensures that the predicted state trajectory is maintained in a compact set, which guarantees existence and uniqueness of the solution of the dynamic model of Eq. 4.3b as well as prevents numerical problems when large (in a norm sense) state trajectories are computed. This issue is particularly relevant when the prediction horizon is long.
- The LEMPC may be modified to handle potentially asynchronous and delayed measurements; see Sect. 4.3.2.
- For certain applications, one may consider driving some of the states of the system to certain set-points and allowing the other states to evolve in a time-varying manner to optimize the economics. For the LEMPC design of Eq. 4.3, this means that part of the system is operated in the first operation mode and part of the system in the second operation mode simultaneously. This is considered in the example of Sect. 4.3.3.
- With the formulated constraints of the LEMPC of Eq. 4.3, the optimization problem is always feasible in the sense that there exists an input trajectory that satisfies the constraints for all $x(t_k) \in \Omega_\rho$. To force the closed-loop state to converge to Ω_ρ when $x(t_k) \notin \Omega_\rho$, one could employ the contractive constraint of Eq. 4.3f in an attempt to decrease the Lyapunov function value or use a constraint, e.g., a terminal constraint, to enforce that the predicted state converges to Ω_ρ. However, no guarantee may be made that the state will converge to Ω_ρ starting from outside Ω_ρ in either case. In the former case, in general, this may not result in a well-defined

problem. If it is well-defined, it may not be possible to decrease the Lyapunov function value for initial states outside of Ω_ρ. In the latter case, feasibility of the resulting problem is not guaranteed.

4.3 Closed-Loop Stability and Robustness Under LEMPC

Closed-loop stability robustness of the closed-loop system of Eq. 4.1 under the LEMPC of Eq. 4.3. Both synchronous (ideal) sampling and asynchronous and delayed sampling are considered. Also, LEMPC is applied to a chemical process example to demonstrate the closed-loop stability and robustness properties.

To complete the closed-loop stability analysis, a few properties of the system of Eq. 4.1 are needed in the analysis. Owing to the fact that f is locally Lipschitz on $\mathbb{X} \times \mathbb{U} \times \mathbb{W}$ and the sets Ω_ρ, $\mathbb{U}$, and $\mathbb{W}$ are compact, there exists a positive constant M such that

$$|f(x, u, w)| \leq M \tag{4.6}$$

for all $x \in \Omega_\rho$, $u \in \mathbb{U}$ and $w \in \mathbb{W}$. By the continuous differentiable property of the Lyapunov function V and the Lipschitz property assumed for the vector field f, there exist positive constants L_x, L_w, L'_x and L'_w such that

$$|f(x, u, w) - f(x', u, 0)| \leq L_x|x - x'| + L_w|w|, \tag{4.7}$$

$$\left|\frac{\partial V(x)}{\partial x} f(x, u, w) - \frac{\partial V(x')}{\partial x} f(x', u, 0)\right| \leq L'_x|x - x'| + L'_w|w| \tag{4.8}$$

for all $x, x' \in \Omega_\rho$, $u \in \mathbb{U}$, and $w \in \mathbb{W}$.

4.3.1 *Synchronous Measurement Sampling*

The stability properties of the LEMPC of Eq. 4.3 for the system of Eq. 4.1 is analyzed under ideal sampling. In order to proceed, preliminary results are presented. First, for a bounded disturbance, the difference between the state of Eq. 4.1 and the nominal state of Eq. 4.1 (the system of Eq. 4.1 with $w \equiv 0$) may be bounded, which is stated in the following proposition.

Proposition 4.1 *Consider the systems*

$$\dot{x}(t) = f(x(t), u(t), w(t)) , \tag{4.9}$$

$$\dot{\hat{x}}(t) = f(\hat{x}(t), u(t), 0) . \tag{4.10}$$

Let $x(t)$ and $\hat{x}(t)$ be the solutions of Eqs. 4.9 and 4.10, respectively, for $t \in [t_0, t_f]$ ($t_f > t_0$) with initial states $x(t_0) = \hat{x}(t_0) \in \Omega_\rho$ and input trajectory $u(t) \in \mathbb{U}$ for all $t \in [t_0, t_f]$ where $u(\cdot)$ is piecewise constant in t. The system of Eq. 4.9 is also forced by some disturbance $w(t) \in \mathbb{W}$ for all $t \in [t_0, t_f]$. If $x(t) \in \Omega_\rho$ and $\hat{x}(t) \in \Omega_\rho$ for all $t \in [t_0, t_f]$, then there exists a function $f_w \in \mathcal{K}$ such that

$$|x(t) - \hat{x}(t)| \leq f_w(t - t_0) \tag{4.11}$$

for all $t \in [t_0, t_f]$.

Proof Consider the difference between the solutions of Eqs. 4.9 and 4.10, which are denoted as $x(t)$ and $\hat{x}(t)$, respectively, and let $e := x - \hat{x}$. From Eq. 4.7 and the fact that $w(t) \in \mathbb{W}$ for all $t \in [t_0, t_f]$, the time-derivative of e may be bounded as follows:

$$|\dot{e}(t)| = \left| f(x(t), u(t), w(t)) - f(\hat{x}(t), u(t), 0) \right| \tag{4.12}$$

$$\leq L_x |x(t) - \hat{x}(t)| + L_w |w(t)| \tag{4.13}$$

$$\leq L_x |e(t)| + L_w \theta \tag{4.14}$$

for $t \in [t_0, t_f]$. Integrating the above bound with respect to time and noting that $e(0) = x(0) - \hat{x}(0) = 0$, the following bound on the error is obtained:

$$|e(t)| \leq f_w(t - t_0) := \frac{L_w \theta}{L_x} (e^{L_x (t - t_0)} - 1) \tag{4.15}$$

for all $t \in [t_0, t_f]$ when $x(t), \hat{x}(t) \in \Omega_\rho$, $u(t) \in \mathbb{U}$, and $w(t) \in \mathbb{W}$ for all $t \in [t_0, t_f]$. It is straightforward to show that $f_w \in \mathcal{K}$ which completes the proof.

Proposition 4.2 bounds the difference between the Lyapunov function values evaluated at two points in Ω_ρ.

Proposition 4.2 *Consider the Lyapunov function V that satisfies Eq. 4.2. There exists a quadratic function f_V such that*

$$V(x_2) - V(x_1) \leq f_V(|x_2 - x_1|) \tag{4.16}$$

for all $x_1, x_2 \in \Omega_\rho$.

Proof Since the Lyapunov function is continuously differentiable and bounded on compact sets, there exists a positive constant $M_V > 0$ such that:

$$V(x_2) \leq V(x_1) + \left| \frac{\partial V(x_1)}{\partial x} \right| |x_2 - x_1| + M_V |x_2 - x_1|^2 \tag{4.17}$$

for all x_1, $x_2 \in \Omega_\rho$. From Eq. 4.2, the partial derivative of V may be bounded as follows:

$$\left| \frac{\partial V(x_1)}{\partial x} \right| \leq \alpha_4(\alpha_1^{-1}(\rho)) \tag{4.18}$$

for all $x_1 \in \Omega_\rho$. From Eqs. 4.17 and 4.18, the existence of a quadratic function $f_V(\cdot)$ that bounds the Lyapunov function values for any two points in Ω_ρ follows:

$$V(x_2) \leq V(x_1) + f_V(|x_2 - x_1|) \tag{4.19}$$

where

$$f_V(s) := \alpha_4(\alpha_1^{-1}(\rho))s + M_V s^2 . \tag{4.20}$$

Theorem 4.1 provides sufficient conditions that guarantee that the state of the closed-loop system of Eq. 4.1 under the LEMPC of Eq. 4.3 is always bounded in Ω_ρ and is ultimately bounded in a small region containing the origin.

Theorem 4.1 *Consider the system of Eq. 4.1 in closed-loop under the LEMPC design of Eq. 4.3 based on the Lyapunov-based controller that satisfies the conditions of Eq. 4.2. Let* $\Delta > 0$, $N \geq 1$, $\varepsilon_w > 0$, *and* $\rho > \rho_e \geq \rho_{\min} > \rho_s > 0$ *satisfy*

$$\rho_e < \rho - f_V(f_w(\Delta)), \tag{4.21}$$

$$-\alpha_3(\alpha_2^{-1}(\rho_s)) + L'_x M \Delta + L'_w \theta \leq -\varepsilon_w/\Delta, \tag{4.22}$$

and

$$\rho_{\min} = \max_{s \in [0,\Delta]} \{V(x(s)) : V(x(0)) \leq \rho_s\}. \tag{4.23}$$

If $x(0) \in \Omega_\rho$, *then the state* $x(t)$ *of the closed-loop system of Eq. 4.1 is always bounded in* Ω_ρ *for all* $t \geq 0$ *and is ultimately bounded in* $\Omega_{\rho_{\min}}$ *if* t_s *is finite.*

Proof The proof is organized into three parts. In Part 1, feasibility of the LEMPC optimization problem of Eq. 4.3 is proved when the state measurement at a specific sampling time is in Ω_ρ. In Part 2, boundedness of the closed-loop state in Ω_ρ is established. In Part 3, ultimate boundedness of the closed-loop state in a small state-space set containing the origin is proved when the switching time is finite.

Part 1: The sample-and-hold input trajectory obtained from the Lyapunov-based controller is a feasible solution to the LEMPC optimization problem of Eq. 4.3 when $x(t_k) \in \Omega_\rho$. Let $\hat{x}(t)$ denote the solution at time t to the nominal sampled-data system:

$$\dot{\hat{x}}(t) = f(\hat{x}(t), h(\hat{x}(\tau_i)), 0) \tag{4.24}$$

for $t \in [\tau_i, \tau_{i+1})$ ($\tau_i := t_k + i\Delta$), $i = 0, 1, \ldots, N-1$ with initial condition $\hat{x}(t_k) = x(t_k) \in \Omega_\rho$. Define $\hat{u}$ as the resulting input trajectory of Eq. 4.24 defined over the interval $[t_k, t_{k+N})$. The input trajectory $\hat{u}$ is a feasible solution to the LEMPC problem.

The input trajectory meets the input constraints by the formulation of the Lyapunov-based controller. For mode 2 operation, the mode 2 contractive constraint of Eq. 4.3f is trivially satisfied with this feasible input trajectory. For mode 1 operation, the region Ω_{ρ_e} is forward invariant under the Lyapunov-based controller applied in a sample-and-hold fashion when $\Omega_{\rho_{\min}} \subseteq \Omega_{\rho_e} \subset \Omega_\rho$ where $\Omega_{\rho_{\min}}$ will be explained further in Part 3.

Part 2: To show that the state is maintained in Ω_ρ when $x(0) \in \Omega_\rho$, mode 1 and mode 2 operation of the LEMPC must be each considered. To show the desired result, it is sufficient to show that if the state at any arbitrary sampling time is such that $x(t_k) \in \Omega_\rho$, then the state at the next sampling time is in Ω_ρ, i.e., $x(t_{k+1}) \in \Omega_\rho$ and the closed-loop state does not come out of Ω_ρ over the sampling period. Recursive application of this result, proves that the closed-loop state is always bounded in Ω_ρ for all $t \geq 0$ when $x(0) \in \Omega_\rho$.

Case 1: If $x(t_k) \in \Omega_{\rho_e}$ and $t_k < t_s$, the LEMPC operates under mode 1 operation. From Part 1, the LEMPC is feasible. Moreover, from the formulation of the LEMPC, the LEMPC computes a control action such that $\tilde{x}(t) \in \Omega_{\rho_e}$ for all $t \in [t_k, t_{k+1})$. Owing to the effect of the bounded disturbances, the closed-loop state does not evolve according to the model of Eq. 4.3b. Nevertheless, if Eq. 4.21 is satisfied, the state at the next sampling time will be contained in Ω_ρ.

To show this result, let ρ_e satisfy Eq. 4.21. The proof proceeds by contradiction. Assume there exists a time $\tau^* \in [t_k, t_{k+1})$ such that $V(x(\tau^*)) > \rho$. The case that $x(t)$ is not defined for some $t \in [t_k, t_{k+1})$ is also covered by this assumption. Define $\tau_1 := \inf\{\tau \in [t_k, t_{k+1}) : V(x(\tau)) > \rho\}$. A standard continuity argument in conjunction with the fact that $V(x(t_k)) \leq \rho_e < \rho$ shows that $\tau_1 \in (t_k, t_{k+1})$, $V(x(t)) \leq \rho$ for all $t \in [t_k, \tau_1]$ with $V(x(\tau_1)) = \rho$, and $V(x(t)) > \rho$ for some $t \in (\tau_1, t_{k+1})$. If ρ_e satisfies Eq. 4.21, then

$$\begin{aligned} \rho = V(x(\tau_1)) &\leq V(\tilde{x}(\tau_1)) + f_V(f_w(\tau_1)) \\ &\leq \rho_e + f_V(f_w(\Delta)) < \rho \end{aligned} \tag{4.25}$$

where the first inequality follows from Propositions 4.1 and 4.2 and the second inequality follows from the fact that $f_V \circ f_w \in \mathscr{K}$ and $\tau_1 \leq \Delta$. Equation 4.25 is a contradiction. Thus, $x(t) \in \Omega_\rho$ for all $t \in [t_k, t_{k+1})$ if Eq. 4.21 is satisfied.

Case 2: Now, consider that the LEMPC under mode 2 operation at an arbitrary sampling time t_k. If $x(t_k) \in \Omega_\rho$, the LEMPC is feasible (Part 1). The LEMPC computes a control action that satisfies

$$\begin{aligned} \frac{\partial V(x(t_k))}{\partial x} f(x(t_k), u^*(t_k|t_k), 0) &\leq \frac{\partial V(x(t_k))}{\partial x} f(x(t_k), h(x(t_k)), 0) \\ &\leq -\alpha_3(|x(t_k)|) \end{aligned} \tag{4.26}$$

for all $x(t_k) \in \Omega_\rho$ where the inequality follows from Eq. 4.2b. Consider the time-derivative of the Lyapunov function for $\tau \in [t_k, t_{k+1})$

$$\begin{aligned}\dot{V}(x(\tau)) &= \frac{\partial V(x(\tau))}{\partial x} f(x(\tau), u^*(t_k|t_k), w(\tau)) \\ &\le -\alpha_3(|x(t_k)|) + \frac{\partial V(x(\tau))}{\partial x} f(x(\tau), u^*(t_k|t_k), w(\tau)) \\ &\quad - \frac{\partial V(x(t_k))}{\partial x} f(x(t_k), u^*(t_k|t_k), 0)\end{aligned} \tag{4.27}$$

for $\tau \in [t_k, t_{k+1})$ where the inequality follows by adding and subtracting the left-hand side of Eq. 4.26 and accounting for the bound of Eq. 4.26. Owing to the Lipschitz bound of Eq. 4.8, Eq. 4.27 may be upper bounded by:

$$\begin{aligned}\dot{V}(x(\tau)) &\le -\alpha_3(|x(t_k)|) + L'_x|x(\tau) - x(t_k)| + L'_w|w(\tau)| \\ &\le -\alpha_3(|x(t_k)|) + L'_x|x(\tau) - x(t_k)| + L'_w\theta\end{aligned} \tag{4.28}$$

for $\tau \in [t_k, t_{k+1})$ where the second inequality follows from the fact that $w(t) \in \mathbb{W} = \{\bar{w} \in \mathbb{R}^l : |\bar{w}| \le \theta\}$ for all $t \ge 0$. Taking into account Eq. 4.6, the continuity of x and the fact that $u^*(t_k|t_k) \in \mathbb{U}$ and $w(\tau) \in \mathbb{W}$, the norm of the difference between the state at $\tau \in [t_k, t_{k+1})$ and the state at t_k scales with Δ. More specifically, the bound of:

$$|x(\tau) - x(t_k)| \le M\Delta \tag{4.29}$$

for $\tau \in [t_k, t_{k+1})$ follows from Eq. 4.6. From Eqs. 4.28 and 4.29, the inequality follows:

$$\dot{V}(x(\tau)) \le -\alpha_3(|x(t_k)|) + L'_x M\Delta + L'_w\theta \tag{4.30}$$

for $\tau \in [t_k, t_{k+1})$.

For any $x(t_k) \in \Omega_\rho \setminus \Omega_{\rho_s}$, it can be obtained that

$$\dot{V}(x(\tau)) \le -\alpha_3(\alpha_2^{-1}(\rho_s)) + L'_x M\Delta + L'_w\theta. \tag{4.31}$$

for $\tau \in [t_k, t_{k+1})$ follows from Eqs. 4.2a and 4.30. If the condition of Eq. 4.22 is satisfied, i.e., $\Delta > 0$ and $\theta > 0$ are sufficiently small, then there exists $\varepsilon_w > 0$ such that the following inequality holds for any $x(t_k) \in \Omega_\rho \setminus \Omega_{\rho_s}$:

$$\dot{V}(x(\tau)) \le -\varepsilon_w/\Delta \tag{4.32}$$

for $\tau \in [t_k, t_{k+1})$. Integrating the bound of Eq. 4.33 for $\tau \in [t_k, t_{k+1})$, one obtains that the following is satisfied:

$$\begin{aligned}V(x(t_{k+1})) &\le V(x(t_k)) - \varepsilon_w \\ V(x(t)) &\le V(x(t_k))\ \forall\, t \in [t_k, t_{k+1})\end{aligned} \tag{4.33}$$

for all $x(t_k) \in \Omega_\rho \setminus \Omega_{\rho_s}$.

If $x(t_k) \in \Omega_\rho \setminus \Omega_{\rho_e}$ where $\Omega_{\rho_e} \supseteq \Omega_{\rho_{\min}}$ (where $\rho_{\min}$ is defined in Eq. 4.23), then using Eq. 4.33 recursively, it follows that the state converges to Ω_{ρ_e} in a finite number of sampling times without coming out of Ω_ρ. If $x(t_k) \in \Omega_\rho \setminus \Omega_{\rho_s}$ and $t_k \geq t_s$, then again, by recursive application of Eq. 4.33, $x(t) \in \Omega_\rho$ for all $t \in [t_k, t_{k+1})$. If $x(t_k) \in \Omega_{\rho_s}$, the state at the next sampling time will be bounded in $\Omega_{\rho_{\min}}$ if $\rho_{\min}$ is defined according to Eq. 4.23. Thus, if $x(t_k) \in \Omega_\rho$, then $x(\tau) \in \Omega_\rho$ for all $\tau \in [t_k, t_{k+1})$ under the LEMPC when the conditions of Eq. 4.21–4.23 are satisfied. Using this result recursively, $x(t) \in \Omega_\rho$ for all $t \geq 0$ when $x(0) \in \Omega_\rho$.

Part 3: If t_s is finite and Eq. 4.23 is satisfied with $\rho > \rho_{\min} > \rho_s > 0$, the closed-loop state is ultimately bounded in $\Omega_{\rho_{\min}}$ owing to the definition of $\rho_{\min}$. In detail, from Part 2, if $x(t_k) \in \Omega_\rho \setminus \Omega_{\rho_s}$ and $t_k \geq t_s$, then $V(x(t_{k+1})) < V(x(t_k))$ until the state converges to Ω_{ρ_s}, which occurs in finite time. Once the closed-loop state converges to $\Omega_{\rho_{\min}}$, it remains inside $\Omega_{\rho_{\min}} \subset \Omega_\rho$ for all times, which follows from the definition of $\rho_{\min}$. This proves that the closed-loop system under the LEMPC of Eq. 4.3 is ultimately bounded in $\Omega_{\rho_{\min}}$ when t_s is finite.

A few notes and remarks on the results on closed-loop stability and robustness under LEMPC are in order:

- The set Ω_ρ is an invariant set for the nominal closed-loop system and is also an invariant set for the closed-loop system subject to bounded disturbances w, i.e., $|w| \leq \theta$, under piecewise constant control action implementation when the conditions stated in Theorem 4.1 are satisfied. This may be interpreted as follows: $\dot{V}$ is negative everywhere in Ω_ρ but the origin when there are no disturbances and the control actions are updated continuously. For sufficiently small disturbances and sampling period, i.e., θ and Δ are sufficiently small, $\dot{V}$ of the closed-loop system will continue to be negative for all $x \in \Omega_\rho \setminus \Omega_{\rho_s}$ where Ω_{ρ_s} is a small set containing the origin.
- Solving the dynamic model of Eq. 4.3b requires a numerical integration method. Therefore, numerical and discretization error will result. Owing to the fact that the error of many numerical integration methods may be bounded by a bound that depends on the integration step size, one may consider the numerical error as a source of bounded disturbance. The integration step size may be selected to be small. i.e., much smaller than the sampling period, to decrease the discretization error. Thus, the stability results remain valid when the discretization error may be bounded with a sufficiently small bound.
- For any state $x(t_k) \in \Omega_\rho$, the LEMPC is feasible where a feasible solution may be readily computed from the Lyapunov-based controller. Moreover, feasibility of the LEMPC, and not optimality of the LEMPC solution, implies closed-loop stability. Both of these issues are important owing to practical computational considerations. While solving the LEMPC to optimality may result in substantial computational time, one could force early termination of the optimization solver and closed-loop stability is still guaranteed (assuming the returned solution is feasible). If the

returned solution is infeasible or at sampling times where the computation time required to solve the LEMPC is significant relative to the sampling period, the control action computed from the Lyapunov-based controller may be applied to the system.

4.3.2 Asynchronous and Delayed Sampling

The design of LEMPC for systems subject to asynchronous and delayed measurements is considered. Specifically, the state of the system of Eq. 4.1 is assumed to be available at asynchronous time instants. The time instances that a state measurement is available are denoted by the time sequence $\{t_a\}_{a\geq 0}$. The sequence $\{t_a\}_{a\geq 0}$ is a random increasing sequence and the interval between two consecutive time instants is not fixed. Moreover, the measurements may also be subject to delays. To model delays in measurements, an auxiliary variable d_a is introduced to indicate the delay corresponding to the measurement received at time t_a, i.e., at time t_a, the measurement $x(t_a - d_a)$ is received.

To study the stability properties in a deterministic framework, two assumptions are made on the time between two successive sampling times and the magnitude of the measurement delay. Specifically, there exist an upper bound T_m on the interval between two successive measurements, i.e., $\sup_{a\geq 0}\{t_{a+1} - t_a\} \leq T_m$, and an upper bound D on the delays, i.e., $d_a \leq D$ for all $a \geq 0$. Both of these assumptions are reasonable from a process control perspective. Because the delays are time-varying, it is possible that at a time instant t_a, the controller may receive a measurement $x(t_a - d_a)$ which does not provide new information, i.e., $t_a - d_a < t_{a-1} - d_{a-1}$. Thus, the maximum amount of time the system might operate in open-loop following t_a is $D + T_m - d_a$. This upper bound will be used in the formulation of LEMPC for systems subject to asynchronous and delayed measurements. The reader may refer to [13] for more discussion on the modeling of asynchronous and delayed measurements.

4.3.2.1 LEMPC Formulation and Implementation

At a sampling time t_a, the MPC is evaluated to obtain the optimal input trajectories based on the received system state value $x(t_a - d_a)$. The optimization problem of the LEMPC for systems subject to asynchronous and delayed measurements at t_a is as follows:

$$\min_{u\in S(\Delta)} \int_{t_a}^{t_a+N\Delta} l_e(\tilde{x}(\tau), u(\tau))\, d\tau \tag{4.34a}$$

$$\text{s.t.} \quad \dot{\tilde{x}}(t) = f(\tilde{x}(t), u(t), 0) \tag{4.34b}$$

$$\tilde{x}(t_a - d_a) = x(t_a - d_a) \tag{4.34c}$$

$$u(t) = u^*(t),\ \forall\, t \in [t_a - d_a, t_a) \tag{4.34d}$$

$$u(t) \in \mathbb{U},\ \forall\, t \in [t_a, t_a + N\Delta) \tag{4.34e}$$

$$V(\tilde{x}(t)) \le \hat{\rho},\ \forall\, t \in [t_a, t_a + N\Delta),$$
$$\text{if } V(x(t_k)) \le \hat{\rho} \text{ and } t_k < t_s \tag{4.34f}$$

$$\dot{\hat{x}}(t) = f(\hat{x}(t), h(\hat{x}(t_a + i\Delta)), 0),$$
$$\forall\, t \in [t_a + i\Delta, t_a + (i+1)\Delta),\ i = 0, \ldots, N-1 \tag{4.34g}$$

$$\hat{x}(t_a) = \tilde{x}(t_a) \tag{4.34h}$$

$$V(\tilde{x}(t)) \le V(\hat{x}(t)),\ \forall\, t \in [t_a, t_a + N_{Da}\Delta),$$
$$\text{if } V(x(t_k)) > \hat{\rho} \text{ or } t_a \ge t_s, \tag{4.34i}$$

where $\tilde{x}$ is the predicted state trajectory with the input trajectory calculated by the LEMPC, $u^*(t)$ is the closed-loop input trajectory applied to the system for $t \in [t_a - d_a, t_a)$, $\hat{x}$ denotes the predicted state trajectory of the system under the Lyapunov-based controller implemented in a sample-and-hold fashion and is computed via the constraints of Eqs. 4.34g and 4.34h, and N_{Da} is the smallest integer that satisfies $T_m + D - d_a \le N_{Da}\Delta$. The integer N_{Da} depends on the current delay d_a, and thus, it may have different values at different time instants. The components of the LEMPC of Eq. 4.34 are explained further below. The optimal solution to the optimization problem is denoted by $u^{a,*}(t|t_a)$, which is defined for $t \in [t_a, t_a + N\Delta)$.

Two calculations are performed in the problem of Eq. 4.34. The first calculation is the estimation of the current state at time t_a. This calculation involves solving the dynamic model of Eq. 4.34b for $t \in [t_a - d_a, t_a)$ with the initial condition of Eq. 4.34c provided by the state measurement received at t_a and with input trajectory of Eq. 4.34d, which are the (closed-loop) input values that have been applied to the system from $t_a - d_a$ to t_a.

Based on the estimate of the current system state, the LEMPC optimization problem is solved to determine the optimal input trajectory that will be applied until the next measurement is received and the LEMPC is resolved. Thus, the second calculation is to evaluate the optimal input trajectory based on $\tilde{x}(t_a)$, which is the estimate of the current state computed by the first calculation. The computed input trajectory must satisfy the input constraint of Eq. 4.34e and the stability constraints of Eqs. 4.34f and 4.34i, which serve a similar purpose as the two Lyapunov-based constraints of the LEMPC of Eq. 4.3.

From the initial time to the switching time t_s, the LEMPC may dictate a time-varying operating policy while ensuring that the closed-loop state trajectory is maintained in Ω_ρ much like the LEMPC of Eq. 4.3. To account for the asynchronous and delayed measurement as well as the disturbance, a subset of the stability region, denoted as $\Omega_{\hat{\rho}}$ with $\hat{\rho} < \rho$, is defined. When a delayed measurement is received at a sampling time, the current state is estimated. If the estimated current state is in the region $\Omega_{\hat{\rho}}$, the LEMPC optimizes the cost function within the region $\Omega_{\hat{\rho}}$; if the estimated current state is in the region $\Omega_\rho / \Omega_{\hat{\rho}}$, the LEMPC, uses the second Lyapunov-based constraint, to compute control actions that drive the state to the region $\Omega_{\hat{\rho}}$. The relation between ρ and $\hat{\rho}$ will be characterized in Eq. 4.40 in Theorem 4.2 below. After the switching time t_s (if t_s is finite), the LEMPC is operated exclusively in the second operation mode to calculate the inputs in a way that the Lyapunov function of the system continuously decreases to steer the state of the system to a neighborhood of the origin.

The control actions applied to the closed-loop system that have been computed by the LEMPC of Eq. 4.34 for systems subject to asynchronous and delayed measurements are defined as follows:

$$u(t) = u^{a,*}(t|t_a) , \tag{4.35}$$

for all $t \in [t_a, t_{a+i})$ and all t_a such that $t_a - d_a > \max_{l<a} t_l - d_l$ and for a given t_a, the variable i denotes the smaller integer that satisfies $t_{a+i} - d_{a+i} > t_a - d_a$. In other words, the input trajectory applied to the closed-loop system for t_a to t_{a+i} is the input trajectory of the LEMPC computed at sampling time t_a. The time t_{a+i} denotes the next sampling time that a measurement is received providing new information about the system in the sense that $t_{a+i} + d_{a+i} > t_a - d_a$.

The implementation strategy of the LEMPC of Eq. 4.34 for systems subject to asynchronous and delayed measurements may be summarized as follows:

1. At the sampling time t_a, the controller receives a delayed state measurement, $x(t_a - d_a)$, and estimates the current state, $\tilde{x}(t_a)$. Go to Step 2.
2. If $t_a < t_s$, go to Step 3. Else, go to Step 3.2.
3. If $\tilde{x}(t_a) \in \Omega_{\hat{\rho}}$, go to Step 3.1. Else, go to Step 3.2.
    3.1. Mode 1 operation of the LEMPC is active, i.e., Eq. 4.34f is imposed in the optimization problem and Eq. 4.34i is inactive. Go to Step 4.
    3.2. Mode 2 operation of the LEMPC is active, i.e., Eq. 4.34i is imposed in the optimization problem and Eq. 4.34f is inactive. Go to Step 4.
4. The LEMPC of Eq. 4.34 is solved to compute an optimal input trajectory $u^{*,a}(t|t_a)$ for $t \in [t_a, t_a + N\Delta)$ and implements the input trajectory $u^{*,a}(t|t_a)$ on the system from t_a to t_{a+i}. Go to Step 5.
5. Go to Step 1 ($a \leftarrow a + 1$).

4.3.2.2 Stability Analysis

The stability properties of the LEMPC of Eq. 4.34 in the presence of asynchronous and delayed measurements are considered. In order to proceed, the following proposition is needed. The proof uses similar arguments as the proof of Theorem 4.1.

Proposition 4.3 *Consider the nominal sampled trajectory $\hat{x}(t)$ of the system of Eq. 4.1 in closed-loop under the Lyapunov-based controller, which satisfies the conditions of Eq. 4.2, obtained by solving recursively*

$$\dot{\hat{x}}(t) = f(\hat{x}(t), h(\hat{x}(t_k)), 0) \tag{4.36}$$

for $t \in [t_k, t_{k+1})$ *where* $t_k = k\Delta$, $k = 0, 1, \ldots$. *Let* $\Delta > 0$, $\varepsilon_s > 0$ *and* $\rho > \rho_s > 0$ *satisfy*

$$-\alpha_3\left(\alpha_2^{-1}(\rho_s)\right) + L'_x M\Delta \le -\varepsilon_s/\Delta. \tag{4.37}$$

Then, if $\hat{x}(0) \in \Omega_\rho$ *and* $\rho_{\min} < \rho$ *where* $\rho_{\min}$ *is defined in Eq. 4.23, the following inequality holds*

$$V(\hat{x}(t)) \le V(\hat{x}(t_k)),\ \forall\, t \in [t_k, t_{k+1}), \tag{4.38}$$

$$V(\hat{x}(t_k)) \le \max\{V(\hat{x}(t_0)) - k\varepsilon_s, \rho_{\min}\}. \tag{4.39}$$

Proposition 4.3 ensures that the state will be ultimately bounded in $\Omega_{\rho_{\min}}$ if the nominal system is controlled by the Lyapunov-based controller $h(x)$ implemented in a sample-and-hold fashion and with an initial condition in Ω_ρ. Theorem 4.2 provides sufficient conditions under which the LEMPC of Eq. 4.34 guarantees that the closed-loop system state is always bounded in Ω_ρ and is ultimately bounded in a small region containing the origin.

Theorem 4.2 *Consider the system of Eq. 4.1 in closed-loop under the LEMPC design of Eq. 4.34 based on the Lyapunov-based controller that satisfies the condition of Eq. 4.2. Let* $\varepsilon_s > 0$, $\Delta > 0$, $\rho > \hat{\rho} > 0$ *and* $\rho > \rho_s > 0$ *satisfy the condition of Eq. 4.37 and satisfy*

$$\hat{\rho} \le \rho - f_V(f_w(N\Delta)) \tag{4.40}$$

and

$$-N_R\varepsilon_s + f_V(f_w(N_D\Delta)) + f_V(f_w(D)) < 0 \tag{4.41}$$

where N_D *is the smallest integer satisfying* $N_D\Delta \ge T_m + D$ *and* N_R *is the smallest integer satisfying* $N_R\Delta \ge T_m$. *If* $N \ge N_D$, $\hat{\rho} \ge \rho_s$, $x(0) \in \Omega_\rho$, $d_0 = 0$, *then the closed-loop state* $x(t)$ *of the system of Eq. 4.1 is always bounded in* Ω_ρ *for* $t \ge 0$. *Moreover, if* t_s *is finite, the closed-loop state is ultimately bounded in* $\Omega_{\rho_a} \subset \Omega_\rho$ *where*

$$\rho_a = \rho_{\min} + f_V(f_W(N_D\Delta)) + f_V(f_W(D)). \tag{4.42}$$

Proof When the state is maintained in the stability region Ω_ρ, the feasibility of the optimization problem of Eq. 4.34 may be proved following the same arguments as in Part 1 of the proof of Theorem 4.1. The remainder of the proof is divided into two parts focusing on proving that the closed-loop state is always bounded in Ω_ρ for $t < t_s$ (Part 1) and is ultimately bounded in Ω_{ρ_a} if t_s is finite (Part 2). In the proof, the worst case scenario from a feedback control point-of-view is considered in the sense that if the state measurement $x(t_a - d_a)$ is received at t_a, then the next asynchronous measurement containing new information is received at t_{a+i} with $t_{a+i} = t_a + T_m$ and $T_m = N\Delta$. The results may then be generalized to the case that $t_{a+i} \le t_a + N\Delta$.

Part 1: Consider $t < t_s$ and two cases need to be considered: if $\tilde{x}(t_a) \in \Omega_{\hat{\rho}}$, then $x(t_{a+i}) \in \Omega_\rho$; and if $\tilde{x}(t_a) \in \Omega_\rho \setminus \Omega_{\hat{\rho}}$, then $V(x(t_{a+i})) < V(x(t_a))$.

When the estimated current state at the sampling time t_a satisfies $\tilde{x}(t_a) \in \Omega_{\hat{\rho}}$, from the constraint of Eq. 4.34f, the predicted state at the next sampling time, $\tilde{x}(t_{a+i})$, is in $\Omega_{\hat{\rho}}$. Applying the similar arguments as that used in Part 2 (Case 1) of the proof of Theorem 4.1, one may show that $x(t) \in \Omega_\rho$ for all $t \in [t_a, t_{a+i})$ and $x(t_{a+i}) \in \Omega_\rho$ when the condition of Eq. 4.40 is satisfied.

When the estimated current state at the sampling time t_a satisfies $\tilde{x}(t_a) \in \Omega_\rho \setminus \Omega_{\hat{\rho}}$, the computed input trajectory is such that the constraint of Eq. 4.34i is satisfied. Again, applying the same contradiction argument as that used in Part 2 (Case 1) of the proof of Theorem 4.1, one may show that when Eq. 4.41 is satisfied the closed-loop state will be maintained in Ω_ρ for all $t \in [t_a, t_{a+i})$ (this essentially requires that the amount of time in open-loop be sufficiently small). Moreover, one may show that the Lyapunov function value will decay for all $x(t_k) \in \Omega_\rho \setminus \Omega_{\hat{\rho}}$.

Specifically, consider any $\tilde{x}(t_a) \in \Omega_\rho/\Omega_{\hat{\rho}}$. From the constraint of Eq. 4.34i imposed on the computed input trajectory, the predicted state trajectory satisfies:

$$V(\tilde{x}(t)) \leq V(\hat{x}(t)) \tag{4.43}$$

for $t \in [t_a, t_a + N_{Da}\Delta)$. By Proposition 4.1 and taking into account that $\hat{\rho} > \rho_s$, the nominal sampled trajectory of Eq. 4.34g satisfies:

$$V(\hat{x}(t_{a+i})) \leq \max\{V(\hat{x}(t_a)) - N_{Da}\varepsilon_s, \rho_{\min}\}. \tag{4.44}$$

By Propositions 4.2 and 4.3, the difference between the Lyapunov function value of the (actual) closed-loop state and the estimated state at t_a may be bounded by:

$$V(x(t_a)) \leq V(\tilde{x}(t_a)) + f_V(f_w(d_a)) = V(\hat{x}(t_a)) + f_V(f_w(d_a)) \tag{4.45}$$

where the equality follows because $\tilde{x}(t_a) = \hat{x}(t_a)$. For $t \in [t_a - d_a, t_{a+i})$, the difference between the Lyapunov function value with respect to the predicted (estimated) state and the value with respect to the (actual) closed-loop state may be bounded by:

$$V(\tilde{x}(t_{a+i})) \leq V(x(t_{a+i})) + f_V(f_w(N_D\Delta)) \tag{4.46}$$

which again follows Propositions 4.2 and 4.3 (again, this requires that $x(t) \in \Omega_\rho$ for all $t \in [t_a - d_a, t_{a+i})$). From the inequalities of Eqs. 4.44–4.46, one obtains that:

$$V(x(t_{a+i})) \leq \max\{V(x(t_a)) - N_{Da}\varepsilon_s, \rho_{\min}\} + f_V(f_w(d_a)) + f_V(f_w(N_D\Delta)). \tag{4.47}$$

Note that in the derivation of the inequality of Eq. 4.47, $N_D\Delta \geq T_m + D - d_a$ for all d_a.

To prove that the Lyapunov function is decreasing between t_a and t_{a+i}, the following inequality must hold

$$N_{Da}\varepsilon_s > f_V(f_w(N_D\Delta)) + f_V(f_w(d_a)) \tag{4.48}$$

for all possible $d_a \leq D$. It is important to point out that $f_V(\cdot)$ and $f_w(\cdot)$ are strictly increasing functions of their arguments, N_{Da} is a decreasing function of the delay d_a and if $d_a = D$ then $N_{Da} = N_R$. If the condition of Eq. 4.41 is satisfied, then the condition of Eq. 4.48 holds for all possible d_a and there exist $\varepsilon_w > 0$ such that the following inequality holds

$$V(x(t_{a+i})) \leq \max\{V(x(t_a)) - \varepsilon_w, \rho_a\} \tag{4.49}$$

which implies that if $x(t_a) \in \Omega_\rho/\Omega_{\hat{\rho}}$, then $V(x(t_{a+i})) < V(x(t_a))$. This also implies that the state converges to $\Omega_{\hat{\rho}}$ in a finite number of sampling times without leaving the stability region.

Part 2: Consider $t \geq t_s$. Following similar steps as in Part 1, the condition of Eq. 4.49 may be derived. Using this condition recursively, it is proved that, if $x(0) \in \Omega_\rho$, then the closed-loop trajectory of the system of Eq. 4.1 under the LEMPC of Eq. 4.34 stay in Ω_ρ and satisfy

$$\limsup_{t\to\infty} V(x(t)) \leq \rho_a. \tag{4.50}$$

This proves the results stated in Theorem 4.2.

4.3.3 Application to a Chemical Process Example

Consider a well-mixed, non-isothermal continuous stirred-tank reactor (CSTR) where an irreversible second-order exothermic reaction $A \to B$ takes place. A is the reactant and B is the desired product. The feed to the reactor consists of A in an inert solvent at flow rate F, temperature T_0 and molar concentration C_{A0}. A jacket is used to provide/remove heat to the reactor. The dynamic equations describing the behavior of the system, obtained through material and energy balances under standard modeling assumptions, are given below:

$$\frac{dC_A}{dt} = \frac{F}{V_R}(C_{A0} - C_A) - k_0 e^{-E/RT} C_A^2 \tag{4.51a}$$

$$\frac{dT}{dt} = \frac{F}{V_R}(T_0 - T) - \frac{\Delta H}{\rho_L C_p} k_0 e^{-E/RT} C_A^2 + \frac{Q}{\rho_L C_p V_R} \tag{4.51b}$$

where C_A denotes the concentration of the reactant A, T denotes the temperature of the reactor, Q denotes the rate of heat input/removal, V_R represents the volume of the reactor, ΔH, k_0, and E denote the enthalpy, pre-exponential constant and activation energy of the reaction, respectively and C_p and ρ_L denote the heat capacity and the density of the fluid in the reactor, respectively. The values of the process parameters used in the simulations are given in Table 4.1. The process model of Eq. 4.51 is

Table 4.1 CSTR parameter values

T_0	300 K	F	$5.0\,\mathrm{m^3 h^{-1}}$
V_R	$1.0\,\mathrm{m^3}$	E	5.0×10^4 kJ kmol^{-1}
k_0	$8.46 \times 10^6\,\mathrm{m^3 kmol^{-1} h^{-1}}$	ΔH	-1.15×10^4 kJ kmol^{-1}
C_p	0.231 kJ kg^{-1}K^{-1}	R	8.314 kJ kmol^{-1}K^{-1}
ρ_L	1000 kg m^{-3}		

numerically simulated using an explicit Euler integration method with integration step $h_c = 1.0 \times 10^{-3}$ h.

The CSTR has two manipulated inputs. One of the inputs is the concentration of A in the inlet to the reactor, C_{A0}, and the other manipulated input is the external heat input/removal, Q. The input vector is given by $u^T = [C_{A0}\ Q]$, and the admissible input values are as follows: $u_1 = C_{A0} \in [0.5, 7.5]$ kmol m^{-3} and $u_2 = Q \in [-50.0, 50.0]$ MJ h^{-1}. The CSTR, described by the equations of Eq. 4.51, has an open-loop asymptotically stable steady-state within the operating range of interest given by $C_{As} = 1.18$ kmol m^{-3} and $T_s = 440.9$ K with corresponding steady-state input values of $C_{A0} = 4.0$ kmol m^{-3} and $Q_s = 0$ MJh^{-1}. The steady-state and the steady-state input are denoted by $x_s^T = [C_{As}\ T_s]$ and $u_s^T = [C_{A0s}\ Q_s]$, respectively. The control objective is to regulate the process in a region around the steady-state to maximize the production rate of B. To accomplish the desired objective, the economic cost function considered in this example is:

$$l_e(x, u) = -k_0 e^{-E/RT} C_A^2 \tag{4.52}$$

which is equal to the negative of the instantaneous production rate of B, i.e., the production rate should be maximized. Also, there is limitation on the amount of reactant material which may be used over the length of operation t_f. Specifically, the input trajectory of u_1 should satisfy the following time-averaged constraint:

$$\frac{1}{t_f} \int_0^{t_f} u_1(t)\, dt = C_{A0,avg} = C_{A0s} \tag{4.53}$$

where t_f denotes the length of operation.

One method to ensure that the average constraint of Eq. 4.53 is satisfied over the entire length of operation is to divide the length of operation into equal-sized operating periods and to construct constraints that are imposed in the EMPC problem to ensure that the average constraint is satisfied over each consecutive operating period. This may be accomplished by using an inventory balance that accounts for the total amount of input energy available over each operating period compared to the total amount of input energy already used in the operating period. The main advantages of enforcing the average constraint in this fashion are that (1) only a limited number of constraints are required to be added to the EMPC, and (2) it ensures that the average constraint is satisfied over the length of operation.

To explain this type of input average constraint implementation, consider a general input average constraint given by:

$$\frac{1}{\tau_M}\int_0^{\tau_M} u(t)\,dt = u_{\text{avg}} \tag{4.54}$$

where τ_M is the operating period length that the average input constraint is imposed, i.e., $M = \tau_M/\Delta$ is the number of sampling periods in the operating period, and u_{avg} is the average input constraint value. Let τ_j denote the jth sampling time of the current operating period where τ_0 and τ_M denote the beginning and ending (time) of the current operating period, respectively. The constraint of Eq. 4.54 is enforced as follows: if the prediction horizon covers the entire operating period, then the average constraint may be enforced directly by imposing the following constraint in the EMPC problem:

$$\sum_{i=j}^{M-1} u(\tau_i) = M u_{\text{avg}} - \sum_{i=0}^{j-1} u^*(\tau_i|\tau_i)\,. \tag{4.55}$$

where $u^*(\tau_i|\tau_i)$ denotes the input computed and applied at sampling time $\tau_i \in [\tau_0, \tau_j)$. The integral of Eq. 4.54 may be converted to a sum in Eq. 4.55 because the input trajectory is piecewise constant.

If the prediction horizon does not cover the entire operating period, then the remaining part of the operating period not covered in the prediction horizon from τ_{j+N} to τ_M should be accounted for in the constraints to ensure feasibility at subsequent sampling times. Namely, at a sampling period $\tau_j \in [\tau_0, \tau_M)$, the following must be satisfied:

$$M u_{\text{avg}} - \sum_{i=j}^{j+N} u(\tau_i) - \sum_{i=0}^{j-1} u^*(\tau_i|\tau_i) \le \max\{M - N - j, 0\} u_{\max}\,, \tag{4.56a}$$

$$M u_{\text{avg}} - \sum_{i=j}^{j+N} u(\tau_i) - \sum_{i=0}^{j-1} u^*(\tau_i|\tau_i) \ge \max\{M - N - j, 0\} u_{\min} \tag{4.56b}$$

where $u_{\min}$ and $u_{\max}$ denote the minimum and maximum admissible input value. Equation 4.56 means that the difference between the total available input energy ($M u_{\text{avg}}$) and the total input energy used from the beginning of the operating period through the end of the prediction horizon must be less/greater than or equal to the total input energy if the maximum/minimum allowable input was applied over the remaining part of the operating period from τ_{k+N} to τ_M, which is the part of the operating period not covered in the prediction horizon.

If the prediction horizon extends over multiple consecutive operating periods, a combination of the constraints of Eqs. 4.55 and 4.56 may be employed. Let $N_{op} = \lceil (j+N)/M \rceil$ denote the number of operating periods covered in the prediction horizon. For the first operating period in the horizon, the constraint is given by:

$$\sum_{i=j}^{M-1} u(\tau_i) + \sum_{i=0}^{j-1} u^*(\tau_i|\tau_i) = M u_{\text{avg}} . \tag{4.57}$$

If $N_{op} > 2$, then, the following set of constraints is imposed:

$$\sum_{i=lM}^{(l+1)M-1} u(\tau_i) = M u_{\text{avg}}, \ \forall\, l \in \{1, 2, \ldots, N_{op} - 2\} . \tag{4.58}$$

For the last operating period covered in the horizon, the following constraint is used:

$$M u_{\text{avg}} - \sum_{j=(N_{op}-1)M}^{(j+N)-(N_{op}-1)M} u(\tau_j) \le \max\{N_{op} M - N - j, 0\} u_{\max} , \tag{4.59a}$$

$$M u_{\text{avg}} - \sum_{j=(N_{op}-1)M}^{(j+N)-(N_{op}-1)M} u(\tau_j) \ge \max\{N_{op} M - N - j, 0\} u_{\min} . \tag{4.59b}$$

The index j is reset to zero at the beginning of each operating period. In the implementation of input average constraints, it may be sufficient, in terms of acceptable closed-loop behavior, to impose the input average constraint over one or two operating periods if the horizon covers multiple operating periods.

An LEMPC is designed and applied to the CSTR model of Eq. 4.51. Since the economic cost does not penalize the use of control energy, the optimal operating strategy with respect to the economic cost of Eq. 4.52 is to operate with the maximum allowable heat rate supplied to the reactor for all time. However, this may lead to a large temperature operating range which may be impractical or undesirable. Therefore, a modified control objective is considered for more practical closed-loop operation of the CSTR under EMPC. The modified control objective is to maximize the reaction rate while feeding a fixed time-averaged amount of the reactant A to the process and while forcing and maintaining operation to/at a set-point temperature of T_s. Additionally, the temperature of the reactor contents must be maintained below the maximum allowable temperature $T \le T_{\max} = 470.0$ K, which is treated as a hard constraint and thus, $\mathbb{X} = \{x \in \mathbb{R}^2 : x_2 \le 470.0\}$.

A stabilizing state feedback controller, i.e., Lyapunov-based controller, is designed for the CSTR. The first input C_{A0} in the stabilizing controller is fixed to the average inlet concentration to satisfy the average input constraint of Eq. 4.53. The second input Q is designed via feedback linearization techniques while accounting for the input constraint. The gain of the feedback linearizing controller is $\gamma = 1.4$ (see [14] for more details regarding the controller design). A quadratic Lyapunov function is considered of the form $V(x) = \bar{x}^T P \bar{x}$ where $\bar{x}$ is the deviation of the states from their corresponding steady-state values and P is the following positive definite matrix:

$$P = \begin{bmatrix} 250 & 5 \\ 5 & 0.2 \end{bmatrix}. \tag{4.60}$$

The stability region of the CSTR under the Lyapunov-based controller is characterized by taking it to be a level set of the Lyapunov function where the time-derivative of the Lyapunov function along the closed-loop state trajectories is negative and is denoted as $\Omega_u = \{x \in \mathbb{R}^2 : V(x) \leq \rho_u\}$ where $\rho_u = 138$. However, $\mathbb{X} \subset \Omega_u$ which is shown in Fig. 4.3. Thus, the set Ω_ρ where $\rho = 84.76$ is defined to account for the state constraint.

Bounded Gaussian process noise is added to the CSTR with a standard deviation of $\sigma = [0.3,\ 5.0]^T$ and bound $\theta = [1.0,\ 20.0]^T$. A random noise vector is generated and applied additively to the right-hand side of the ODEs of Eq. 4.51 over the sampling period ($\Delta = 0.01$ h) and the bounds are given for each element of the noise vector ($|w_i| \leq \theta_i$ for $i = 1,\ 2$). Through extensive closed-loop simulations of the CSTR under the Lyapunov-based controller and under the LEMPC (described below) and with many realizations of the process noise, the set Ω_{ρ_e} is determined to be $\rho_e = 59.325$.

The first differential equation of Eq. 4.51 (C_A) is input-to-state-stable (ISS) with respect to T. Therefore, a contractive Lyapunov-based constraint may be applied to the LEMPC to ensure that the temperature converges to a neighborhood of the optimal steady-state temperature value. Namely, a Lyapunov function for the temperature ordinary differential equation (ODE) of Eq. 4.51b is defined and is given by: $V_T(T) := (T - T_s)^2$. The LEMPC formulation is given by:

$$\min_{u \in S(\Delta)} \quad -\int_{t_k}^{t_{k+N}} k_0 e^{-E/R\tilde{T}(\tau)} \tilde{C}_A^2(\tau)\, d\tau \tag{4.61a}$$

$$\text{s.t.} \quad \dot{\tilde{C}}_A(t) = \frac{F}{V}(u_1(t) - \tilde{C}_A(t)) - k_0 e^{-E/R\tilde{T}(t)} \tilde{C}_A^2(t) \tag{4.61b}$$

$$\dot{\tilde{T}}(t) = \frac{F}{V}(T_0 - \tilde{T}(t)) - \frac{\Delta H k_0}{\rho_L C_p} e^{-E/R\tilde{T}(t)} \tilde{C}_A^2(t) + \frac{u_2(t)}{\rho_L C_p V_R} \tag{4.61c}$$

$$\tilde{C}_A(t_k) = C_A(t_k),\ \tilde{T}(t_k) = T(t_k) \tag{4.61d}$$

$$u_1(t) \in [0.5, 7.5]\,\text{kmol m}^{-3},\ \forall\, t \in [t_k, t_{k+N}) \tag{4.61e}$$

$$u_2(t) \in [-50.0, 50.0]\,\text{MJ h}^{-1},\ \forall\, t \in [t_k, t_{k+N}) \tag{4.61f}$$

$$\sum_{i=j}^{M-1} u_1(\tau_i) + \sum_{i=0}^{j-1} u_1^*(\tau_i|\tau_i) = M C_{A0s} \tag{4.61g}$$

$$MC_{A0s} - \sum_{j=M}^{j+N-M} u_1(\tau_j) \le \max\{2M - N - j, 0\}u_{1,\max} \tag{4.61h}$$

$$MC_{A0s} - \sum_{j=M}^{j+N-M} u_1(\tau_j) \ge \max\{2M - N - j, 0\}u_{1,\min} \tag{4.61i}$$

$$\tilde{T}(t) \le T_{\max} \tag{4.61j}$$

$$V(\tilde{x}(t)) \le \rho_e, \ \forall\, t \in [t_k, t_{k+N}) \tag{4.61k}$$

$$\frac{\partial V_T(T(t_k))}{\partial T} f_2(\tilde{x}(t_k), u(t_k), 0) \le \frac{\partial V_T(T(t_k))}{\partial T} f_2(\tilde{x}(t_k), h(\tilde{x}(t_k)), 0) \tag{4.61l}$$

where $f_2(\cdot)$ is the right-hand side of the ODE of Eq. 4.51b. The CSTR is initialized at many states distributed throughout state-space including some cases where the initial state is outside Ω_u. The LEMPC described above is applied to the CSTR with an operating period over which to enforce the average input constraint of $M = 20$ ($\tau_M = 0.2$ h) and a prediction horizon of $N = 20$.

Several simulations of 50.0 h length of operation are completed. In all cases, the LEMPC is able to force the system to Ω_ρ and maintain operation inside Ω_ρ without violating the state constraint. The closed-loop state trajectories over the first 1.0 h are shown in Fig. 4.3 for one initial condition starting inside Ω_ρ and one starting outside Ω_u. From Fig. 4.3, the LEMPC forces the temperature to a neighborhood of the temperature set-point where it maintains the temperature thereafter. The reactant

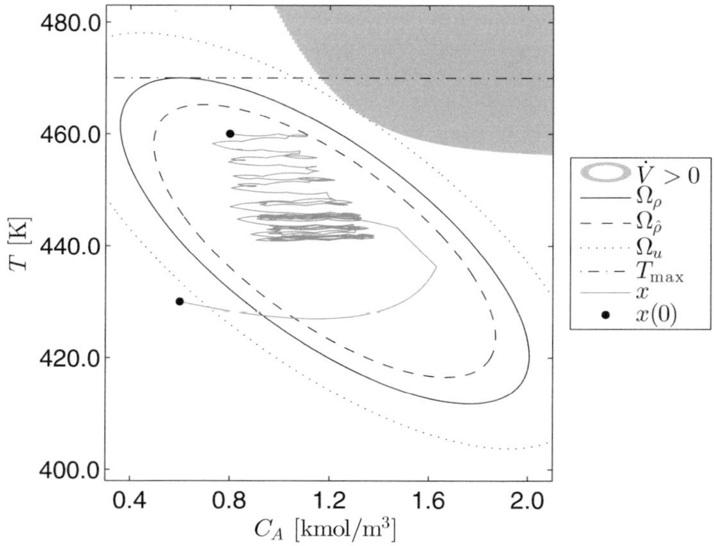

Fig. 4.3 Two closed-loop state trajectories under the LEMPC in state-space

Table 4.2 Average economic cost over several simulations under the LEMPC, the Lyapunov-based controller applied in a sample-and-hold fashion, and the constant input equal to u_s

$\bar{J}_E$ under LEMPC	$\bar{J}_E$ under $k(x)$	$\bar{J}_E$ under u_s
14.17	14.10	14.09
14.18	14.11	14.09
14.17	14.10	14.08
14.17	14.09	14.06
14.18	14.10	14.10
14.17	14.09	14.08
14.18	14.09	14.10
14.18	14.08	14.08
14.19	14.08	14.10
14.18	14.07	14.07
14.18	14.11	14.11
14.18	14.08	14.07
14.17	14.06	0.36*
14.19	14.06	14.10

For the case denoted with a "*", the system under the constant input u_s settled at a different steady-state, i.e., not x_s

concentration trajectory varies with time and never settles at a steady-state owing to a periodic-like forcing of the inlet concentration.

The CSTR was simulated with the same realization of the process noise and same initial condition under the Lyapunov-based controller applied in a sample-and-hold fashion and under a constant input equal to the steady-state input. To evaluate the average economic cost under LEMPC and under the other two control strategies, the following index is defined:

$$\bar{J}_e := \frac{1}{t_f} \int_0^{t_f} l_e(x(t), u(t))\, dt \tag{4.62}$$

where x and u denote the closed-loop state and input trajectories. The average economic cost over each of these simulations is reported in Table 4.2. Owing to the fact that the simulations are performed over many operating periods, the average closed-loop performance index is essentially a measure of the asymptotic average performance. The transient performance is also discussed below. From these results, an average of 0.6 % closed-loop performance benefit is observed with the LEMPC over the Lyapunov-based controller and the constant input u_s^*. It is important to note that for one of the simulations that was initialized outside Ω_u the CSTR under the constant input u_s^* settled on an offsetting steady-state which is denoted with an asterisk in Table 4.2.

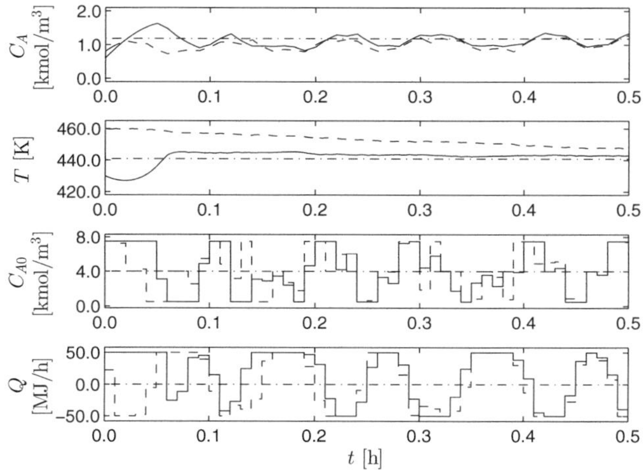

Fig. 4.4 The closed-loop state and input trajectories of the CSTR under the LEMPC of Eq. 4.61 for two initial conditions (*solid* and *dashed trajectories*) and the steady-state is the *dashed-dotted line*

Figure 4.4 gives the state and input trajectories of the CSTR under the LEMPC of Eq. 4.61 for two initial conditions: (0.6 kmol m^{-3}, 430.0 K), referred to as the low temperature initial condition, and (0.8 kmol m^{-3}, 460.0 K), referred to as the high temperature initial condition. One of the most interesting results of these simulations is the asymmetric responses about the temperature set-point dictated by the LEMPC. For the low temperature initial condition, the heat rate input computed by the LEMPC is initially at the maximum admissible value to force the temperature to the set-point as fast as possible. This behavior is desirable with respect to the economic cost because the production rate is monotonically increasing with temperature. On the other hand, for the high temperature initial condition, the rate at which the computed input trajectory of the LEMPC forces the temperature to the set-point is much slower rate than that of the case with the lower initial temperature. Again, this type of behavior is desirable owing to the fact that the production rate of B is greater at higher temperature.

To quantify the difference in the transient closed-loop performance under LEMPC relative to the transient closed-loop performance achieved under the Lyapunov-based controller and under the constant input equal to u_s, the closed-loop economic performance over the first operating period is used, i.e., the index $\bar{J}_e$ with $t_f = 0.2$ h. The performance benefit under the LEMPC relative to that achieved under the Lyapunov-based controller and under the constant input is 12.15 % and 20.79 %, respectively, and a clear advantage on transient performance is realized under LEMPC. Moreover, the asymmetric response dictated by the LEMPC is a unique property of EMPC that addresses a potential drawback of tracking MPC methodologies. In particular, the stage cost of tracking MPC typically penalize positive and negative deviation from the set-point equally even though from an economic stand-point this may not be appropriate.

4.4 Closed-Loop Performance Under LEMPC

Owing to the availability of the explicit stabilizing (Lyapunov-based) controller, the corresponding Lyapunov function, and the stability region used to design LEMPC, a terminal equality constraint may be readily designed for the LEMPC problem. The terminal equality constraint allows for guaranteed closed-loop performance properties, while maintaining the unique recursive feasibility property of LEMPC. Nominally operated systems are considered, i.e., the system of Eq. 4.1 with $w \equiv 0$ and with slight abuse of notation, the nominally operated system will be written as

$$\dot{x} = f(x, u) \tag{4.63}$$

where $x \in \mathbb{X}$ and $u \in \mathbb{U}$. The system of Eq. 4.63 satisfies all of the relevant assumptions as the system of Eq. 4.1.

An economic cost $l_e : \mathbb{X} \times \mathbb{U} \to \mathbb{R}$ is assumed for the system of Eq. 4.63 which is continuous on $\mathbb{X} \times \mathbb{U}$. The existence of a steady-state and steady-state input pair, denoted as (x_s^*, u_s^*), that minimizes the economic cost in the sense that the minimum of l_e is attained at the pair (x_s^*, u_s^*) is assumed. For simplicity, the minimizing pair is assumed to be unique. With these assumptions, the minimizing steady-state pair is given by:

$$(x_s^*, u_s^*) = \underset{x_s \in \mathbb{X}, u_s \in \mathbb{U}}{\arg\min} \left\{ l_e(x_s, u_s) : f(x_s, u_s) = 0 \right\} .$$

Without loss of generality, the minimizing pair will be taken to be the origin of Eq. 4.63, i.e., $f(0, 0) = 0$.

4.4.1 Stabilizability Assumption

A terminal equality constraint imposed in the LEMPC optimization problem will be computed at each sampling time based on the solution of the sampled-data system consisting of the continuous-time system of Eq. 4.63 with an explicit controller applied in a sample-and-hold fashion. To consider infinite-time closed-loop economic performance (to be made precise below), a stronger assumption is considered on the explicit controller (Lyapunov-based controller). A relaxation of the assumption is discussed in a remark below.

Assumption 4.1 There exists a locally Lipschitz feedback controller $h : \mathbb{X} \to \mathbb{U}$ with $h(0) = 0$ for the system of Eq. 4.1 that renders the origin of the closed-loop system under continuous implementation of the controller locally exponentially stable. More specifically, there exist constants $\rho > 0$, $c_i > 0$, $i = 1, 2, 3, 4$ and a continuously differentiable Lyapunov function $V : \mathbb{X} \to \mathbb{R}_+$ such that the following inequalities hold:

$$c_1 |x|^2 \leq V(x) \leq c_2 |x|^2 , \tag{4.64a}$$

$$\frac{\partial V(x)}{\partial x} f(x, h(x)) \leq -c_3 |x|^2, \tag{4.64b}$$

$$\left| \frac{\partial V(x)}{\partial x} \right| \leq c_4 |x|, \tag{4.64c}$$

for all $x \in \Omega_\rho \subseteq \mathbb{X}$.

When the controller of Assumption 4.1 is applied in a sample-and-hold (zeroth-order hold) fashion, i.e., the controller is an emulation controller, the resulting closed-loop system is a sampled-data system. To distinguish from the state and input trajectories of the system under the emulation controller and the state and input trajectories of the closed-loop system under LEMPC, z and v will be used for the former, and x and u^* will be used for the latter, respectively. The sampled-data system consisting of the system of Eq. 4.63 under the sample-and-hold implementation of the explicit controller is given by:

$$\begin{aligned} \dot{z}(t) &= f(z(t), v(t)) \\ v(t) &= h(z(t_k)) \end{aligned} \tag{4.65}$$

for $t \in [t_k, t_{k+1})$ where $t_k = k\Delta$ and $k = 0, 1, \ldots$ with initial condition $z(0) = z_0 \in \Omega_\rho$. From Corollary 2.2, the origin of the closed-loop system of Eq. 4.65 is exponentially stable under sufficiently fast sampling, i.e., the sampling period, Δ, is sufficiently small. Moreover, from the proof of Corollary 2.2, V is a Lyapunov function for the closed-loop sampled-data system in the sense that there exists a constant $\hat{c}_3 > 0$ such that

$$\frac{\partial V(z(t))}{\partial z} f(z(t), h(z(t_k))) \leq -\hat{c}_3 |z(t)|^2 \tag{4.66}$$

for all $t \in [t_k, t_{k+1})$ and integers $k \geq 0$ where $z(t)$ is the solution of Eq. 4.65 at time t with initial condition $z(0) = x(0) \in \Omega_\rho$ where $x(0)$ denotes a state measurement of the system of Eq. 4.63 at the initial time.

4.4.2 *Formulation and Implementation of the LEMPC with a Terminal Equality Constraint*

The solution of Eq. 4.65 may be leveraged in the design of a terminal equality constraint. Specifically, the system of Eq. 4.65 is initialized with a state measurement of the system of Eq. 4.63 at $t = 0$. Using forward simulation of the system of Eq. 4.65, the state may be computed at t_{k+N}. The computed state is then used as a terminal constraint in LEMPC in the sense that the predicted state of the LEMPC must converge to the state of Eq. 4.65, i.e., $\tilde{x}(t_{k+N}) = z(t_{k+N})$. Using this design principle, the formulation of the LEMPC with a terminal equality constraint formulated based on z is given by the problem:

$$\min_{u \in S(\Delta)} \int_{t_k}^{t_{k+N}} l_e(\tilde{x}(\tau), u(\tau))\, d\tau \tag{4.67a}$$

$$\text{s.t.} \quad \dot{\tilde{x}}(t) = f(\tilde{x}(t), u(t)) \tag{4.67b}$$

$$\tilde{x}(t_k) = x(t_k) \tag{4.67c}$$

$$\tilde{x}(t_{k+N}) = z(t_{k+N}) \tag{4.67d}$$

$$u(t) \in \mathbb{U},\ \forall\, t \in [t_k, t_{k+N}) \tag{4.67e}$$

$$V(\tilde{x}(t)) \leq \rho_e,\ \forall\, t \in [t_k, t_{k+N}) \quad \text{if } V(x(t_k)) \leq \rho_e \text{ and } t_k < t_s \tag{4.67f}$$

$$\frac{\partial V(x(t_k))}{\partial x} f(x(t_k), u(t_k)) \leq \frac{\partial V(x(t_k))}{\partial x} f(x(t_k), h(x(t_k))) \quad \text{if } V(x(t_k)) > \rho_e \text{ or } t_k \geq t_s \tag{4.67g}$$

where all the components of the LEMPC optimization problem of Eq. 4.67 are similar to that of the problem of Eq. 4.3 except for the additional constraint of Eq. 4.67d. The terminal constraint of Eq. 4.67d enforces that the computed input trajectory steers the predicted state trajectory to the state $z(t_{k+N})$ at the end of the prediction horizon. The terminal constraint of Eq. 4.67d differs from traditional terminal equality constraints in the sense that it is not necessarily a steady-state. However, it is important to note that the terminal constraint in the LEMPC of Eq. 4.67 exponentially converges to the steady-state owing to the stability properties of the explicit controller.

The implementation of the LEMPC of Eq. 4.67 is similar to that of the LEMPC of Eq. 4.3. Before the optimization problem of Eq. 4.67 is solved, the terminal constraint, $z(t_{k+N})$, is computed by recursively solving the system of Eq. 4.65 over t_{k+N-1} to t_{k+N} and is initialized with $z(t_{k+N-1})$, which corresponds to the terminal constraint at the previous sampling time. At the initial time ($t = 0$), $z(t_N)$ is computed by first initializing the system of Eq. 4.65 with $x(0)$ and recursively solving this system from the initial time to $t_N = N\Delta$. For added robustness, especially to numerical and discretization error, one may reinitialize the system of Eq. 4.65 with a state measurement at each sampling time, i.e., initialize the system of Eq. 4.65 with $x(t_k)$ and use forward simulation to compute $z(t_{k+N})$. However, in the nominal scenario considered here, numerical error is not considered.

4.4.3 Closed-Loop Performance and Stability Analysis

The closed-loop stability properties of the LEMPC follows from the results of the previous section and are stated in the following corollary.

Corollary 4.1 *Let the conditions of Theorem 4.1 be satisfied. If $x(0) \in \Omega_\rho$, then the closed-loop state trajectory of the system of Eq. 4.63 under the LEMPC of Eq. 4.67 based on the Lyapunov-based controller that satisfies Eq. 4.2 is always bounded in Ω_ρ for all $t \geq 0$. Moreover, the LEMPC problem of Eq. 4.67 is recursively feasible.*

Proof If similar conditions are satisfied as that of Theorem 4.1 and the LEMPC problem of Eq. 4.67 is feasible, it follows from Theorem 4.1 that the closed-loop state is bounded in Ω_ρ. Initial feasibility at $t = 0$ follows from the fact that the sample-and-hold input trajectory used to compute the terminal constraint, $z(t_N)$, is a feasible solution to the optimization problem because (1) it forces the predicted state to the terminal constraint, (2) satisfies the input constraint, (3) maintains the state in Ω_{ρ_e} if $x(0) \in \Omega_{\rho_e}$ or trivially satisfies the contractive constraint of Eq. 4.67g. For the subsequent sampling time ($t = \Delta$), the piecewise defined input trajectory consisting of $u^*(t|0)$ for $t \in [\Delta, N\Delta)$ and $u(t) = h(z(t_N))$ for $t \in [N\Delta, (N+1)\Delta)$ is a feasible solution to the optimization problem at $t = \Delta$. Applying this result recursively for all future sampling times, recursive feasibility follows.

Finite-time and infinite-time economic performance is considered. The analysis follows closely that of [15], which analyzes closed-loop performance of EMPC formulated with an equality terminal constraint equal to x_s^*. Let $J_e^*(x(t_k))$ denote the optimal value of Eq. 4.67a at time t_k given the state measurement $x(t_k)$. The first result gives the finite-time average performance under the LEMPC of Eq. 4.67. Without loss of generality, take $l_e(x, u) \geq 0$ for all $x \in \Omega_\rho$ and $u \in \mathbb{U}$.

Theorem 4.3 *Consider the system of Eq. 4.65 under the LEMPC of Eq. 4.67 based on the Lyapunov-based controller that satisfies Eq. 4.2. For any strictly positive finite integer, T, the closed-loop average economic performance is bounded by:*

$$\int_0^{T\Delta} l_e(x(t), u^*(t))\, dt \leq \int_0^{(T+N)\Delta} l_e(z(t), v(t))\, dt \tag{4.68}$$

where x and u^* denote the closed-loop state and input trajectories and z and v denote the resulting state and input trajectories of the system of Eq. 4.65.

Proof Let $u^*(t|t_k)$ for $t \in [t_k, t_{k+N})$ be the optimal input trajectory of Eq. 4.67 at t_k. The piecewise defined input trajectory consisting of $u^*(t|t_k)$ for $t \in [t_{k+1}, t_{k+N})$ and $u(t) = h(z(t_{k+N}))$ for $t \in [t_{k+N}, t_{k+N+1})$ is a feasible solution to the optimization problem at the next sampling time (t_{k+1}). Utilizing the feasible solution to the problem of Eq. 4.67 at t_{k+1}, the difference between the optimal value of Eq. 4.67a at any two successive sampling times may be bounded as follows:

$$\begin{aligned} & J_e^*(x(t_{k+1})) - J_e^*(x(t_k)) \\ & \leq \int_{t_{k+N}}^{t_{k+N+1}} l_e(z(t), h(z(t_{k+N})))\, dt - \int_{t_k}^{t_{k+1}} l_e(x(t), u^*(t_k|t_k))\, dt\,. \end{aligned} \tag{4.69}$$

Let T be any strictly positive integer. Summing the difference of the optimal value of Eq. 4.67a at two subsequent sampling times, the sum may be lower bounded by:

$$\begin{aligned}\sum_{k=0}^{T-1}\left[J_e^*(x(t_{k+1})) - J_e^*(x(t_k))\right] &= J_e^*(x(T\Delta)) - J_e^*(x(0)) \\ &\geq -J_e^*(x(0)) \end{aligned} \tag{4.70}$$

where the inequality follows from the fact that $l_e(x, u) \geq 0$ for all $x \in \Omega_\rho$ and $u \in \mathbb{U}$. At the initial time, the optimal value of Eq. 4.67a may be bounded by the cost under the explicit controller over the prediction horizon:

$$J_e^*(x(0)) \leq \int_0^{t_N} l_e(z(t), v(t))\, dt\ . \tag{4.71}$$

Moreover, the left-hand side of Eq. 4.70 may be upper bounded as follows:

$$\begin{aligned}&\sum_{k=0}^{T-1}\left[J_e^*(x(t_{k+1})) - J_e^*(x(t_k))\right] \\ &\quad\leq \sum_{k=0}^{T-1}\left(\int_{t_{k+N}}^{t_{k+N+1}} l_e(z(t), h(z(t_{k+N})))\, dt - \int_{t_k}^{t_{k+1}} l_e(x(t), u^*(t_k|t_k))\, dt\right) \\ &\quad= \int_{t_N}^{(T+N)\Delta} l_e(z(t), v(t))\, dt - \int_0^{T\Delta} l_e(x(t), u^*(t))\, dt \end{aligned} \tag{4.72}$$

which follows from Eq. 4.69. From Eqs. 4.70 and 4.72, the closed-loop economic performance over the initial time to $T\Delta$ is no worse than the closed-loop performance under the explicit control from initial time to $(T + N)\Delta$:

$$\begin{aligned}\int_0^{T\Delta} l_e(x(t), u^*(t))\, dt &\leq J_e^*(x(0)) + \int_{t_N}^{T\Delta+N\Delta} l_e(z(t), v(t))\, dt \\ &\leq \int_0^{(T+N)\Delta} l_e(z(t), v(t))\, dt \end{aligned} \tag{4.73}$$

where the second inequality follows from Eq. 4.71. This proves the bound of Eq. 4.68.

The upper limit of integration of the right-hand side of Eq. 4.68, i.e., $(T + N)\Delta$, arises from the fact that a fixed prediction horizon is used in the LEMPC of Eq. 4.67. If, instead, $T\Delta$ represents the final operating time of a given system, one could employ a shrinking horizon from $(T - N)\Delta$ to $T\Delta$ in the LEMPC and the upper limit of integration of the right-hand side of Eq. 4.68 would be $T\Delta$. Also, as a consequence of the performance bound of Eq. 4.68, the average finite-time economic performance may be bounded as follows:

$$\frac{1}{T\Delta}\int_0^{T\Delta} l_e(x(t), u^*(t))\,dt \le \frac{1}{T\Delta}\int_0^{T\Delta} l_e(z(t), v(t))\,dt + \frac{1}{T\Delta}\int_{T\Delta}^{(T+N)\Delta} l_e(z(t), v(t))\,dt \tag{4.74}$$

for any integer $T > 0$ (and for the fixed horizon case). From the right-hand side of Eq. 4.74, the second term of the right-hand side is less significant as T gets large.

To consider asymptotic average closed-loop economic performance of the system under the LEMPC of Eq. 4.67, the asymptotic average closed-loop performance under the explicit controller needs to be considered. Because the state and input trajectory asymptotically converge to (x_s^*, u_s^*), it is straightforward to show that the asymptotic average economic performance under the explicit controller is no worse than the economic cost at the optimal steady-state pair, which is stated in the lemma below.

Lemma 4.1 *The asymptotic average economic cost of the system of Eq. 4.65 under the Lyapunov-based controller that satisfies Assumption 4.1 for any initial condition* $z(0) \in \Omega_\rho$ *is*

$$\lim_{T\to\infty}\frac{1}{T\Delta}\int_0^{T\Delta} l_e(z(t), v(t))\,dt = l_e(x_s^*, u_s^*) \tag{4.75}$$

where $\Delta \in (0, \Delta^*)$ *(*$\Delta^* > 0$ *is defined according to Corollary 2.2) and* z *and* v *denote the state and input trajectories of the system of Eq. 4.65.*

Proof Recall, the economic stage cost function l_e is continuous on the compact set $\Omega_\rho \times \mathbb{U}$ and $z(t) \in \Omega_\rho$ and $v(t) \in \mathbb{U}$ for all $t \ge 0$. Thus, the integral:

$$\frac{1}{T\Delta}\int_0^{T\Delta} l_e(z(t), v(t))\,dt < \infty \tag{4.76}$$

for any integer $T > 0$. Since $z(t)$ and $v(t)$ exponentially converge to the optimal steady-state pair (x_s^*, u_s^*) as $t \to \infty$, the limit of the integral of Eq. 4.76 as T tends to infinity exists and is equal to $l_e(x_s^*, u_s^*)$. To prove the limit, it is sufficient to show that for any $\varepsilon > 0$, there exists a T^* such that for $T > T^*$, the following holds:

$$\left|\frac{1}{T\Delta}\int_0^{T\Delta} l_e(z(t), v(t))\,dt - l_e(x_s^*, u_s^*)\right| < \varepsilon \tag{4.77}$$

To simplify the presentation, define $I(T_1, T_2)$ as the following integral:

$$I(T_1, T_2) := \int_{T_1\Delta}^{T_2\Delta} l_e(z(t), v(t))\,dt \tag{4.78}$$

where the arguments of I are integers representing the integers of the lower and upper limits of integration, respectively. Since $z(t)$ and $v(t)$ converge to x_s^* and u_s^* as t tends to infinity, respectively, $l_e(x(t), v(t)) \to l_e(x_s^*, u_s^*)$ as t tends to infinity.

Furthermore, $z(t) \in \Omega_\rho$ and $v(t) \in \mathbb{U}$ for all $t \geq 0$, so for every $\varepsilon > 0$, there exists an integer $\tilde{T} > 0$ such that

$$|l_e(z(t), v(t)) - l_e(x_s^*, u_s^*)| < \varepsilon/2 \tag{4.79}$$

for $t \geq \tilde{T}\Delta$. For any $T > \tilde{T}$,

$$\begin{aligned} |I(0,T) - T\Delta l_e(x_s^*, u_s^*)| &= |I(0,\tilde{T}) + I(\tilde{T},T) - T\Delta l_e(x_s^*, u_s^*)| \\ &\leq \int_0^{\tilde{T}\Delta} |l_e(z(t), v(t)) - l_e(x_s^*, u_s^*)|\, dt \\ &\quad + \int_{\tilde{T}\Delta}^{T\Delta} |l_e(z(t), v(t)) - l_e(x_s^*, u_s^*)|\, dt \\ &\leq \tilde{T}\tilde{M} + (T - \tilde{T})\varepsilon/2 \end{aligned} \tag{4.80}$$

where

$$\tilde{M} := \sup_{t \in [0, \tilde{T}\Delta]} \left\{ |l_e(z(t), v(t)) - l_e(x_s^*, u_s^*)| \right\} .$$

For any $T > T^* = 2\tilde{T}(\tilde{M} - \varepsilon/2)/\varepsilon$ (which implies $(\tilde{M} - \varepsilon/2)\tilde{T}/T < \varepsilon/2$), the following inequality is satisfied:

$$\begin{aligned} |I(0,T)/T - l_e(x_s^*, u_s^*)| &\leq \tilde{T}\tilde{M}/T + (1 - \tilde{T}/T)\varepsilon/2 \\ &= (\tilde{M} - \varepsilon/2)\tilde{T}/T + \varepsilon/2 < \varepsilon \end{aligned} \tag{4.81}$$

which proves the limit of Eq. 4.75.

With Lemma 4.1, the asymptotic average closed-loop economic performance under the LEMPC is no worse than the closed-loop performance at the economically optimal steady-state.

Theorem 4.4 *Consider the system of Eq. 4.65 under the LEMPC of Eq. 4.67 based on the Lyapunov-based controller that satisfies Assumption 4.1. Let* $\Delta \in (0, \Delta^*)$ *(*$\Delta^* > 0$ *is defined according to Corollary 2.2). The closed-loop asymptotic average economic performance is no worse than the economic cost at steady-state, i.e., the following bound holds:*

$$\limsup_{T \to \infty} \frac{1}{T\Delta} \int_0^{T\Delta} l_e(x(t), u^*(t))\, dt \leq l_e(x_s^*, u_s^*) \tag{4.82}$$

Proof From Theorem 4.3, for any $T > 0$:

$$\frac{1}{T\Delta} \int_0^{T\Delta} l_e(x(t), u^*(t))\, dt \leq \frac{1}{T\Delta} \int_0^{(T+N)\Delta} l_e(z(t), v(t))\, dt . \tag{4.83}$$

As T increases, both sides of the inequality of Eq. 4.83 remain finite owing to the fact that l_e is continuous and the state and input trajectories are bounded in compact sets. The limit of the right-hand side as $T \to \infty$ is equal to $l_e(x_s^*, u_s^*)$ (Lemma 4.1). Therefore, one may readily obtain that:

$$\limsup_{T\to\infty} \frac{1}{T\Delta} \int_0^{T\Delta} l_e(x(t), u^*(t))\, dt \le l_e(x_s^*, u_s^*) \tag{4.84}$$

which proves the desired result.

While the finite-time performance results of Theorem 4.3 require that the explicit Lyapunov-based controller be designed such that Eq. 4.2 is satisfied, the infinite-time average performance results of Theorem 4.4 require that the Lyapunov-based controller satisfies the stronger assumption (Assumption 4.1), which is required to obtain the performance bound of Eq. 4.82. A Lyapunov-based controller that satisfies Eq. 4.2 when implemented in a sample-and-hold fashion with a sufficiently small sampling period will force the state to a small compact set containing the steady-state. When the LEMPC of Eq. 4.67 is designed with a Lyapunov-based controller that only satisfies Eq. 4.2, a weaker result on the asymptotic average economic closed-loop performance is obtained. Namely, the closed-loop asymptotic average performance may be bounded as follows:

$$\limsup_{T\to\infty} \frac{1}{T\Delta} \int_0^{T\Delta} l_e(x(t), u^*(t))\, dt \le \max_{x,z\in\Omega_{\rho_{\min}}} l_e(x, h(z)) \tag{4.85}$$

where $\Omega_{\rho_{\min}}$ is defined as in Theorem 4.1. Note that the size of set $\Omega_{\rho_{\min}}$ may be made arbitrarily small by making the sampling period arbitrarily small.

Remark 4.3 For systems with average constraints e.g., like that imposed in the example of Sect. 4.3.3, the average constraint design methodologies for asymptotic average constraints [15] and for transient average constraints [16], which were presented for EMPC with a terminal equality constraint equal to x_s^*, may be extended to the LEMPC of Eq. 4.67 when the average constraint is satisfied under the explicit controller. The methods of [15, 16] go beyond imposing the average constraint over successive operating periods, which is the method employed in Sect. 4.3.3.

Remark 4.4 The performance results of this section hold for any prediction horizon size even when $N = 1$. The use of a short horizon may be computationally advantageous for real-time application. Also, owing to the fact that the terminal equality constraint of Eq. 4.67d may be a point in the state-space away from the steady-state, it is expected that the feasible region of the LEMPC of Eq. 4.67 would be larger than the feasible region of EMPC with a terminal equality constraint equal to the steady-state for most cases especially when a short prediction horizon is used in the EMPC formulation.

4.5 LEMPC with a Time-Varying Stage Cost

One of the unique advantages of EMPC relative to other control methodologies is the integration of economic objectives directly within a control framework. For stability purposes, most of the EMPC schemes use a steady-state to impose constraints in the EMPC optimization problem to ensure closed-loop stability in their formulations, e.g., the EMPC formulated with a terminal constraint described in Chap. 3 and the Lyapunov-based constraints of Eqs. 4.3e and 4.3f used in the LEMPC of Eq. 4.3. Also, these EMPC schemes have been formulated with time-invariant economic cost functions. However, when the time-scale or update frequency of time-varying economic considerations arising from, for example, variable energy pricing or product demand changes is comparable to the time-scale of the process/system dynamics, it may be desirable to formulate the EMPC scheme with a time-dependent cost function.

In this section, an LEMPC scheme is developed that may accommodate an explicitly time-varying economic cost function. First, the formulation of the LEMPC scheme is presented. Second, closed-loop stability, in the sense of boundedness of the closed-loop state, is proven through a theoretical treatment of the LEMPC scheme. The LEMPC scheme is applied to a chemical process example to demonstrate that the LEMPC with time-varying economic cost achieves closed-loop stability and results in improved closed-loop economic performance over a conventional approach to optimization and control.

4.5.1 Class of Economic Costs and Stabilizability Assumption

Consider the class of systems described by the system of Eq. 4.1 with all of the relevant assumptions. Instead of the time-invariant economic cost, the system of Eq. 4.1 is assumed to be equipped with a time-dependent economic cost function, which has the following form $l_e(t, x, u)$ (the function l_e maps time and the state and input vectors to a scalar that is a measure of the economics, i.e., $l_e : [0, \infty) \times \mathbb{X} \times \mathbb{U} \rightarrow \mathbb{R}$). No restriction on the form of the cost function is required for stability. However, some limitations to the cost function that may be considered must be made to solve the optimization problem. From a practical point-of-view, many of the cost functions that would be used to describe the economics of a system may be piecewise continuous functions of time and continuous with respect to the state and input vectors.

In a traditional or conventional approach, if the economics change results in a change in the optimal operating steady-state, the optimal steady-state is updated and the updated optimal steady-state is sent down to a feedback controller, e.g., tracking MPC, to force the system to operate at the updated optimal steady-state. To account for the various potential operating steady-states, the existence of a set of steady-states for the system of Eq. 4.1, which is denoted as $\Gamma = \{x_s \in \mathbb{R}^n : \exists\, u_s \in \mathbb{U} \text{ s.t. } f(x_s, u_s, 0) = 0\} \subset \mathbb{X}$, is assumed for the system of Eq. 4.1. An additional assumption is made on the set Γ to ensure that the acceptable operating region is non-empty which is stated below. For a given system, the equilibrium manifold Γ may be taken as the set of admissible operating steady-states, i.e., the set of possible operating points under the conventional approach to optimization and control.

As in the case for the LEMPC formulated with a time-invariant economic stage cost, a stabilizability assumption is needed. For each $x_s \in \Gamma$, the existence of a Lyapunov-based controller that renders x_s of the nominal system of Eq. 4.1 asymptotically stable under continuous implementation is assumed. For simplicity of notation, the notation $h(x; x_s)$ where x_s is a parameter which is used to denote the Lyapunov-based controller with respect to $x_s \in \Gamma$, i.e., the control law $h(x; x_s)$ renders the steady-state x_s asymptotically stable for the nominally operated closed-loop system. Also, for two points in Γ, e.g., $x_{s,1}, x_{s,2} \in \Gamma$, no relationship is assumed between the two controllers $h(x; x_{s,1})$ and $h(x; x_{s,2})$ other than the former renders the steady-state $x_{s,1}$ asymptotically stable and the latter renders the steady-state $x_{s,2}$ asymptotically stable. Thus, the two controllers may be designed utilizing different techniques.

Using converse theorems, the existence of Lyapunov functions $V(\cdot; x_s)$ for all $x_s \in \Gamma$ follows from the stabilizability assumption. The Lyapunov functions satisfy the following conditions:

$$\alpha_1(|x - x_s|; x_s) \leq V(x; x_s) \leq \alpha_2(|x - x_s|; x_s) \tag{4.86a}$$

$$\frac{\partial V(x; x_s)}{\partial x} f(x, h(x; x_s), 0) \leq -\alpha_3(|x - x_s|; x_s) \tag{4.86b}$$

$$\left| \frac{\partial V(x; x_s)}{\partial x} \right| \leq \alpha_4(|x - x_s|; x_s) \tag{4.86c}$$

for $x \in D(x_s)$ and each $x_s \in \Gamma$ where $\alpha_i(\cdot; x_s)$, $i = 1, 2, 3, 4$ are class $\mathscr{K}$ function and $D(x_s)$ is an open neighborhood of x_s. Owing to the fact that there exists a Lyapunov function for each x_s, different class $\mathscr{K}$ functions exist for each Lyapunov function. This is captured by the parameterization of the functions α_i, $i = 1, 2, 3, 4$, and $\alpha_i(\cdot; x_s)$ denotes the ith class $\mathscr{K}$ function for the Lyapunov function with respect to the steady-state x_s.

For each $x_s \in \Gamma$, the stability region $\Omega_{\rho(x_s)}$ may be characterized for the closed-loop system of Eq. 4.1 with the Lyapunov-based controller $h(x; x_s)$. The symbol $\Omega_{\rho(x_s)}$ where $x_s \in \Gamma \subset \mathbb{R}^n$ is a fixed parameter denotes a level set of the Lyapunov function with respect to x_s, i.e., $\Omega_{\rho(x_s)} = \{x \in \mathbb{R}^n : V(x; x_s) \leq \rho(x_s)\}$ where $\rho(x_s)$ depends on x_s. The union of the stability regions is denoted as $\mathscr{X} = \bigcup_{x_s \in \Gamma} \Omega_{\rho(x_s)}$ and it is assumed to be a non-empty, compact, and connected set.

4.5.2 *The Union of the Stability Regions*

A simple demonstration of the construction of the set $\mathscr{X}$ is provided to embellish the concept of the union set $\mathscr{X}$. The stability region of a closed-loop system under an explicit stabilizing control law may be estimated for a steady-state in Γ through the off-line computation described below. After the stability regions of sufficiently many steady-states in Γ are computed, the union of these sets may be described algebraically through various mathematical techniques, e.g., curve fitting and convex optimization techniques. The basic algorithm is

1. For $j = 1$ to J (if Γ consists of an infinite number of points, J is a sufficiently large positive integer).
   1.1 Select a steady-state, $x_{s,j}$, in the set Γ.
   1.2 Partition the state-space near $x_{s,j}$ into I discrete points (I is a sufficiently large positive integer).
   1.3 Initialize $\rho(x_{s,j}) := \infty$.
   1.4 For $i = 1$ to I:
      1.4.1 Compute $\dot{V}(x_i; x_{s,j})$ where x_i denotes the i discrete point from the partitioning of the state-space. If $\dot{V}(x_i; x_{s,j}) \geq 0$, go to Step 1.4.2. Else, go to Step 1.4.3.
      1.4.2 If $V(x_i; x_{s,j}) < \rho(x_{s,j})$, set $\rho(x_{s,j}) := V(x_i; x_{s,j})$. Go to Step 1.4.3.
      1.4.3 If $i + 1 \leq I$, go to Step 1.4.1 and $i \leftarrow i + 1$. Else, go to Step 2.
2. Save $\rho(x_{s,j})$ (if necessary, reduce $\rho(x_{s,j})$ such that the set $\Omega_{\rho(x_{s,j})}$ only includes points where the time-derivative of the Lyapunov function is negative).
3. If $j + 1 \leq J$, go to Step 1 and $j \leftarrow j + 1$. Else, go to Step 4.
4. Approximate the union set with analytic algebraic expressions (constraints) using appropriate techniques.

If Γ consists of a finite number of points, then J could be taken as the number of points in Γ. If the number of points in Γ is large or infinite, J could be a sufficiently large integer. From a practical stand-point, these numbers need to be small enough such that this type of calculation may be implemented. Figure 4.5 gives an illustration of the construction of $\mathcal{X}$ using this procedure. The following example provides a tractable illustration of the construction of $\mathcal{X}$ for a scalar system.

Example 4.2 Consider the nonlinear scalar system described by

$$\dot{x} = x - 2x^2 + xu \tag{4.87}$$

with admissible inputs in the set $\mathbb{U} = [-100, 100]$ and with the set of admissible operating steady-states defined as $\Gamma = \{x_s \in [-25, 25]\}$. The steady-states in Γ are open-loop unstable. For any $x_s \in \Gamma$, the system of Eq. 4.87 may be written in the following input-affine form:

$$\dot{\bar{x}} = f(\bar{x}) + g(\bar{x})\bar{u} \tag{4.88}$$

where $\bar{x} = x - x_s$ and $\bar{u} = u - u_s$. Consider a quadratic Lyapunov function of the form:

$$V(x; x_s) = \frac{1}{2}(x - x_s)^2 \tag{4.89}$$

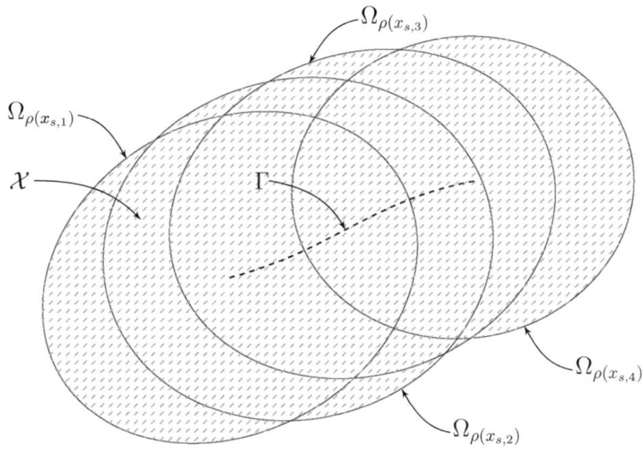

Fig. 4.5 An illustration of the construction of the stability region $\mathscr{X}$. The *shaded region* corresponds to the set $\mathscr{X}$

for the closed system of Eq. 4.87 under the following Lyapunov-based feedback control law [17]:

$$\hat{h}(x; x_s) = \begin{cases} -\dfrac{L_f V + \sqrt{L_f V^2 + L_g V^4}}{L_g V} & \text{if } L_g V \neq 0 \\ 0 & \text{if } L_g V = 0 \end{cases} \tag{4.90}$$

for a $x_s \in \Gamma$ where $L_f V$ and $L_g V$ are the Lie derivatives of the function V with respect to f and g, respectively (these functions are similarly parameterized for each x_s). To account for the bound on the available control energy, the controller is formulated as

$$h(x; x_s) = 100 \text{ sat}\left(\frac{\hat{h}(x; x_s)}{100}\right) \tag{4.91}$$

where sat$(\cdot)$ denotes the standard saturation function.

For this particular case, the stability region of the system of Eq. 4.87 with the stabilizing controller of Eq. 4.91 for the minimum and maximum steady-state in the set Γ are used to approximate the set $\mathscr{X}$. For the steady-state $x_{s,1} = -25$ with corresponding steady-state input $u_{s,1} = -51$, the largest level set of the Lyapunov function where the Lyapunov function is decreasing along the state trajectory with respect to the steady-state $x_{s,1}$ is $\Omega_{\rho(x_{s,1})} = \{x \in \mathbb{R} : V(x; -25) \leq 300.25\}$, i.e., $\rho(x_{s,1}) = 300.25$. For the steady-state $x_{s,2} = 25$ and $u_{s,2} = 49$, the level set is $\Omega_{\rho(x_{s,2})} = \{x \in \mathbb{R} : V(x, 25) \leq 2775.49\}$, i.e., $\rho(x_{s,2}) = 2775.49$. Therefore, the union of the stability region is described as $\mathscr{X} = \{x \in [-49.5, 99.5]\}$.

4.5.3 Formulation of LEMPC with Time-Varying Economic Cost

The formulation of the LEMPC with the time-varying economic stage cost is given in this subsection. First, the overall methodology of employing the set $\mathscr{X}$ in the design of the LEMPC is described. As a consequence of the construction method used for $\mathscr{X}$, any state in $\mathscr{X}$ is in a stability region of at least one steady-state. This means that there exists an input trajectory that satisfies the input constraint and that maintains operation in $\mathscr{X}$. The ability to maintain the state in $\mathscr{X}$ is guaranteed because the input trajectory obtained from the Lyapunov-based controller with respect to the steady-state x_s such that the current state $x(t_k) \in \Omega_{\rho(x_s)}$ will maintain the state in $\Omega_{\rho(x_s)} \subset \mathscr{X}$. The stability properties of $\mathscr{X}$ make it an attractive choice to use in the formulation of a LEMPC. Namely, use $\mathscr{X}$ to formulate a region constraint that is imposed in the optimization problem of EMPC to ensure that $\mathscr{X}$ is an invariant set.

In any practical setting, the closed-loop system is subjected to disturbances and uncertainties causing the closed-loop state trajectory to deviate from the predicted (open-loop) nominal trajectory. Enforcing that the predicted state to be in $\mathscr{X}$ is not sufficient for maintaining the closed-loop state trajectory in $\mathscr{X}$ because disturbances may force the state out of $\mathscr{X}$. To make $\mathscr{X}$ an invariant set, a subset of $\mathscr{X}$ is defined and is denoted as $\hat{\mathscr{X}}$. The set $\hat{\mathscr{X}}$ is designed such that any state starting in $\hat{\mathscr{X}}$, which may be forced outside of $\hat{\mathscr{X}}$ by the disturbances, will be maintained in $\mathscr{X}$ over the sampling period when the computed control action is such that the predicted state is maintained in $\hat{\mathscr{X}}$.

Any state $x(t_k) \in \mathscr{X} \setminus \hat{\mathscr{X}}$ where $x(t_k)$ denotes a measurement of the state at sampling time t_k may be forced back into the set $\hat{\mathscr{X}}$. This statement holds as a result of the method used to construct $\hat{\mathscr{X}}$ and $\mathscr{X}$. Specifically, a steady-state $\hat{x}_s \in \Gamma$ may be found such that $x(t_k) \in \Omega_{\rho(\hat{x}_s)}$. Then, a contractive Lyapunov-based constraint like that of Eq. 4.3f is imposed in the formulation of the LEMPC to ensure that the computed control action decreases the Lyapunov function by at least the rate given by the Lyapunov-based controller. This guarantees that the closed-loop state will converge to $\hat{\mathscr{X}}$ in finite-time. Here, $\hat{\mathscr{X}}$ and $\mathscr{X}$ are analogous to Ω_{ρ_e} and Ω_ρ in the LEMPC design of Eq. 4.3 with a time-invariant economic cost.

Given the overview and purposes of the sets $\mathscr{X}$ and $\hat{\mathscr{X}}$, the sets Γ, $\mathscr{X}$, and $\hat{\mathscr{X}}$ are summarized. First, the set Γ is the set of points in state-space that satisfies the steady-state model equation for some $u_s \in \mathbb{U}$, i.e., $f(x_s, u_s, 0) = 0$. Second, $\mathscr{X}$, which is the union of the stability regions $\Omega_{\rho(x_s)}$ constructed for each steady-state in Γ, is assumed to be a non-empty, compact, and connected set. Third, the set $\hat{\mathscr{X}}$ is assumed to be a non-empty, compact, and connected set with $\hat{\mathscr{X}} \subset \mathscr{X}$, and it is further clarified in Sect. 4.5.5.

Using the sets Γ, $\mathscr{X}$, and $\hat{\mathscr{X}}$, the LEMPC formulation with an explicitly time-varying cost is given by the following optimization problem:

$$\min_{u \in S(\Delta)} \int_{t_k}^{t_{k+N}} l_e(\tau, \tilde{x}(\tau), u(\tau))\, d\tau \tag{4.92a}$$

$$\text{s.t. } \dot{\tilde{x}}(t) = f(\tilde{x}(t), u(t), 0) \tag{4.92b}$$

$$\tilde{x}(t_k) = x(t_k) \tag{4.92c}$$

$$u(t) \in \mathbb{U}, \ \forall\, t \in [t_k, t_{k+N}) \tag{4.92d}$$

$$\tilde{x}(t) \in \hat{\mathscr{X}}, \ \forall\, t \in [t_k, t_{k+N}) \text{ if } x(t_k) \in \hat{\mathscr{X}} \tag{4.92e}$$

$$\tilde{x}(t) \in \mathscr{X}, \ \forall\, t \in [t_k, t_{k+N}) \text{ if } x(t_k) \in \mathscr{X} \setminus \hat{\mathscr{X}} \tag{4.92f}$$

$$\frac{\partial V(x(t_k); \hat{x}_s)}{\partial x} f(x(t_k), u(t_k), 0) \le \frac{\partial V(x(t_k); \hat{x}_s)}{\partial x} f(x(t_k), h(x(t_k); \hat{x}_s), 0)$$
$$\text{if } x(t_k) \notin \hat{\mathscr{X}}, \ x(t_k) \in \Omega_{\rho(\hat{x}_s)} \text{ with } \hat{x}_s \in \Gamma \tag{4.92g}$$

where all of the notation used is similar to that used in the LEMPC formulation of Eq. 4.3. The optimal solution of this optimization problem is denoted as $u^*(t|t_k)$ and it is defined for $t \in [t_k, t_{k+N})$. The control action computed for the first sampling period of the prediction horizon is denoted as $u^*(t_k|t_k)$. In the optimization problem of Eq. 4.92, Eq. 4.92a defines the time-dependent economic cost functional to be minimized over the prediction horizon. The constraint of Eq. 4.92b is the nominal model of the system of Eq. 4.1 which is used to predict the evolution of the system with input trajectory $u(t)$ computed by the LEMPC. The dynamic model is initialized with a measurement of the current state (Eq. 4.92c). The constraint of Eq. 4.92d restricts the input trajectory take values within the admissible input set.

Similar to the LEMPC design of Eq. 4.3, the LEMPC of Eq. 4.92 is a dual-mode controller. The constraint of Eq. 4.92e defines mode 1 operation of the LEMPC and is active when the state at the current sampling time $x(t_k) \in \hat{\mathscr{X}}$. It enforces that the predicted state trajectory be maintained in $\hat{\mathscr{X}}$. The constraint of Eq. 4.92g defines mode 2 operation of the LEMPC and is active when the state is outside $\hat{\mathscr{X}}$. It is used to force the state back into the $\hat{\mathscr{X}}$ which is guaranteed for any $x(t_k) \in \mathscr{X}$. The constraint of Eq. 4.92f is active when $x(t_k) \in \mathscr{X} \setminus \hat{\mathscr{X}}$ and ensures the predicted state be contained in the set $\mathscr{X}$. Although Eq. 4.92f is not needed for stability, it is used to ensure that the LEMPC optimizes the input trajectory with knowledge that the state must be contained in $\mathscr{X}$, and potentially improves the closed-loop economic performance when the LEMPC is operating under mode 2 operation compared to not imposing such a constraint.

Figure 4.6 illustrates the sets and different operation modes of the closed-loop system under the LEMPC of Eq. 4.92. Over the first sampling period, the LEMPC, operating in mode 1, computes a control action that maintains the predicted state $\tilde{x}(t_{k+1})$ inside $\hat{\mathscr{X}}$. However, the closed-loop state at the next sampling time $x(t_{k+1})$ is driven outside of $\hat{\mathscr{X}}$ by disturbances. The LEMPC, operating in mode 2, ensures that the computed control action decreases the Lyapunov function based on the steady-state $\hat{x}_s$ over the next sampling period to force the state back into $\hat{\mathscr{X}}$.

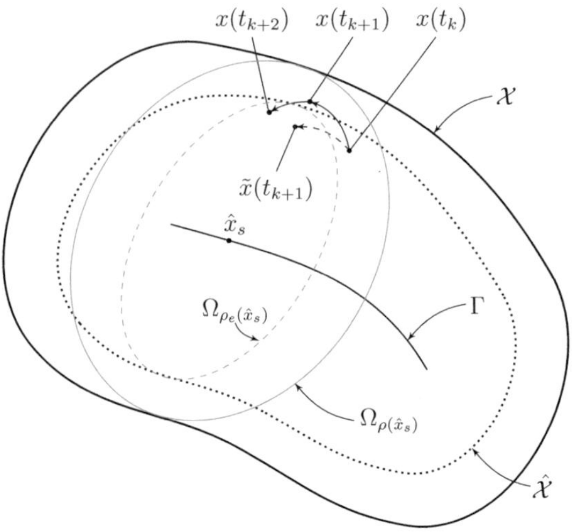

Fig. 4.6 The illustration gives the state evolution under the LEMPC of Eq. 4.92 with a time-varying economic stage cost over two sampling periods

4.5.4 Implementation Strategy

The LEMPC of Eq. 4.92 is implemented in a receding horizon fashion. The optimization problem is repeatedly solved every sampling time after receiving state feedback from the system. The implementation strategy may be summarized as follows:

1. At sampling time t_k, the LEMPC receives a state measurement $x(t_k)$ from the sensors.
2. If $x(t_k) \in \hat{\mathcal{X}}$, go to Step 2.1. Else, go to Step 2.2.
   - 2.1 LEMPC operates in mode 1: the constraint of Eq. 4.92e is active and the constraints of Eqs. 4.92f and 4.92g are inactive, go to Step 3.
   - 2.2 Find $\hat{x}_s \in \Gamma$ such that $x(t_k) \in \Omega_{\rho(\hat{x}_s)}$, go to Step 2.3.
   - 2.3 LEMPC operates in mode 2: the constraint of Eq. 4.92e is inactive and the constraints of Eqs. 4.92f and 4.92g are active, go to Step 3.
3. The LEMPC computes the optimal input trajectory $u^*(t|t_k)$ for $t \in [t_k, t_{k+N})$, go to Step 4.
4. The LEMPC sends the control action, $u^*(t_k|t_k)$, computed for the first sampling period of the prediction horizon to the control actuators to apply to the system in a sample-and-hold fashion from t_k to t_{k+1}. Go to Step 5.
5. Set $k \leftarrow k + 1$. Go to Step 1.

4.5.5 Stability Analysis

In this subsection, Theorem 4.5 provides sufficient conditions for closed-loop stability in the sense of boundedness of the closed-loop system state inside the set $\mathscr{X}$ under the LEMPC of Eq. 4.92 for any initial condition $x(0) \in \mathscr{X}$. It follows the ideas of the analysis of Sect. 4.3. The assumption on the set $\hat{\mathscr{X}}$ that is needed to ensure closed-loop stability is given below.

Assumption 4.2 Let $\hat{\mathscr{X}} \subset \mathscr{X}$ be a non-empty, compact, and connected set such that if $x(0) \in \hat{\mathscr{X}}$ and the constant control $\hat{u} \in \mathbb{U}$ is such that $\tilde{x}(t) \in \hat{\mathscr{X}}$ for all $t \in [0, \Delta]$ where $\tilde{x}(t)$ is the solution to

$$\dot{\tilde{x}}(t) = f(\tilde{x}(t), \hat{u}, 0) \tag{4.93}$$

for $t \in [0, \Delta]$ and $\tilde{x}(0) = x(0)$, then $x(\Delta) \in \mathscr{X}$ where $x(\Delta)$ denotes the closed-loop state of Eq. 4.3 under the constant control $\hat{u}$.

Assumption 4.2 is satisfied for the case that instead of using the mode 1 constraint of Eq. 4.92e, the constraint $\tilde{x}(t) \in \Omega_{\rho_e(x_s)}$ for $t \in [t_k, t_{k+N})$ where $\Omega_{\rho_e(x_s)}$ is designed according to a similar condition as in Eq. 4.21 for some $x_s \in \Gamma$ such that $x(t_k) \in \Omega_{\rho_e(x_s)}$. For this case, $\hat{\mathscr{X}}$ may be constructed by taking the union of sets $\Omega_{\rho_e(x_s)}$ for all $x_s \in \Gamma$ where $\Omega_{\rho_e(x_s)}$ is similar to the set Ω_{ρ_e} (for each $x_s \in \Gamma$) in the LEMPC of Eq. 4.3. Nevertheless, Assumption 4.2 is needed to cover the more general case with the mode 1 constraint of Eq. 4.92e.

To avoid introducing convoluted notation, the sufficient conditions of the Theorem are stated as similar conditions as Eqs. 4.22 and 4.23 must hold for each $x_s \in \Gamma$. This means that there exists positive constants: ρ, $\rho_{\min}$, ρ_s, L'_x, L'_w, M, and ε_w that satisfy similar conditions for each $x_s \in \Gamma$. Moreover, all of these parameters depend on x_s.

Theorem 4.5 *Consider the system of Eq. 4.1 in closed-loop under the LEMPC design of Eq. 5 based on the set of controllers that satisfy the conditions of Eq. 4.86 for each $x_s \in \Gamma$. Let $\varepsilon_w(x_s) > 0$, $\Delta > 0$, $\rho(x_s) > \rho_e(x_s) \geq \rho_{\min}(x_s) > \rho_s(x_s) > 0$ for all $x_s \in \Gamma$ satisfy a similar condition as Eq. 4.22 for each $x_s \in \Gamma$ and let $\mathscr{X} = \cup_{x_s \in \Gamma} \Omega_{\rho(x_s)}$ be a non-empty, compact, and connected set and $\hat{\mathscr{X}}$ satisfy Assumption 4.2. If $x(0) \in \mathscr{X}$ and $N \geq 1$, then the state $x(t)$ of the closed-loop system is always bounded in $\mathscr{X}$ for all $t \geq 0$.*

Proof The proof of Theorem 4.5 consists of the following parts: first, the feasibility of the optimization problem of Eq. 4.92 is proven for any state $x(t_k) \in \mathscr{X}$. Second, boundedness of the closed-loop state trajectory $x(t) \in \mathscr{X}$ for all $t \geq 0$ is proven for any initial state starting in $\mathscr{X}$.

Part 1: Owing to the construction of $\mathscr{X}$, any state $x(t_k) \in \mathscr{X}$ is in the stability region $\Omega_{\rho(x_s)}$ of the Lyapunov-based controller designed for some steady-state $x_s \in \Gamma$. This implies that there exists an input trajectory that is a feasible solution because the input trajectory obtained from the Lyapunov-based controller is a feasible solution to the optimization of Eq. 4.92 as it satisfies the constraints (refer to Theorem 4.1, Part 1 on how this input trajectory is obtained). The latter claim is guaranteed by the closed-loop stability properties of the Lyapunov-based controller ($h(\cdot; x_s)$).

Part 2: If $x(t_k) \in \mathscr{X} \setminus \hat{\mathscr{X}}$, then the LEMPC of Eq. 4.92 operates in mode 2. Since $x(t_k) \in \mathscr{X}$, a steady-state $\hat{x}_s \in \Gamma$ may be found such that the current state $x(t_k) \in \Omega_{\rho(\hat{x}_s)}$. Utilizing the Lyapunov-based controller $h(\cdot; \hat{x}_s)$, the LEMPC computes a control action that satisfies the constraint of Eq. 4.92g:

$$\frac{\partial V(x(t_k); \hat{x}_s)}{\partial x} f(x(t_k), u^*(t_k|t_k), 0) \leq \frac{\partial V(x(t_k); \hat{x}_s)}{\partial x} f(x(t_k), h(x(t_k); \hat{x}_s), 0) \tag{4.94}$$

for some $\hat{x}_s \in \Gamma$ where $u^*(t_k|t_k)$ is the optimal control action computed by the LEMPC to be applied in a sample-and-hold fashion to the system of Eq. 4.1 for $t \in [t_k, t_{k+1})$. From Eq. 4.86b, the term in the right-hand side of the inequality of Eq. 4.94 may be upper bounded by a class $\mathscr{K}$ function as follows:

$$\frac{\partial V(x(t_k), \hat{x}_s)}{\partial x} f(x(t_k), u^*(t_k), 0) \leq -\alpha_3(|x(t_k) - \hat{x}_s|; \hat{x}_s) \tag{4.95}$$

for all $x(t_k) \in \mathscr{X}$ and for some $\hat{x}_s \in \Gamma$. Following similar steps as that used in Theorem 4.5, Part 2, one may show that the Lyapunov function value, i.e., the Lyapunov function for the steady-state $\hat{x}_s$, will decay over the sampling period when a similar condition to Eq. 4.22 is satisfied for each $x_s \in \Gamma$.

If $x(t_k) \in \hat{\mathscr{X}}$, then $x(t_{k+1}) \in \mathscr{X}$ owing to the construction of $\hat{\mathscr{X}}$, i.e., if Assumption 4.2 is satisfied. If $x(t_k) \in \mathscr{X} \setminus \hat{\mathscr{X}}$, then $x(t_{k+1}) \in \mathscr{X}$ because the state is forced to a smaller level set of the Lyapunov function with respect to the steady-state $\hat{x}_s \in \Gamma$ over the sampling period. Both of these results together imply that $x(t_{k+1}) \in \mathscr{X}$ for all $x(t_k)$ under the LEMPC of Eq. 4.92. Using this result recursively, the closed-loop state is always bounded in $\mathscr{X}$ when the initial state is in $\mathscr{X}$.

Remark 4.5 The LEMPC of Eq. 4.92 does not have a switching time like the LEMPC of Eq. 4.3 whereby the mode 2 constraint is exclusively imposed after the switching time to enforce the closed-loop state to a specific steady-state. To ensure there exists a feasible path from any state in $\mathscr{X}$ to the desired operating steady-state more conditions are needed. The interested reader may refer to [18] that provides some conditions that accomplish such a goal. Additionally, no guarantees are made that the closed-loop state will converge to $\hat{\mathscr{X}}$ when the state is in $\mathscr{X} \setminus \hat{\mathscr{X}}$ owing to the fact that the mode 2 constraint could be formulated with respect to a different steady-state at each sampling time. However, enforcing convergence to $\hat{\mathscr{X}}$ may be readily accomplished through implementation by enforcing a mode 2 constraint formulated with respect to the same steady-state at each sampling time until the state converges to $\hat{\mathscr{X}}$.

4.5.6 Application to a Chemical Process Example

Consider a non-isothermal CSTR where three parallel reactions take place. The reactions are elementary irreversible exothermic reactions of the form: $A \to B$, $A \to C$, and $A \to D$. The desired product is B; while, C and D are byproducts. The feed of the reactor consists of the reactant A in an inert solvent and does not contain any of the products. Using first principles and standard modeling assumptions, a nonlinear dynamic model of the process is obtained:

$$\frac{dC_A}{dt} = \frac{F}{V_R}(C_{A0} - C_A) - \sum_{i=1}^{3} k_{0,i} e^{-E_i/RT} C_A \tag{4.96a}$$

$$\frac{dT}{dt} = \frac{F}{V_R}(T_0 - T) - \frac{1}{\rho_L C_p} \sum_{i=1}^{3} \Delta H_i k_{0,i} e^{-E_i/RT} C_A + \frac{Q}{\rho_L C_p V_R} \tag{4.96b}$$

where C_A is the concentration of the reactant A, T is the temperature of the reactor, Q is the rate of heat supplied or removed from the reactor, C_{A0} and T_0 are the reactor feed reactant concentration and temperature, respectively, F is a constant volumetric flow rate through the reactor, V_R is the constant liquid hold-up in the reactor, ΔH_i, $k_{0,i}$, and E_i, $i = 1, 2, 3$ denote the enthalpy changes, pre-exponential constants and activation energies of the three reactions, respectively, and C_p and ρ_L denote the heat capacity and the density of the fluid in the reactor. The process parameters are given in Table 4.3. The CSTR has two manipulated inputs: the inlet concentration C_{A0} with available control energy $0.5\ \text{kmol}\ \text{m}^{-3} \le C_{A0} \le 7.5\ \text{kmol}\ \text{m}^{-3}$ and the heat rate to/from the vessel Q with available control energy $-1.0 \times 10^5\ \text{kJ}\ \text{h}^{-1} \le Q \le 1.0 \times 10^5\ \text{kJ}\ \text{h}^{-1}$. The state vector is $x^T = [C_A\ T]$ and the input vector is $u^T = [C_{A0}\ Q]$.

4.5.6.1 Stability Region Construction

Supplying or removing significant amount of thermal energy to/from the reactor (nonzero Q) is considered to be undesirable from an economic perspective. Therefore, the set $\mathscr{X}$ is constructed considering steady-states with a steady-state reactant inlet concentration of $C_{A0s} \in [2.0,\ 6.0]\ \text{kmol}\ \text{m}^{-3}$ and no heat rate supplied/removed from the reactor, i.e., $Q_s = 0.0\ \text{kJ}\ \text{h}^{-1}$. The corresponding steady-states in the desired operating range form a set denoted as Γ of admissible operating steady-states. Several of these steady-states have been verified to be open-loop unstable, i.e., the eigenvalues of the linearization around the steady-states corresponding to the minimum and maximum steady-state inlet concentrations are $\lambda_{1,\min} = -1.00$, $\lambda_{2,\min} = 2.73$ and $\lambda_{1,\max} = -1.00$, $\lambda_{2,\max} = 2.10$, respectively. The set Γ covers approximately a temperature range of 50 K.

Table 4.3 CSTR process parameters

Feedstock volumetric flow rate	$F = 5.0\,\text{m}^3\text{h}^{-1}$
Feedstock temperature	$T_0 = 300\,\text{K}$
Reactor volume	$V_R = 5.0\,\text{m}^3$
Pre-exponential factor for reaction 1	$k_{01} = 6.0 \times 10^5\,\text{h}^{-1}$
Pre-exponential factor for reaction 2	$k_{02} = 6.0 \times 10^5\,\text{h}^{-1}$
Pre-exponential factor for reaction 3	$k_{03} = 6.0 \times 10^5\,\text{h}^{-1}$
Reaction enthalpy change for reaction 1	$\Delta H_1 = -5.0 \times 10^4\,\text{kJ kmol}^{-1}$
Reaction enthalpy change for reaction 2	$\Delta H_2 = -5.2 \times 10^4\,\text{kJ kmol}^{-1}$
Reaction enthalpy change for reaction 3	$\Delta H_3 = -5.4 \times 10^4\,\text{kJ kmol}^{-1}$
Activation energy for reaction 1	$E_1 = 5.0 \times 10^4\,\text{kJ kmol}^{-1}$
Activation energy for reaction 2	$E_2 = 7.53 \times 10^4\,\text{kJ kmol}^{-1}$
Activation energy for reaction 3	$E_3 = 7.53 \times 10^4\,\text{kJ kmol}^{-1}$
Heat capacity	$C_p = 0.231\,\text{kg m}^{-3}$
Density	$\rho_L = 1000\,\text{kJ kg}^{-1}\text{K}^{-1}$
Gas constant	$R = 8.314\,\text{kJ kmol}^{-1}\text{K}^{-1}$

A set of two proportional controllers with saturation to account for the input constraints is used in the design of the Lyapunov-based controller:

$$h(x; x_s) = \begin{bmatrix} 3.5\ \text{sat}\left(\dfrac{K_1(x_{s,1} - x_1) + u_{1,s} - 4.0}{3.5}\right) + 4.0 \\ 10^5\ \text{sat}\left(\dfrac{K_2(x_{s,2} - x_2) + u_{2,s}}{10^5}\right) \end{bmatrix} \tag{4.97}$$

where $K_1 = 10$ and $K_2 = 8000$ are the gains of each proportional controller. The proportional controller gains have been tuned to give the largest estimate of the stability region for a given steady-state. A quadratic Lyapunov function of the form:

$$V(x; x_s) = (x - x_s)^T P(x - x_s) \tag{4.98}$$

where P is a positive definite matrix is used to estimate the stability regions of many steady-states in the set Γ, i.e., the stability region for a given steady-state in Γ is taken to be a level set of the Lyapunov function where the Lyapunov function is decreasing along the state trajectory. To estimate $\mathscr{X}$, the procedure outlined in Sect. 4.5.2 is employed. To obtain the largest estimate of the region $\mathscr{X}$, several P matrices were used. The results of this procedure are shown in Fig. 4.7. The union of these regions

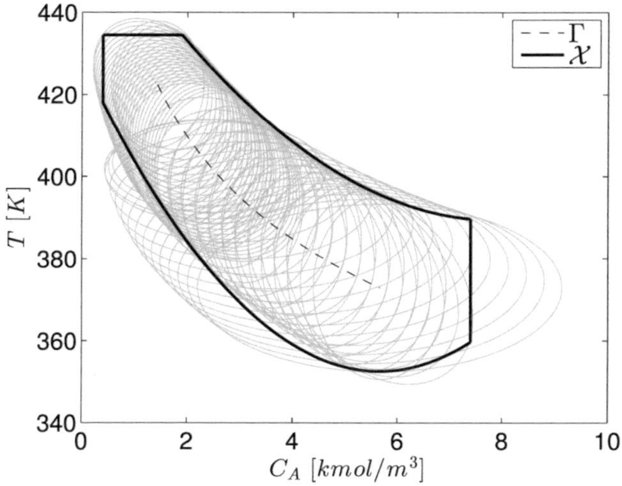

Fig. 4.7 The construction of the set $\mathscr{X}$ for the CSTR of Eq. 4.96

$\mathscr{X}$ was approximated with two quadratic polynomial inequalities and three linear state inequalities given by:

$$\begin{aligned} 1.26x_1^2 - 19.84x_1 + 467.66 - x_2 &\geq 0 \\ 2.36x_1^2 - 26.72x_1 + 428.26 - x_2 &\leq 0 \\ 0.4 \leq x_1 &\leq 7.4 \\ x_2 &\leq 434.5 \end{aligned} \tag{4.99}$$

which will be used in the formulation of the LEMPC to ensure that the state trajectories are maintained inside $\mathscr{X}$ (note that x_2 is lower bounded by the second inequality).

4.5.6.2 Closed-Loop Simulation Results

The control objective of this chemical process example is to operate the CSTR in an economically optimal manner while accounting for changing economic factors and maintaining the system operation inside a bounded set. For this chemical process example, the economic measure being considered is

$$l_e(t, x, u) = A_1(t)u_2^2 + A_2(t)u_1 - A_3 r_1(x) + A_4(x_2 - 395)^2 \tag{4.100}$$

where $r_1(x)$ is the reaction rate of the first reaction that produces the desired product:

$$r_1(x) = k_{01} e^{-E_1/Rx_2} x_1 \ . \tag{4.101}$$

The economic measure of Eq. 4.100 penalizes energy usage/removal, penalizes reactant material consumption, credits the production rate of the desired product, and penalizes the deviation of the operating temperature from the median operating temperature. The fourth term of the economic cost is used to prevent the LEMPC from operating the CSTR at the boundary of the allowable operating range for long periods of time which is considered undesirable from a practical perspective. In this fashion, the economic cost consists of terms that are associated with the operating cost/profit (economic terms) as well as terms that ensure that the LEMPC operates the CSTR in a practical and safe fashion.

For this study, the weights A_1 and A_2 are considered to vary with time; while, $A_3 = 278$ and $A_4 = 0.4$ are constants over the 5.0 h simulated operation of the CSTR under the LEMPC. The weight A_1 is equal to 4.0×10^{-6} for $t = 0.0$–4.0 h and 5.0×10^{-6} for 4.0–5.0 h, and the time-dependent weight A_2 is given by the following piecewise constant relationship:

$$A_2(t) = \begin{cases} 333 & 0.0\,\text{h} \le t < 1.0\,\text{h} \\ 167 & 1.0\,\text{h} \le t < 2.0\,\text{h} \\ 83 & 2.0\,\text{h} \le t < 3.0\,\text{h} \\ 17 & 3.0\,\text{h} \le t < 4.0\,\text{h} \\ 167 & 4.0\,\text{h} \le t < 5.0\,\text{h} \end{cases} \tag{4.102}$$

Since the economic cost is considered to account for more than just the operating cost/profit of the CSTR, the weights may be considered to account for more than the price of a particular resource. For instance, the variation of the weight A_2 may be caused by supply and/or demand changes of the reactant A. While these weights may come from a higher level information technology system, careful tuning of these weights is critical to achieve both practical operation with LEMPC and economically optimal (with respect to the actual operating cost) operation. For this particular study, the economic cost has been chosen to vary on a time-scale comparable to the one of the process dynamics.

In the first set of simulations, nominal operation ($w \equiv 0$) is considered to understand the operation of the CSTR under the LEMPC operating in mode 1 only. The formulation of the LEMPC with explicitly time-varying cost function used to accomplish the desired control objective is

$$\begin{aligned} \min_{u \in S(\Delta)} \; & \int_{t_k}^{t_{k+N}} l_e(\tau, \tilde{x}(\tau), u(\tau))\, d\tau \\ \text{s.t. } & \dot{\tilde{x}}(t) = f(\tilde{x}(t), u(t), 0) \\ & \tilde{x}(t_k) = x(t_k) \\ & u(t) \in \mathbb{U}, \ \forall\, t \in [t_k, t_{k+N}) \\ & 1.26\tilde{x}_1^2(t) - 19.84\tilde{x}_1(t) + 467.66 - \tilde{x}_2(t) \ge 0, \ \forall\, t \in [t_k, t_{k+N}) \\ & 2.36\tilde{x}_1^2(t) - 26.72\tilde{x}_1(t) + 428.26 - \tilde{x}_2(t) \le 0, \ \forall\, t \in [t_k, t_{k+N}) \end{aligned} \tag{4.103}$$

$$0.4 \leq \tilde{x}_1(t) \leq 7.4, \ \forall \, t \in [t_k, t_{k+N})$$
$$\tilde{x}_2(t) \leq 434.5, \ \forall \, t \in [t_k, t_{k+N})$$

where the economic measure l_e is given in Eq. 4.100. Since no disturbances or uncertainties are present, the set $\hat{\mathscr{X}}$ is taken to be $\mathscr{X}$ ($\hat{\mathscr{X}} = \mathscr{X}$). The sampling period and the prediction horizon of the LEMPC is $\Delta = 0.1$ h and $N = 10$, respectively. These parameters have been chosen through extensive simulations such that the total prediction horizon is sufficiently long to yield good economic performance of the closed-loop system. To solve the LEMPC optimization problem at each sampling period, the open-source interior point solver Ipopt [19] was used. A fourth-order Runge-Kutta method with integration step of 5.0×10^{-4} h was used to numerically solve the nonlinear ODEs of Eq. 4.96. To assess the total economic performance of each simulation, the total economic measure over the simulated operation of the CSTR is defined as

$$\sum_{j=0}^{M-1} \left[A_1(t_j) u_2^2(t_j) + A_2(t_j) u_1(t_j) - A_3 r_1(x(t_j)) + A_4 (x_2(t_j) - 395)^2 \right] \quad (4.104)$$

where M is the number of integration steps over the entire simulated time t_f and t_j denotes an integration time step.

Since the exact future values of the cost weights may not be known exactly in a practical setting, two cases were simulated: (1) the LEMPC of Eq. 4.103 is formulated with a perfect forecast of time-varying economic weights and (2) the LEMPC of Eq. 4.103 is formulated with constant A_1 and A_2 (no forecasting). The two cases are denoted as LEMPC-1 and LEMPC-2, respectively. For LEMPC-2, the previously obtained weights A_1 and A_2 are used in the optimization problem until the LEMPC receives new weight values which are obtained at the time instance in which the weights change.

The CSTR is initialized at the initial condition of $C_A(0) = 2.0\,\text{kmol}\,\text{m}^{-3}$ and $T(0) = 410.0$ K. The results of two simulations are shown in Figs. 4.8 and 4.9 under LEMPC-1 and LEMPC-2, respectively. The total economic cost of the CSTR under LEMPC-1 is 2.37×10^4; while, the economic cost of the CSTR under LEMPC-2 is 2.91×10^4. The key factor that contributes to the performance degradation of the second simulation (as depicted in Figs. 4.8 and 4.9) may be observed in the input trajectories that the two LEMPC schemes compute. For the LEMPC-1 simulation, the LEMPC knows that the cost of the reactant material decreases at the beginning of each of the first four hours of operation so it waits to utilize this resource until the beginning of each of these hours when the price is less than in the previous hour. For the LEMPC-2 simulation, the LEMPC uses the minimum amount of reactant material at the beginning of each of these four hours. Also, the cost of the thermal energy Q increases over the last hour of the simulated operation. In the first case, the LEMPC utilizes the thermal energy before the price increases to increase the reactor temperature, and then, uses less energy thereafter. In the second case, the LEMPC supplies heat to the reactor when the cost of thermal energy has already increased.

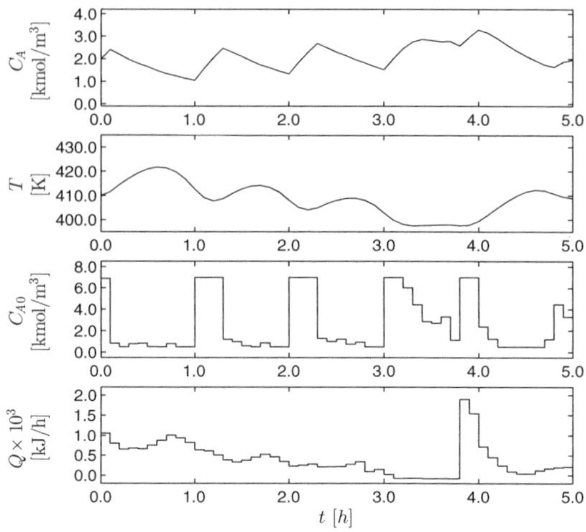

Fig. 4.8 The states and inputs of the nominally operated CSTR under LEMPC-1 (mode 1 operation only) initialized at $C_A(0) = 2.0\,\text{kmol}\,\text{m}^{-3}$ and $T(0) = 410.0\,\text{K}$

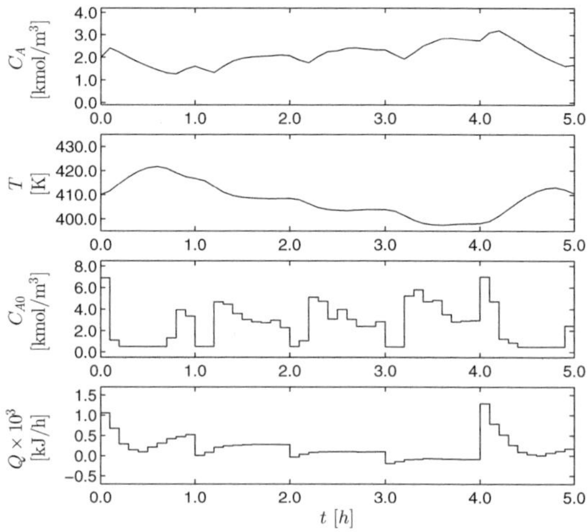

Fig. 4.9 The states and inputs of the nominally operated CSTR under LEMPC-2 (mode 1 operation only) initialized at $C_A(0) = 2.0\,\text{kmol}\,\text{m}^{-3}$ and $T(0) = 410.0\,\text{K}$

Comparing the evolution of the states in both cases, the regions of operation in state-space between the two cases are similar.

Over the course of both of the simulations, the LEMPC schemes operate the CSTR in a time-varying (transient) fashion. If the economic weights become fixed or if a significant time-scale separation between economic cost change and the process dynamics existed, steady-state operation would become optimal for this particular economic cost and nonlinear model. Also, the LEMPC in this example is not formulated with any periodic, average, or integral input constraints, and is not formulated with any stabilizing constraints to enforce convergence to the economically optimal steady-state. Therefore, the reason for the time-varying operation is due to

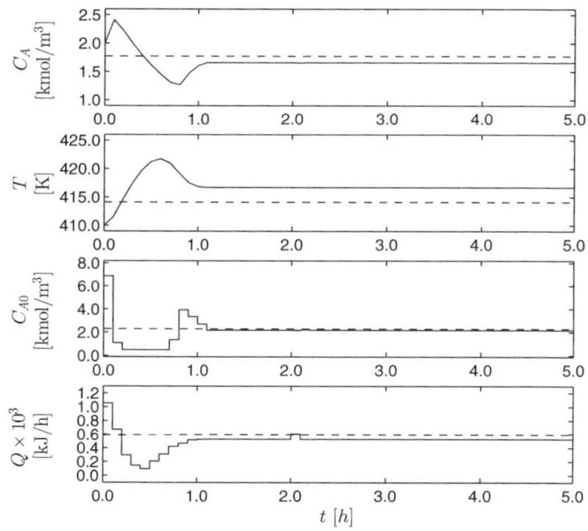

Fig. 4.10 The states and inputs of the CSTR under the LEMPC of Eq. 4.103 (mode 1 operation only) when the economic cost weights are constant with time (*solid line*) with the economically optimal steady-state (*dashed line*)

the economic cost changing with time on a time-scale comparable to the process dynamics. To demonstrate this point, Fig. 4.10 shows the state and input trajectories under LEMPC of Eq. 4.103 (mode 1 operation of the controller only) where the economic cost weights are constant with time. Recalling the LEMPC does not have any constraints that enforce convergence to the steady-state, the CSTR under the LEMPC with a prediction horizon of $N = 10$ settles on an offsetting steady-state from the economically optimal steady-state, i.e., the steady-state in Γ that optimizes the economic cost.

To assess the economic performance of the CSTR under the LEMPC, a comparison between the CSTR under the LEMPC and under a conventional approach to optimization and control, i.e., steady-state optimization with tracking MPC, was carried out. The CSTR is simulated under a Lyapunov-based MPC (LMPC), formulated with a quadratic cost, where the LMPC works to drive the system to the economically optimal steady-state which is the minimizer of

$$\begin{aligned} \min_{(x_s, u_s)} \; & l_e(t, x_s, u_s) \\ \text{s.t.} \; & f(x_s, u_s, 0) = 0 \\ & u_s \in \mathbb{U}, \; x_s \in \Gamma \end{aligned} \tag{4.105}$$

for a fixed t. The optimal steady-state at a given time t is denoted as $x_s^*(t)$ and the optimal steady-state with time for the economic weights is given in Table 4.4.

Table 4.4 The optimal steady-state variation with respect to the time-varying economic weights

t	$C^*_{A,s}$	T^*_s	$C^*_{A0,s}$	Q^*_s
$0.0\,\text{h} \leq t < 1.0\,\text{h}$	1.77	414.1	2.30	591.5
$1.0\,\text{h} \leq t < 2.0\,\text{h}$	2.12	407.5	2.61	300.9
$2.0\,\text{h} \leq t < 3.0\,\text{h}$	2.40	402.9	2.87	151.2
$3.0\,\text{h} \leq t < 4.0\,\text{h}$	2.75	398.1	3.20	−80.0
$4.0\,\text{h} \leq t < 5.0\,\text{h}$	2.12	407.5	2.61	240.6

The formulation of LMPC is as follows:

$$
\begin{aligned}
\min_{u \in S(\Delta)} & \int_{t_k}^{t_{k+N}} \left(\left| \tilde{x}(\tau) - x_s^*(\tau) \right|_{Q_c} + \left| u(\tau) - u_s^*(\tau) \right|_{R_c} \right) d\tau \\
\text{s.t.} & \dot{\tilde{x}}(t) = f(\tilde{x}(t), u(t), 0) \\
& \tilde{x}(t_k) = x(t_k) \\
& u(t) \in \mathbb{U}, \; \forall \, t \in [t_k, t_{k+N}) \\
& \frac{\partial V(x(t_k); x_s^*(t_k))}{\partial x} f(x(t_k), u(t_k), 0) \\
& \leq \frac{\partial V(x(t_k); x_s^*(t_k))}{\partial x} f(x(t_k), h(x(t_k); x_s^*(t_k)), 0)
\end{aligned}
\tag{4.106}
$$

where the cost function is a quadratic cost function that penalizes the deviation of states and inputs from the optimal (time-varying) steady-state. The sampling period and prediction horizon of the LMPC are chosen to be the same as the ones of the LEMPC. The weighting matrices are $Q_c = \text{diag}([2788.0, 0.6])$ and $R_c = \text{diag}([27.8, 5.0 \times 10^{-7}])$. A quadratic Lyapunov function of the form given in Eq. 4.98 with a positive definite matrix $P = \text{diag}([280.0, 9.0])$ is considered. The Lyapunov-based controller used in the formulation of the Lyapunov-based constraint is a set of proportional controllers (P-controllers) like that of Eq. 4.97 with gains $K_1 = 1$ and $K_2 = 6000$. The P-controllers have been tuned less aggressively compared to the P-controllers used in the construction of the set $\mathcal{X}$ to allow the LMPC more freedom in the optimization of the control action.

The CSTR is initialized at several states in state-space and is simulated with three control strategies: (1) LEMPC-1, (2) LEMPC-2, and (3) LMPC. The total economic cost of each simulation is given in Table 4.5. The operating trajectories of a simulation under LMPC are also given in Fig. 4.11 to demonstrate the differences in achievable trajectories with the conventional MPC formulation working to track the economically optimal steady-state. Clearly, the operating trajectories of the LEMPC cannot be obtained by a tracking MPC, regardless of the tuning of the weighting matrices. From the results of Table 4.5, the economic performance of the system under both of the LEMPC schemes is better than the performance under the tracking LMPC.

Table 4.5 The total economic cost of the closed-loop reactor over several simulations with different initial states

Initial state		Total economic cost ($\times 10^5$) and performance improvement				
$x_1(0)$	$x_2(0)$	LMPC	LEMPC-1	Improvement (%)	LEMPC-2	Improvement (%)
2.0	410.0	0.908	0.237	73.9	0.291	68.0
2.0	425.0	2.325	0.456	80.4	0.507	78.2
4.0	370.0	4.274	1.234	71.1	1.075	74.8
4.0	395.0	2.744	0.152	94.4	0.192	93.0
5.0	370.0	4.164	0.634	84.8	0.643	84.6
6.0	360.0	5.370	1.375	74.4	1.225	77.2

The performance improvement is relative to the economic performance under LMPC

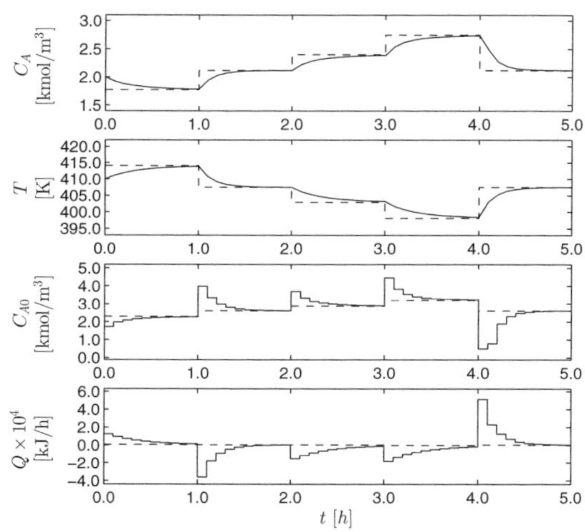

Fig. 4.11 The states and inputs of the CSTR under the LMPC of Eq. 4.106 used to track the economically optimal steady-state (*dashed line*)

For two of the initial conditions, the economic performance was better with LEMPC-2 compared to LEMPC-1 (Table 4.5). The closed-loop evolution of the CSTR with the two LEMPC schemes for one of these simulations is shown in Figs. 4.12 and 4.13. This is a result of not having a sufficiently long prediction horizon for these two initial conditions. More specifically, this behavior is caused by initializing the CSTR far away from the economically optimal region to operate the process. For this prediction horizon ($N = 10$), the LEMPC cannot simulate enough of the future evolution of the process to recognize that there is an economically better region to operate the process. As a result, the state is maintained away from this optimal region at the beginning of both simulations. For the LEMPC-2 simulation, the maximum allowable amount of reactant concentration is fed to the process from 0.0 to 1.8 h. This causes the rates of the three reactions to increase. Since the heat rate supplied/removed from the reactor is penalized in the cost and the LEMPC does not know that the price of the reactant material will decrease at 2.0 h, Q and C_{A0} decrease up to about 2.0h to maintain stability. This decrease in Q and C_{A0} decreases

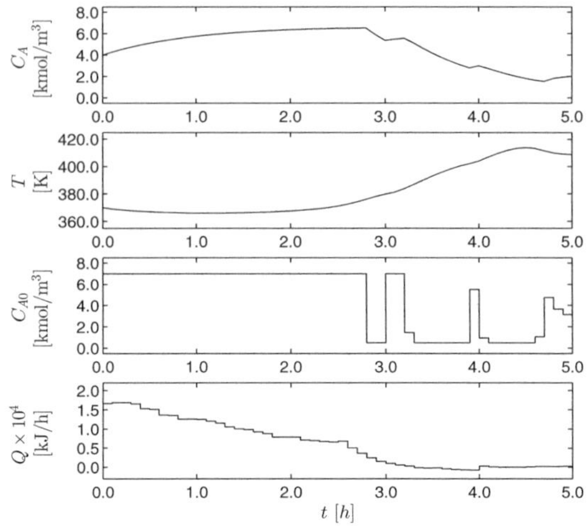

Fig. 4.12 The states and inputs of the nominally operated CSTR under LEMPC-1 initialized at $C_A(0) = 4.0\,\text{kmol}\,\text{m}^{-3}$ and $T(0) = 370.0\,\text{K}$

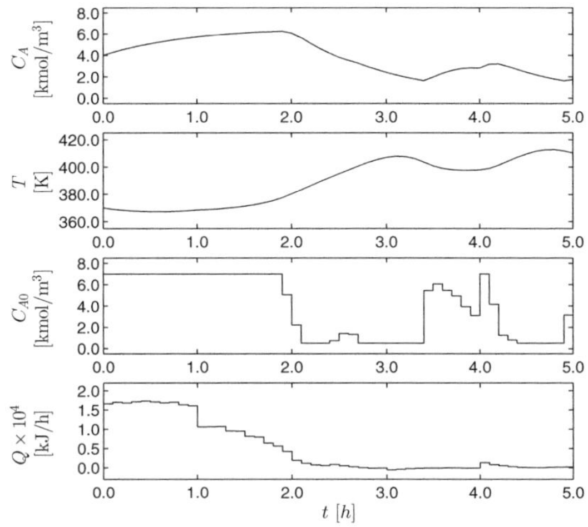

Fig. 4.13 The states and inputs of the nominally operated CSTR under LEMPC-2 initialized at $C_A(0) = 4.0\,\text{kmol}\,\text{m}^{-3}$ and $T(0) = 370.0\,\text{K}$

the reactant concentration in the reactor while increasing the temperature bringing the states closer to the economically optimal region of operation. The LEMPC is then able to observe the economically optimal region of operation along its prediction horizon. Thus, it forces the states to this region. For LEMPC-1, the LEMPC knows that the reactant price will decrease at the beginning of each of the first four hours. Therefore, it maintains feeding the maximum allowable reactant material to maximize the reaction rate of the first reaction, and it supplies less heat to the reactor compared to LEMPC-2. As a result of this behavior, process operation is maintained far enough away from the optimal region of operation.

To assess the stability and robustness properties of the LEMPC of Eq. 4.92, the size where the LEMPC is able to operate the system in a time-varying manner to optimize the process economic cost is reduced and the two-mode control strategy is employed. Process noise is added to the closed-loop system and is modeled as bounded Gaussian white noise on the inlet reactant concentration and inlet temperature which has zero mean and the following standard deviation and bound: 0.5 and 1.0 kmol m^{-3}, respectively, for the inlet concentration noise and 3.0 and 10.0 K, respectively, for the inlet temperature noise. To simulate the noise, a new random number is generated and used to add noise in the process model over each integration step. The region $\hat{\mathscr{X}}$ is approximated through the following constraints:

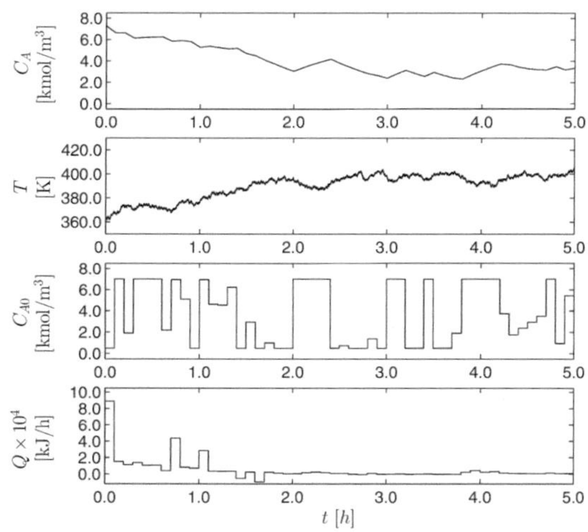

Fig. 4.14 The states and inputs of the CSTR under the two-mode LEMPC with added process noise; evolution with respect to time

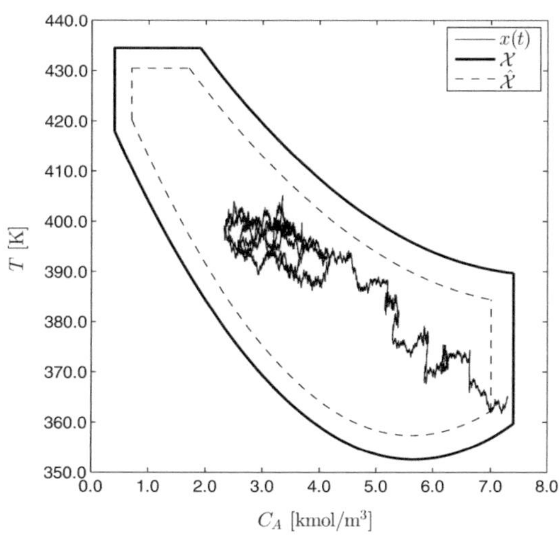

Fig. 4.15 The states and inputs of the CSTR under the two-mode LEMPC with added process noise; state-space plot

$$\begin{aligned} 1.20x_1^2 - 19.17x_1 + 460.61 - x_2 &\geq 0 \\ 2.59x_1^2 - 29.14x_1 + 438.36 - x_2 &\leq 0 \\ 0.7 \leq x_1 &\leq 7.1 \\ x_2 &\leq 431.5 \end{aligned} \tag{4.107}$$

which has been estimated through extensive simulations with the given process model, economic cost, and process noise as the region whereby closed-loop stability may be maintained. The results of a closed-loop simulation of the CSTR are displayed in Figs. 4.14 and 4.15. The LEMPC does maintain the process inside the region $\mathscr{X}$ for the duration of the simulation as observed in Figs. 4.14 and 4.15.

4.6 Conclusions

In this chapter, various LEMPC designs were developed, which are capable of optimizing closed-loop performance with respect to general economic considerations for nonlinear systems. Numerous issues arising in the context of chemical process control were considered including closed-loop stability, robustness, closed-loop performance, asynchronous and delayed sampling, and explicitly time-varying economic cost functions. The formulations of the LEMPC schemes were provided as well as rigorous theoretical treatments of the schemes were carried out. Closed-loop stability, in the sense of boundedness of the closed-loop state, under the LEMPC designs was proven. Additionally, when desirable, the LEMPC designs may be used to enforce convergence of the closed-loop state to steady-state. Under a specific terminal constraint design, the closed-loop system under the resulting LEMPC scheme was shown to achieve at least as good closed-loop performance as that achieved under an explicit stabilizing controller. Demonstrations of the effectiveness of the LEMPC schemes on chemical process examples were also provided. Moreover, the closed-loop properties of these examples under the LEMPC schemes were compared with respect to existing approaches to optimization and control. In all cases considered, the closed-loop economic performance under the LEMPC designs was better relative to the conventional approaches.

References

1. Scokaert POM, Mayne DQ, Rawlings JB (1999) Suboptimal model predictive control (feasibility implies stability). IEEE Trans Autom Control 44:648–654
2. Massera JL (1956) Contributions to stability theory. Ann Math 64:182–206
3. Krasovskii NN (1963) Stability of motion. Stanford University Press, Stanford
4. Hahn W (1967) Stability of motion. Springer, New York
5. Khalil HK (2002) Nonlinear systems, 3rd edn. Prentice Hall, Upper Saddle River
6. Lin Y, Sontag E, Wang Y (1996) A smooth converse Lyapunov theorem for robust stability. SIAM J Control Optim 34:124–160

7. Teel AR, Praly L (1999) Results on converse Lyapunov functions from class-KL estimates. In: Proceedings of the 38th IEEE conference on decision and control, vol 3. Phoenix, AZ, pp 2545–2550
8. Mhaskar P, El-Farra NH, Christofides PD (2006) Stabilization of nonlinear systems with state and control constraints using Lyapunov-based predictive control. Syst Control Lett 55:650–659
9. Muñoz de la Peña D, Christofides PD (2008) Lyapunov-based model predictive control of nonlinear systems subject to data losses. IEEE Trans Autom Control 53:2076–2089
10. Christofides PD, Liu J, Muñoz de la Peña D (2011) Networked and distributed predictive control: methods and nonlinear process network applications. Advances in industrial control series. Springer, London, England
11. Lee EB, Markus L (1967) Foundations of optimal control theory., SIAM series in applied mathematics. Wiley, New York
12. Vinter R (2010) Optimal control., Modern Birkhäuser classicsBirkhäuser, Boston
13. Liu J Muñoz de la Peña D, Christofides PD, Davis JF (2009) Lyapunov-based model predictive control of nonlinear systems subject to time-varying measurement delays. Int J Adapt Control Signal Process 23:788–807
14. Heidarinejad M, Liu J, Christofides PD (2012) Economic model predictive control of nonlinear process systems using Lyapunov techniques. AIChE J 58:855–870
15. Angeli D, Amrit R, Rawlings JB (2012) On average performance and stability of economic model predictive control. IEEE Trans Autom Control 57:1615–1626
16. Müller MA, Angeli D, Allgöwer F (2014) Transient average constraints in economic model predictive control. Automatica 50:2943–2950
17. Sontag ED (1989) A 'universal' construction of Artstein's theorem on nonlinear stabilization. Syst Control Lett 13:117–123
18. Lao L, Ellis M, Christofides PD (2014) Smart manufacturing: handling preventive actuator maintenance and economics using model predictive control. AIChE J 60:2179–2196
19. Wächter A, Biegler LT (2006) On the implementation of an interior-point filter line-search algorithm for large-scale nonlinear programming. Math Program 106:25–57

Chapter 5
State Estimation and EMPC

5.1 Introduction

In the previous chapters, all EMPC schemes are developed under the assumption of state feedback. However, this assumption does not hold in many practical applications. To address this issue, in this chapter, we introduce two output feedback EMPC schemes based on state estimation.

First, working with a class of full-state feedback linearizable nonlinear systems, a high-gain observer-based output feedback EMPC scheme is presented. In this scheme, a high-gain observer is used to estimate the system state using output measurements and the EMPC uses the observer state estimates. Sufficient conditions for the stability of the closed-loop system are derived using singular perturbation arguments. A chemical process example is used to demonstrate the ability of the high-gain observer-based EMPC to achieve time-varying operation that leads to a superior economic performance compared to the performance achieved under steady-state operation.

To improve the robustness of the closed-loop system especially to plant/model mismatch and uncertainties and to reduce the sensitivity of the state observer to measurement noise, a robust moving horizon estimation (RMHE) based output feedback EMPC design is subsequently presented. Bounded process and measurement noise is considered. To achieve fast convergence of the state estimates to the actual state (inducing an effective separation principle between the state observer and controller designs), a deterministic (high-gain) observer is first applied for a small time period with continuous output measurements to drive the estimation error to a small value. After the initial time period, a RMHE designed based on the deterministic observer is used to provide more accurate and smooth state estimates to the EMPC and thus, improves the robustness of the closed-loop system to noise. In the RMHE design, the deterministic observer is used to calculate a reference estimate and a confidence region for the state estimate. The confidence region is subsequently used as a constraint in the RMHE problem. Closed-loop stability is rigorously analyzed, and

M. Ellis et al., *Economic Model Predictive Control*,
Advances in Industrial Control, DOI 10.1007/978-3-319-41108-8_5

conditions that ensure closed-loop stability are derived. Extensive simulations based on a chemical process example illustrate the effectiveness of the second scheme.

5.1.1 System Description

We consider nonlinear systems described by the following state-space model:

$$\begin{aligned}\dot{x}(t) &= f(x(t)) + g(x(t))u(t) + l(x(t))w(t) \\ y(t) &= h(x(t)) + v(t)\end{aligned} \tag{5.1}$$

where $x(t) \in \mathbb{R}^n$ denotes the state vector, $u(t) \in \mathbb{R}^p$ denotes the control (manipulated) input vector, $w(t) \in \mathbb{R}^m$ denotes the disturbance vector, $y(t) \in \mathbb{R}^q$ denotes the measured output vector and $v(t) \in \mathbb{R}^q$ is the measurement noise vector. The control input vector is restricted to a nonempty convex set $\mathbb{U} \subseteq \mathbb{R}^p$ such that $\mathbb{U} := \{u \in \mathbb{R}^p : |u| \leq u^{\max}\}$ where $u^{\max}$ is the magnitude of the input constraint. It is assumed that the noise vectors are bounded such as $w(t) \in \mathbb{W}$ and $v(t) \in \mathbb{V}$ for all $t \geq 0$ where

$$\begin{aligned}\mathbb{W} &:= \{w \in \mathbb{R}^m : |w| \leq \theta_w,\ \theta_w > 0\} \\ \mathbb{V} &:= \{v \in \mathbb{R}^q : |v| \leq \theta_v,\ \theta_v > 0\}\end{aligned}$$

where θ_w and θ_v are positive real numbers. Moreover, it is assumed that the output measurement vector y of the system is continuously available at all times. It is further assumed that f, g, l and h are sufficiently smooth functions and $f(0) = 0$ and $h(0) = 0$.

5.1.2 Stabilizability Assumption

We assume that there exists a state feedback controller $u = k(x)$, which renders the origin of the nominal closed-loop system asymptotically and locally exponentially stable while satisfying the input constraints for all the states x inside a given stability region. Using converse Lyapunov theorems, this assumption implies that there exist class $\mathcal{K}$ functions $\alpha_i(\cdot)$, $i = 1, 2, 3, 4$ and a continuously differentiable Lyapunov function $V : D \rightarrow \mathbb{R}$ for the closed-loop system, that satisfy the following inequalities:

$$\alpha_1(|x|) \leq V(x) \leq \alpha_2(|x|) \tag{5.2a}$$

$$\frac{\partial V(x)}{\partial x}(f(x) + g(x)k(x)) \leq -\alpha_3(|x|) \tag{5.2b}$$

$$\left|\frac{\partial V(x)}{\partial x}\right| \leq \alpha_4(|x|) \tag{5.2c}$$

$$k(x) \in \mathbb{U} \tag{5.2d}$$

for all $x \in D \subseteq \mathbb{R}^n$ where D is an open neighborhood of the origin. We denote the region $\Omega_\rho \subseteq D$ as the stability region of the closed-loop system under the controller $k(x)$. Using the smoothness assumed for f and g, and taking into account that the manipulated input is bounded, there exists a positive constant M such that

$$|f(x) + g(x)u| \leq M \tag{5.3}$$

for all $x \in \Omega_\rho$ and $u \in \mathbb{U}$. In addition, by the continuous differentiable property of the Lyapunov function V and the smoothness of f and g, there exist positive constants L_x, L_u, C_x, $C_{g'}$ and C_g such that

$$\begin{aligned}
\left|\frac{\partial V(x)}{\partial x}f(x) - \frac{\partial V(x')}{\partial x}f(x')\right| &\leq L_x|x - x'| \\
\left|\frac{\partial V(x)}{\partial x}g(x) - \frac{\partial V(x')}{\partial x}g(x')\right| &\leq L_u|x - x'| \\
|f(x) - f(x')| &\leq C_x|x - x'| \\
|g(x) - g(x')| &\leq C_{g'}|x - x'| \\
\left|\frac{\partial V(x)}{\partial x}g(x)\right| &\leq C_g
\end{aligned} \tag{5.4}$$

for all $x, x' \in \Omega_\rho$ and $u \in \mathbb{U}$.

5.2 High-Gain Observer-Based EMPC Scheme

To simplify the presentation but without loss of generality, we restrict our consideration to single-input single-output nonlinear systems in this section. Moreover, we consider systems without process disturbances and measurement noise. The later assumption is relaxed in the subsequent section where robustness is explicitly addressed. The system in Eq. 5.1 reduces to the following system:

$$\begin{aligned}
\dot{x} &= f(x) + g(x)u \\
y &= h(x)
\end{aligned} \tag{5.5}$$

where $u \in \mathbb{R}$ and $y \in \mathbb{R}$. The presented approach may be extended to multi-input multi-output systems in a conceptually straightforward manner.

It is assumed that the system in Eq. 5.5 is full-state feedback linearizable. Thus, the relative degree of the output with respect to the input is n. Assumption 5.1 below states this requirement.

Assumption 5.1 There exists a set of coordinates

$$z = \begin{bmatrix} z_1 \\ z_2 \\ \vdots \\ z_n \end{bmatrix} = T(x) = \begin{bmatrix} h(x) \\ L_f h(x) \\ \vdots \\ L_f^{n-1} h(x) \end{bmatrix} \tag{5.6}$$

such that the system of Eq. 5.1 may be written as:

$$\begin{aligned} \dot{z}_1 &= z_2 \\ &\vdots \\ \dot{z}_{n-1} &= z_n \\ \dot{z}_n &= L_f^n h(T^{-1}(z)) + L_g L_f^{n-1} h(T^{-1}(z))u \\ y &= z_1 \end{aligned}$$

where $L_g L_f^{n-1} h(x) \neq 0$ for all $x \in \mathbb{R}^n$ ($L_f h(x)$ and $L_g h(x)$ denote Lie derivatives of the function h with respect to f and g, respectively).

Using Assumption 5.1, the system of Eq. 5.5 may be rewritten in the following compact form:

$$\begin{aligned} \dot{z} &= Az + B[L_f^n h(T^{-1}(z)) + L_g L_f^{n-1} h(T^{-1}(z))u] \\ y &= Cz \end{aligned}$$

where

$$A = \begin{bmatrix} 0_{n-1} & I_{n-1} \\ 0 & 0_{n-1}^T \end{bmatrix}, \quad B = \begin{bmatrix} 0_{n-1} \\ 1 \end{bmatrix}, \quad C = \begin{bmatrix} 1 \\ 0_{n-1} \end{bmatrix}^T,$$

0_{n-1} denotes a $n-1$ dimensional vector with all elements equal to zero, and I_{n-1} denotes the $n-1$ by $n-1$ identity matrix.

Remark 5.1 Assumption 5.1 imposes certain practical restrictions on the applicability of the method. However, this should be balanced with the nature of the results achieved by the output feedback controller in the sense that for a sufficiently large observer gain, the closed-loop system under the output feedback controller approaches the closed-loop stability region and performance of the state feedback controller. Essentially, a nonlinear separation-principle is achieved because of Assumption 5.1 and the use of a high-gain observer (please see Theorem 5.1). This is an assumption imposed in most previous works that use high-gain observers for state estimation, starting from the early work of Khalil and co-workers [1]. With respect to practical restrictions, our example demonstrates that the method is applicable to a class of chemical reactor models. The requirement of full state linearizability may be relaxed to input/output linearizability where the relative degree r is smaller than the system dimension n. For the input/output linearizability case, an additional observer is required to estimate the state of the inverse dynamics; please see [2] for a detailed development of this case.

5.2.1 State Estimation via High-Gain Observer

The state estimation-based EMPC developed in this section takes advantage of a high-gain observer [1, 3], which obtains estimates of the output derivatives up to order $n-1$ and consequently, computes estimates of the transformed state z. From the estimated transformed state, the system state may be estimated through the inverse transformation $T^{-1}(\cdot)$. The state estimate is denoted by $\hat{x}$. Proposition 5.1 below defines the high-gain observer equations and establishes precise conditions under which the combination of the high-gain observer and of the controller $k(x)$ together with appropriate saturation functions to eliminate wrong estimates enforce asymptotic stability of the origin in the closed-loop system for sufficiently large observer gain. The proof of the proposition follows from the results in [2, 4].

Proposition 5.1 *Consider the nonlinear system of Eq. 5.5 for which Assumption 5.5 holds. Also, assume that there exists a $k(x)$ for which Eq. 5.2 holds and it enforces local exponential stability of the origin in the closed-loop system. Consider the nonlinear system of Eq. 5.5 under the output feedback controller*

$$u = k(\hat{x}) \tag{5.7}$$

where

$$\hat{x} = T^{-1}(sat(\hat{z})) \tag{5.8}$$

and

$$\dot{\hat{z}} = A\hat{z} + L(y - C\hat{z}) \tag{5.9}$$

with

$$L = \left[\frac{a_1}{\varepsilon} \; \frac{a_2}{\varepsilon^2} \; \cdots \; \frac{a_n}{\varepsilon^n}\right]^T,$$

and the parameters a_i are chosen such that the roots of

$$s^n + a_1 s^{n-1} + \cdots + a_{n-1}s + a_n = 0 \tag{5.10}$$

are in the open left-half of the complex plane. Then given δ, there exists ε^* such that if $\varepsilon \in (0, \varepsilon^*]$, $|\hat{z}(0)| \leq z_m$, $x(0) \in \Omega_\delta$ with z_m being the maximum of the vector $\hat{z}$ for $|\hat{z}| \leq \beta_z(\delta_z, 0)$ where β_z is a class $\mathcal{KL}$ function and $\delta_z = \max\{|T(x)|, x \in \Omega_\delta\}$; the origin of the closed-loop system is asymptotically stable. This stability property implies that for $\varepsilon \in (0, \varepsilon^*]$ and given some positive constant $e_m > 0$ there exists positive real constant $t_b > 0$ such that if $x(0) \in \Omega_\delta$ and $|\hat{z}(0)| \leq z_m$, then $|x(t) - \hat{x}(t)| \leq e_m$ for all $t \geq t_b$.

Remark 5.2 In Proposition 5.1, the saturation function, $sat(\cdot)$, is used to eliminate the peaking phenomenon associated with the high-gain observer, see for example [1]. Also, the estimated state $\hat{x}$ is considered to have converged to the actual state x when the estimation error $|x - \hat{x}|$ is less than a given bound e_m. The time needed to converge, is given by t_b which is proportional to the observer gain $1/\varepsilon$. During this transient, the value of the Lyapunov function $V(x)$ may increase.

5.2.2 *High-Gain Observer-Based EMPC*

In this section, we consider the design of an estimation-based LEMPC for nonlinear systems. We assume that the LEMPC is evaluated at synchronous time instants $\{t_{k\geq 0}\}$ with $t_k = t_0 + k\Delta,\ k = 0, 1, \ldots$ where $t_0 = 0$ is the first time that LEMPC is evaluated and $\Delta > 0$ is the LEMPC sampling time.

5.2.2.1 Implementation Strategy

The high-gain observer of Eq. 5.9 continuously receives output measurements and computes estimated system states. At each sampling time t_k, the LEMPC obtains the estimated system state, which is denoted by $\hat{x}(t_k)$, from the observer. Based on $\hat{x}(t_k)$, the LEMPC uses the system model of Eq. 5.5 to predict the future evolution of the system over a finite prediction horizon while minimizing an economic cost function.

The two-mode operation paradigm presented in Chap. 4 is adopted in the design of the LEMPC. From the initial time t_0 up to a specific time t_s the LEMPC operates in the first operation mode to optimize the economic cost function while maintaining the closed-loop system state in the stability region Ω_ρ. Without loss of generality, t_s is assumed to be a multiple of LEMPC sampling time. In the first operation mode, a subset of the stability region, denoted by Ω_{ρ_e} with $\rho_e < \rho$, is defined in order to account for the high-gain observer effect, i.e., there is a discrepancy between the estimated state and the actual state. If the estimated state is in the region Ω_{ρ_e}, the LEMPC minimizes the cost function while constraining the predicted state trajectory to be within the region Ω_{ρ_e} over the prediction horizon. If the estimated state is in the region $\Omega_\rho \backslash \Omega_{\rho_e}$, the LEMPC computes control actions that optimize the economic cost subject to a condition that ensures that the control actions drive the system state to the region Ω_{ρ_e}. After time t_s, the LEMPC operates in the second operation mode and calculates the inputs in a way that the state of the closed-loop system is driven to a neighborhood of the desired steady-state.

The above described implementation strategy of the LEMPC may be summarized as follows:

Algorithm 5.1 High-gain observer-based LEMPC implementation algorithm

1. Based on the output measurements $y(t)$, the high-gain observer continuously estimates the state $\hat{x}(t)$ (for all $t \geq t_0 = 0$). The LEMPC receives the estimated state at a sampling time t_k from the observer.

2. If $t_k < t_s$, go to Step 3. Else, go to Step 4.
3. If $\hat{x}(t_k) \in \Omega_{\rho_e}$, go to Step 3.1. Else, go to Step 3.2.
   3.1. The controller optimizes the economic cost function while constraining the predicted state trajectory to lie within Ω_{ρ_e}. Go to Step 5.
   3.2. The controller optimizes the economic cost function while ensuring the computed control actions drive the state to the region Ω_{ρ_e}. Go to Step 5.
4. The controller computes control actions that drive the state to a small neighborhood of the origin.
5. Go to Step 1 ($k \leftarrow k + 1$).

5.2.2.2 LEMPC Formulation

The LEMPC is evaluated at each sampling time to obtain the future input trajectories based on estimated state $\hat{x}(t_k)$ provided by the high-gain observer. Specifically, the optimization problem of the LEMPC is as follows:

$$\min_{u \in S(\Delta)} \int_{t_k}^{t_{k+N}} l_e(\tilde{x}(\tau), u(\tau))\, d\tau \tag{5.11a}$$

$$\text{s.t.}\quad \dot{\tilde{x}}(t) = f(\tilde{x}(t)) + g(\tilde{x}(t))u(t) \tag{5.11b}$$

$$\tilde{x}(t_k) = \hat{x}(t_k) \tag{5.11c}$$

$$u(t) \in \mathbb{U},\ \forall\, t \in [t_k, t_{k+N}) \tag{5.11d}$$

$$V(\tilde{x}(t)) \leq \rho_e,\ \forall\, t \in [t_k, t_{k+N}),\quad \text{if } V(\hat{x}(t_k)) \leq \rho_e \text{ and } t_k < t_s \tag{5.11e}$$

$$L_g V(\hat{x}(t_k))u(t_k) \leq L_g V(\hat{x}(t_k))k(\hat{x}(t_k)),\quad \text{if } V(\hat{x}(t_k)) > \rho_e \text{ or } t_k \geq t_s \tag{5.11f}$$

where $\tilde{x}$ is the predicted trajectory of the system with control inputs calculated by the LEMPC. The notation used in the LEMPC of Eq. 5.11 is similar to that of the LEMPC of Eq. 4.3. The constraint of Eq. 5.11b is the system model used to predict the future evolution of the system. The model is initialized with the estimated state $\hat{x}(t_k)$ computed by the high-gain observer. The constraint of Eq. 5.11d accounts for the inputs constraints. The constraint of Eq. 5.11e is associated with the mode 1 operation of the LEMPC, which restricts the predicted system state to be in the set Ω_{ρ_e}, while the constraint of Eq. 5.11f is associated with the mode 2 operation of the LEMPC. The latter constraint restricts the control input for the first sampling period of the prediction horizon so that the amount of reduction of the Lyapunov function value is at least at the same level as that achieved by applying the controller $k(x)$. The constraint of Eq. 5.11f is used when $\hat{x}(t_k) \notin \Omega_{\rho_e}$ or when $t_k \geq t_s$.

The optimal solution to the optimization problem is denoted by $u^*(t|t_k)$ for $t \in [t_k, t_{k+N})$. The control actions computed by the LEMPC that are applied to the system are defined as follows:

$$u(t) = u^*(t|t_k), \ \forall\, t \in [t_k, t_{k+1}) \tag{5.12}$$

which is computed at each sampling time.

5.2.3 Closed-Loop Stability Analysis

In this subsection, the closed-loop stability of the output feedback EMPC is analyzed and a set of sufficient conditions is derived. In order to present the results, we need the following proposition, which states the closed-loop stability properties under the LEMPC with full state feedback. The proposition is a slight variation of Theorem 4.1, and therefore, its proof is omitted.

Proposition 5.2 (Theorem 4.1) *Consider the system of Eq. 5.5 in closed-loop under the LEMPC of Eq. 5.11 with state feedback, i.e., $\tilde{x}(t_k) = x(t_k)$, based on a controller $k(\cdot)$ that satisfies the conditions of Eq. 5.2. Let $\varepsilon_w > 0$, $\Delta > 0$ and $\rho > \rho_s > 0$ satisfy the following constraint:*

$$-\alpha_3(\alpha_2^{-1}(\rho_s)) + L_x M \Delta \leq -\varepsilon_w/\Delta. \tag{5.13}$$

If $x(0) \in \Omega_\rho$, then $x(t) \in \Omega_\rho$ for all $t \geq 0$. Furthermore, there exists a class $\mathscr{KL}$ function β and a class $\mathscr{K}$ function γ such that

$$|x(t)| \leq \beta(|x(t^*)|, t - t^*) + \gamma(\rho^*) \tag{5.14}$$

with $\rho^* = \max\{V(x(t+\Delta)) : V(x(t)) \leq \rho_s\}$, for all $x(t^*) \in B_\delta \subset \Omega_\rho$ and for all $t \geq t^* > t_s$ where t^* is such that $x(t^*) \in B_\delta$.

Theorem 5.1 below provides sufficient conditions such that the state estimation-based LEMPC of Eq. 5.11 with the high-gain observer of Eq. 5.9 guarantees that the state of the closed-loop system of Eq. 5.5 is always bounded and is ultimately bounded in a small region containing the origin. To this end, let:

$$e_i = \frac{1}{\varepsilon^{n-i}}(y^{(i-1)} - \hat{z}_i), \ i = 1, \ldots, n, \tag{5.15}$$

$$e^T = [e_1\, e_2\, \cdots\, e_n] \tag{5.16}$$

and

$$A^* = \begin{bmatrix} -a_1 & 1\,0\cdots 0\,0 \\ \vdots & \vdots\,\vdots\,\ddots\,\vdots\,\vdots \\ -a_{n-1} & 0\,0\cdots 0\,1 \\ -a_n & 0\,0\cdots 0\,0 \end{bmatrix}, \ b = \begin{bmatrix} 0 \\ \vdots \\ 0 \\ 1 \end{bmatrix} \tag{5.17}$$

where $y^{(i-1)}$ is the $(i-1)$-th derivative of the output measurement y and $\hat{z}_i$ is the i-th component of $\hat{z}$.

Theorem 5.1 *Consider the closed-loop system of Eq. 5.5 with the state estimation-based LEMPC of Eq. 5.11 based on a feedback controller $k(\cdot)$ that satisfies the conditions of Eq. 5.2. Let Assumption 5.1, Eqs. 5.13, 5.15–5.17 hold and choose the parameters a_i $(i = 1, \ldots, n)$ such that the roots of Eq. 5.10 are in the open left-half of the complex plane. Then there exist a class $\mathscr{KL}$ function β, a class $\mathscr{K}$ function γ, a pair of positive real numbers (δ_x, d_x), $0 < \rho_e < \rho$, $\varepsilon^* > 0$ and $\Delta^* > 0$ such that if* $\max\{|x(0)|, |e(0)|\} \leq \delta_x$, $\varepsilon \in (0, \varepsilon^*]$, $\Delta \in (0, \Delta^*]$,

$$-\alpha_3(\alpha_1^{-1}(\rho_s)) + (M\Delta + e_m)(L_x + L_u u^{\max}) < 0 \tag{5.18}$$

and

$$\rho_e \leq \rho - \alpha_4(\alpha_1^{-1}(\rho))M \max\{t_b(\varepsilon), \Delta\} \tag{5.19}$$

with t_b defined in Proposition 5.1, then $x(t) \in \Omega_\rho$ for all $t \geq 0$. Furthermore, for all $t \geq t^* > t_s$, the following bound holds:

$$|x(t)| \leq \beta(|x(t^*)|, t - t^*) + \gamma(\rho^*) + d_x\ . \tag{5.20}$$

Proof When the control action applied to the closed-loop system of Eq. 5.5 is obtained from the state estimation-based LEMPC of Eq. 5.11, the closed-loop system takes the following singularly perturbed form:

$$\begin{aligned} \dot{x} &= f(x) + g(x)u^*(\hat{x}) \\ \varepsilon\dot{e} &= A^*e + \varepsilon b L_f^n h(T^{-1}(z)) + \varepsilon b L_g L_f^{n-1} h(T^{-1}(z))u^*(\hat{x}) \end{aligned} \tag{5.21}$$

where the notation $u^*(\hat{x})$ is used to emphasize that the control action computed by the state estimation-based LEMPC is a function of the estimated state.

First, we compute the reduced-order slow and fast closed-loop subsystems related to Eq. 5.21 and prove the closed-loop stability of the slow and fast subsystems. Setting $\varepsilon = 0$ in Eq. 5.21, we obtain the corresponding slow subsystem as follows:

$$\dot{x} = f(x) + g(x)u^*(\hat{x}) \tag{5.22a}$$

$$A^*e = 0 \tag{5.22b}$$

Taking into account the fact that A^* is non-singular and $e = [0\ 0\ \cdots\ 0]^T$ is the unique solution of Eq. 5.22b, we may obtain $\hat{z}_i = y^{(i-1)}$, $i = 1, \ldots, n$ and $x(t) = \hat{x}(t)$. This means that the closed-loop slow subsystem is reduced to the one studied in Proposition 5.2 under state feedback. According to Proposition 5.2, if $x(0) \in$

$B_\delta \subset \Omega_\rho$, then $x(t) \in \Omega_\rho$ for all $t \geq 0$ and for all $t \geq t^* > t_s$, the following bound holds:

$$|x(t)| \leq \beta(|x(t^*)|, t - t^*) + \gamma(\rho^*) \tag{5.23}$$

where ρ^* and t^* have been defined in Proposition 5.2.

Introducing the fast time scale $\bar{\tau} = t/\varepsilon$ and setting $\varepsilon = 0$, the closed-loop fast subsystem may be represented as follows:

$$\frac{de}{d\bar{\tau}} = A^* e \, . \tag{5.24}$$

Since A^* is Hurwitz, the closed-loop fast subsystem is also stable. Moreover, there exist $k_e \geq 1$ and $a_e > 0$ such that:

$$|e(\bar{\tau})| \leq k_e |e(0)| e^{-a_e \bar{\tau}} \tag{5.25}$$

for all $\bar{\tau} \geq 0$.

Next, we consider $t \in (0, \max\{\Delta, t_b\}]$ and $t \geq \max\{\Delta, t_b\}$ separately and prove that if conditions stated in Theorem 1 are satisfied, boundedness of the state is ensured. Note that t_b decreases as ε decreases. When $x(0) \in B_{\delta_x} \subset \Omega_{\rho_e} \subset \Omega_\rho$, and $\delta_x < \delta$, considering the closed-loop system state trajectory:

$$\dot{x}(t) = f(x(t)) + g(x(t))u^*(\hat{x}(0))$$

for $t \in (0, \max\{\Delta, t_b\}]$ and using Eqs. 5.2 and 5.3, we obtain that for all $t \in (0, \max\{\Delta, t_b\}]$:

$$\begin{aligned} V(x(t)) &= V(x(0)) + \int_0^t \dot{V}(x(\tau))d\tau \\ &= V(x(0)) + \int_0^t \frac{\partial V(x(\tau))}{\partial x} \dot{x}(\tau) d\tau \\ &\leq \rho_e + M \max\{\Delta, t_b(\varepsilon)\} \alpha_4(\alpha_1^{-1}(\rho)) \end{aligned} \tag{5.26}$$

Since t_b decreases as ε decreases, there exist Δ_1 and ε_1 such that if $\Delta \in (0, \Delta_1]$ and $\varepsilon \in (0, \varepsilon_1]$, Eq. 5.19 holds and thus,

$$V(x(t)) < \rho \tag{5.27}$$

for all $t \in (0, \max\{\Delta, t_b\}]$.

For $t \geq \max\{\Delta, t_b\}$, we have that $|x(t) - \hat{x}(t)| \leq e_m$ (this follows from Proposition 1 and e_m decreases as ε decreases), and we may write the time derivative of the Lyapunov function along the closed-loop system state of Eq. 5.5 under the state estimation-based LEMPC of Eqs. 5.11f, 5.9 and 5.11 for all $t \in [t_k, t_{k+1})$ (assuming without loss of generality that $t_k = \max\{\Delta, t_b\}$) as follows

$$\dot{V}(x(t)) = \frac{\partial V(x(t))}{\partial x}(f(x(t)) + g(x(t))u^*(\hat{x}(t_k))) \,. \tag{5.28}$$

Adding and subtracting the term $\partial V(\hat{x}(t_k))/\partial x(f(\hat{x}(t_k)) + g(\hat{x}(t_k))u^*(\hat{x}(t_k)))$ to/from the above inequality and from Eqs. 5.2 and 5.11f, we obtain

$$\begin{aligned}\dot{V}(x(t)) \le & -\alpha_3(\alpha_1^{-1}(\rho_s)) \\ & + \frac{\partial V(x)}{\partial x}\left(f(x(t)) - f(\hat{x}(t_k)) + (g(x(t)) - g(\hat{x}(t_k))u^*(\hat{x}(t_k)))\right)\end{aligned} \tag{5.29}$$

Using the smoothness properties of V, f, g and Eq. 5.4, we may obtain

$$\dot{V}(x(t)) \le -\alpha_3(\alpha_1^{-1}(\rho_s)) + (L_x + L_u u^{\max})|x(t) - \hat{x}(t_k)| \tag{5.30}$$

From the triangle inequality, Eq. 5.3, and the fact that the estimation error is bounded by e_m for $t \ge \max\{\Delta, t_b\}$,

$$|x(t) - \hat{x}(t_k)| \le |x(t) - x(t_k)| + |x(t_k) - \hat{x}(t_k)| \le M\Delta + e_m$$

for $x(t), x(t_k), \hat{x}(t_k) \in \Omega_\rho$ where $|x(t_k) - \hat{x}(t_k)| \le e_m$. Thus,

$$\dot{V}(x(t)) \le -\alpha_3(\alpha_1^{-1}(\rho_s)) + (L_x + L_u u^{\max})(M\Delta + e_m) \tag{5.31}$$

Picking ε_2 and Δ_2 such that for all $\varepsilon \in (0, \varepsilon_2]$ and for all $\Delta \in (0, \Delta_2]$, Eq. 5.18 is satisfied, the closed-loop system state $x(t)$ is bounded in Ω_ρ, for all $t \ge \max\{\Delta, t_b\}$. Finally, using similar arguments to the proof of Theorem 1 in [5], we have that there exist class $\mathscr{KL}$ function β, positive real numbers (δ_x, d_x) (note that the existence of $\delta_x < \delta$ such that $|x(0)| \le \delta_x$ follows from the smoothness of V), and $0 < \varepsilon^* < \min\{\varepsilon_1, \varepsilon_2\}$ and $0 < \Delta^* < \min\{\Delta_1, \Delta_2\}$ such that if $\max\{|x(0)|, |e(0)|\} \le \delta_x$, $\varepsilon \in (0, \varepsilon^*]$ and $\Delta \in (0, \Delta^*]$, then, the bound of Eq. 5.20 holds for all $t \ge 0$.

Remark 5.3 Under the state feedback LEMPC, the closed-loop system state is always bounded in Ω_ρ for both mode 1 and mode 2 operation; however, for mode 2 operation, after time t^* the closed-loop system state enters the ball B_δ, and the closed-loop system state may be bounded by Eq. 5.23. On the other hand, in state estimation-based LEMPC, the closed-loop system state is always bounded in Ω_ρ, if the initial system state belongs in $B_{\delta_x} \subset \Omega_{\rho_e} \subset \Omega_\rho$.

Remark 5.4 The major motivation for taking advantage of the nonlinear controller $k(x)$ arises from the need for formulating an a priori feasible economic MPC problem for a well-defined set of initial conditions. The control action of $k(x)$ is always a feasible candidate for the LEMPC design (even though the LEMPC via optimization is free to choose a different control action) and the LEMPC may take advantage of $k(x)$ to characterize its own corresponding stability region. In addition, the closed-loop system state is always bounded in the invariant stability region of $k(x)$.

5.2.4 Application to a Chemical Process Example

Consider a well-mixed, non-isothermal continuous stirred tank reactor (CSTR) where an irreversible, second-order, endothermic reaction $A \rightarrow B$ takes place, where A is the reactant and B is the desired product. The feedstock of the reactor consists of the reactant A in an inert solvent with flow rate F, temperature T_0 and molar concentration C_{A0}. Due to the non-isothermal nature of the reactor, a jacket is used to provide heat to the reactor. The dynamic equations describing the behavior of the reactor, obtained through material and energy balances under standard modeling assumptions, are given below:

$$\frac{dC_A}{dt} = \frac{F}{V_L}(C_{A0} - C_A) - k_0 e^{-E/RT} C_A^2 \tag{5.32a}$$

$$\frac{dT}{dt} = \frac{F}{V_L}(T_0 - T) + \frac{-\Delta H}{\rho_L C_p} k_0 e^{-E/RT} C_A^2 + \frac{Q_s}{\rho_L C_p V_L} \tag{5.32b}$$

where C_A denotes the concentration of the reactant A, T denotes the temperature of the reactor, Q_s denotes the steady-state rate of heat supply to the reactor, V_L represents the volume of the reactor, ΔH, k_0, and E denote the enthalpy, pre-exponential constant and activation energy of the reaction, respectively, and C_p and ρ_L denote the heat capacity and the density of the fluid in the reactor, respectively. The values of the process parameters used in the simulations are given in Table 5.1. The process model of Eq. 5.32 is numerically simulated using an explicit Euler integration method with integration step $h_c = 1.0 \times 10^{-3}$ h.

The process model has one stable steady-state in the operating range of interest. The control objective is to economically optimize the process in a region around the stable steady-state (C_{As}, T_s) to maximize the average production rate of B through manipulation of the concentration of A in the inlet to the reactor, C_{A0}. The steady-state C_{A0} value associated with the steady-state point is denoted by C_{A0s}. The process model of Eq. 5.32 belongs to the following class of nonlinear systems:

$$\dot{x}(t) = f(x(t)) + g(x(t))u(t)$$

Table 5.1 CSTR model parameter values

$T_0 = 300$	K	$F = 5.0$	$m^3\ h^{-1}$
$V_L = 1.0$	m^3	$E = 5.0 \times 10^3$	kJ $kmol^{-1}$
$k_0 = 13.93$	$m^3\ kmol^{-1}\ h^{-1}$	$\Delta H = 1.15 \times 10^4$	kJ $kmol^{-1}$
$C_p = 0.231$	kJ $kg^{-1} K^{-1}$	$R = 8.314$	kJ $kmol^{-1} K^{-1}$
$\rho_L = 1000$	kg m^{-3}	$C_{As} = 2.0$	kmol m^{-3}
$T_s = 350$	K	$C_{A0s} = 4.0$	kmol m^{-3}
$Q_s = 1.73 \times 10^5$	kJ h^{-1}		

where $x^T = [x_1\ x_2] = [C_A - C_{As}\ T - T_s]$ is the state, $u = C_{A0} - C_{A0s}$ is the input, $f = [f_1\ f_2]^T$ and $g_i = [g_{i1}\ g_{i2}]^T$ $(i = 1, 2)$ are vector functions. The input is subject to constraint as follows: $|u| \leq 3.5\,\text{kmol m}^{-3}$. There is an economic measure considered in this example as follows:

$$\frac{1}{t_f}\int_0^{t_f} k_0 e^{-E/RT(\tau)} C_A^2(\tau)\, d\tau \tag{5.33}$$

where t_f is the time duration of the reactor operation. The economic objective function of Eq. 5.33 describes the average production rate over the entire process operation. We also consider that there is a limitation on the amount of reactant material which may be used over a specific period $t_p = 1.0\,\text{h}$. Specifically, $u = C_{A0} - C_{A0s}$ should satisfy the following constraint:

$$\frac{1}{t_p}\int_0^{t_p} u(\tau)\, d\tau = 1.0\,\text{kmol m}^{-3}. \tag{5.34}$$

It should be emphasized that due to the second-order dependence of the reaction rate on the reactant concentration, the production rate may be improved through switching between the upper and lower bounds of the manipulated input, as opposed to steady-state operation via uniform in time distribution of the reactant in the feed; refer to the discussion contained in Sect. 1.3.2 for further explanation of this point.

In this section, we will design a state estimation-based LEMPC to manipulate the C_{A0} subject to the material constraint. In the first set of simulations, we assume that state feedback information is available at synchronous time instants while in the second set of simulations, we take advantage of a high-gain observer to estimate the reactant concentration from temperature measurements.

In terms of the Lyapunov-based controller, we use a proportional controller (P-controller) of the form $u = -\gamma_1 x_1 - \gamma_2 x_2$ subject to input constraints where $\gamma_1 = 1.6$ and $\gamma_2 = 0.01$ and a quadratic Lyapunov function $V(x) = x^T P x$ where

$$P = \text{diag}([110.11, 0.12])\ ,$$

and $\rho = 430$. It should be emphasized that Ω_ρ has been estimated through evaluation of $\dot{V}$ when we apply the proportional controller. We assume that the full system state $x = [x_1\ x_2]^T$ is measured and sent to the LEMPC at synchronous time instants $t_k = k\Delta$, $k = 0, 1, \ldots$, with $\Delta = 0.01\,\text{h} = 36\,\text{s}$ in the first set of simulations. For the output feedback LEMPC (second set of simulations), only temperature (x_2) is measured and a high-gain observer is utilized to estimate the reactant concentration from temperature measurements.

Considering the material constraint which needs to be satisfied through one period of process operation, a decreasing LEMPC horizon sequence $N_0, \ldots, N_{99}$ where $N_i = 100 - i$ and $i = 0, \ldots, 99$ is utilized at the different sampling times. At each sampling time t_k, the LEMPC with horizon N_k takes into account the leftover amount of reactant material and adjusts its horizon to predict future system state up to time

$t_p = 1.0\,\text{h}$ to maximize the average production rate. Since the LEMPC is evaluated at discrete-time instants during the closed-loop simulation, the material constraint is enforced as follows:

$$\sum_{i=0}^{M-1} u(t_i) = \frac{t_p}{\Delta} 1.0\,\text{kmol}\,\text{m}^{-3} \tag{5.35}$$

where $M = 100$. As LEMPC proceeds at different sampling times, the constraint is adjusted according to the optimal manipulated input at previous sampling times.

The state feedback LEMPC formulation for the chemical process example in question has the following form:

$$\max_{u \in S(\Delta)} \frac{1}{N_k \Delta} \int_{t_k}^{t_{k+N_k}} [k_0 e^{-E/R(\tilde{x}_2(\tau)+T_s)} (\tilde{x}_1(\tau) + C_{As})^2] \, d\tau \tag{5.36a}$$

$$\text{s.t.} \quad \dot{\tilde{x}}(t) = f(\tilde{x}(t)) + g(\tilde{x}(t))u(t), \ \forall\, t \in [t_k, t_{k+N_k}) \tag{5.36b}$$

$$\tilde{x}(t_k) = x(t_k) \tag{5.36c}$$

$$u(t) \in \mathbb{U}, \ \forall\, t \in [t_k, t_{k+N_k}) \tag{5.36d}$$

$$\sum_{i=k}^{k+N_k-1} u(t_i|t_k) = \zeta_k \tag{5.36e}$$

$$V(\tilde{x}(t)) \leq \rho, \ \forall\, t \in [t_k, t_{k+N_k}) \tag{5.36f}$$

where $x(t_k)$ is the process state measurement at sampling time t_k and the predicted system state along the LEMPC horizon is restricted to lie within the invariant set Ω_ρ through enforcement of the constraint of Eq. 5.36f subject to the manipulated input constraint of Eq. 5.36d. The constraint of Eq. 5.36e implies that the optimal values of u along the prediction horizon should be chosen to satisfy the material constraint where the explicit expression of ζ_k may be computed based on Eq. 5.35 and the optimal manipulated input values prior to sampling time t_k. In other words, this constraint indicates the amount of the remaining reactant material at each sampling time. Thus, it ensures that the material constraint is enforced through one period of process operation.

In terms of the initial guess for solving the optimization problem of Eq. 5.36, at the first sampling time we take advantage of the Lyapunov-based controller while for the subsequent sampling times, a shifted version of the optimal solution of the previous sampling time is utilized. The simulations were carried out using Java programming language in a Pentium 3.20 GHz computer and the optimization problems were solved using the open source interior point optimizer Ipopt [6]. The purpose of the following set of simulations is to demonstrate that: (I) the LEMPC design subject to state and output feedback restricts the system state in an invariant set; (II) the LEMPC design maximizes the economic measure of Eq. 5.36a; and (III) the LEMPC design achieves a higher objective function value compared to steady-state operation with equal distribution in time of the reactant material. We have also performed simulations for the case that the constraint of Eq. 5.36f is not included in the LEMPC design of

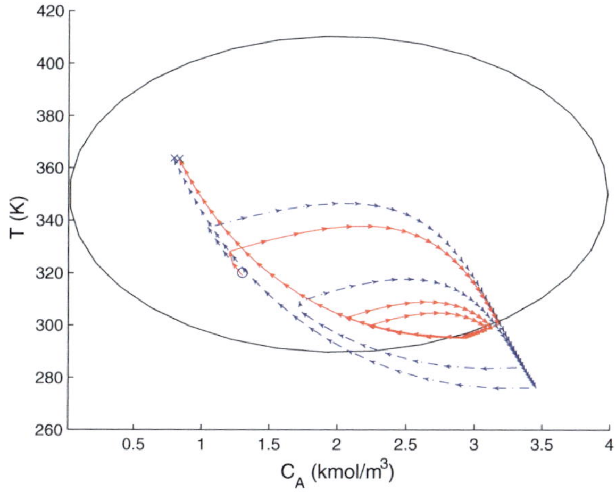

Fig. 5.1 The stability region Ω_ρ and the state trajectories of the process under the LEMPC design of Eq. 5.36 with state feedback and initial state $(C_A(0), T(0)) = (1.3\,\text{kmol}\,\text{m}^{-3}, 320\,\text{K})$ for one period of operation with (*solid line*) and without (*dash-dotted line*) the constraint of Eq. 5.36f. The symbols ○ and × denote the initial ($t = 0.0$ h) and final ($t = 1.0$ h) state of these closed-loop system trajectories, respectively

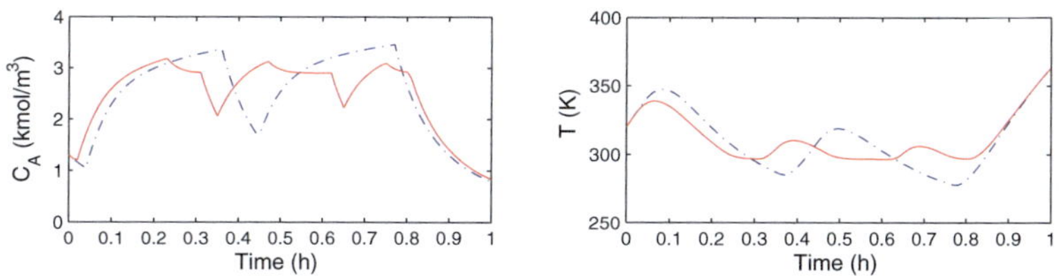

Fig. 5.2 State trajectories of the process under the LEMPC design of Eq. 5.36 with state feedback and initial state $(C_A(0), T(0)) = (1.3\,\text{kmol}\,\text{m}^{-2}, 320\,\text{K})$ for one period of operation with (*solid line*) and without (*dash-dotted line*) the constraint of Eq. 5.36f

Eq. 5.36. In this case, the process state is not constrained to be in a specific invariant set.

In the first set of simulations, we take the CSTR operation time $t_f = t_p = 1.0$ h. Figures 5.1, 5.2 and 5.3 illustrate the process state profile in state space (temperature T versus concentration C_A) considering the stability region Ω_ρ, the time evolution of process state and the manipulated input profile for the LEMPC formulation of Eq. 5.36 with and without the state constraint of Eq. 5.36f, respectively. In both cases, the initial process state is $(1.3\,\text{kmol}\,\text{m}^{-3}, 320\,\text{K})$. For both cases, the material constraint is satisfied while in the unconstrained state case, there is more freedom to compute the optimal input trajectory to maximize the average production rate. It needs to be emphasized that the process state trajectory under the LEMPC design of Eq. 5.36 subject to the constraint of Eq. 5.36f never leaves the invariant level set Ω_ρ when this constraint is enforced.

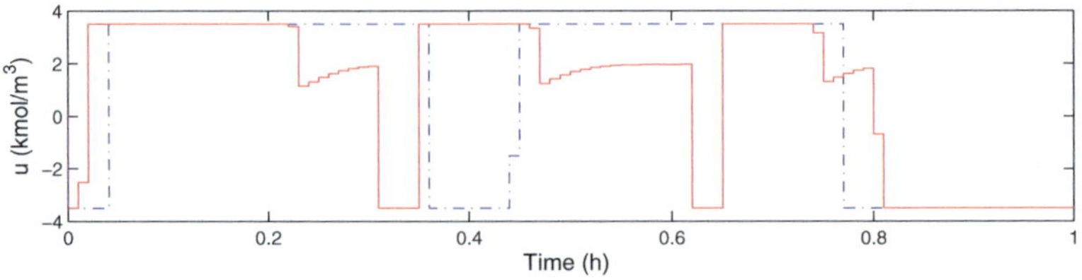

Fig. 5.3 Manipulated input trajectory under the LEMPC design of Eq. 5.36 with state feedback and initial state $(C_A(0), T(0)) = (1.3\,\text{kmol m}^{-3}, 320\,\text{K})$ for one period of operation with (*solid line*) and without (*dash-dotted line*) the constraint of Eq. 5.36f

We have also compared the time-varying operation through LEMPC of Eq. 5.36 to steady-state operation where the reactant material is uniformly distributed in the feed to the reactor over the process operation time of 1 h from a closed-loop performance point of view. To carry out this comparison, we have computed the total cost of each operating scenario based on an index of the following form:

$$J = \frac{1}{t_M} \sum_{i=0}^{M-1} [k_0 e^{-\frac{E}{RT(t_i)}} C_A^2(t_i)]$$

where $t_0 = 0.0\,\text{h}$, $t_M = 1.0\,\text{h}$ and $M = 100$. To be consistent in comparison, both of the simulations have been initialized from the steady-state point $(2.0\,\text{kmol m}^{-3}, 350\,\text{K})$. We find that through time-varying LEMPC operation, there is approximately 7 % improvement with respect to steady-state operation. Specifically, in the case of LEMPC operation with $\rho = 430$ the cost is 13.48, in the case of LEMPC operation with $\rho = \infty$ (LEMPC of Eq. 5.36 without the state constraint of Eq. 5.36f) the cost is 13.55 and in the case of steady-state operation the cost is 12.66.

We have also performed closed-loop simulation with the state estimation-based LEMPC (again, $t_f = t_p = 1.0\,\text{h}$). For this set of simulation, the high-gain observer parameters are $\varepsilon = 0.01$, $a_1 = a_2 = 1$, $\rho_e = 400$ and $z_m = 1685$; the high-gain observer is of the form of Eq. 5.9 with $n = 2$. In this case, the LEMPC formulation at each sampling time is initialized by the estimated system state $\hat{x}(t_k)$ while the output (temperature) measurement is continuously available to the high-gain observer. To ensure that the actual system state is restricted in Ω_ρ, we set $\rho_e = 400$. Figures 5.4, 5.5 and 5.6 illustrate the process state profile in state-space (temperature T versus concentration C_A) considering the stability region Ω_ρ, the time evolution of process states and the manipulated input profile for the LEMPC formulation of Eq. 5.36 using high-gain observer and with the state constraint of Eq. 5.36f, respectively. Similar to the state feedback case, the initial process state is $(1.3\,\text{kmol m}^{-3}, 320\,\text{K})$. Through LEMPC implementation, the material constraint is satisfied while the closed-loop system state is restricted inside the stability region Ω_ρ. The cost is 12.98 which is greater than the one for steady-state operation (12.66).

Also, we performed a set of simulations to compare LEMPC with the Lyapunov-based controller from an economic closed-loop performance point of view for opera-

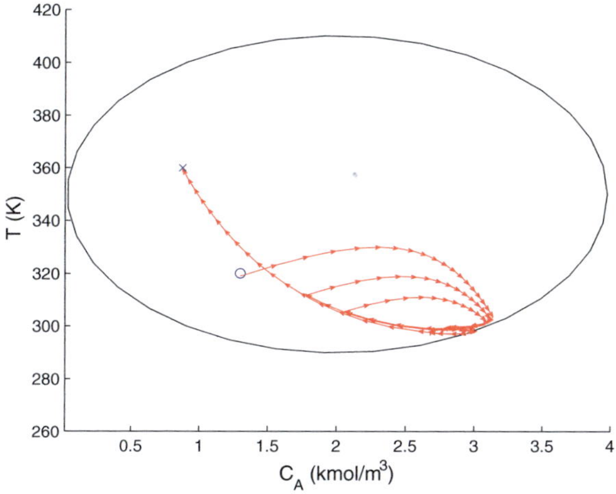

Fig. 5.4 The stability region Ω_ρ and the state trajectories of the process under the state under the state estimation-based LEMPC and initial state $(C_A(0), T(0)) = (1.3\,\text{kmol}\,\text{m}^{-3}, 320\,\text{K})$ for one period of operation subject to the constraint of Eq. 5.36f. The symbols $\circ$ and $\times$ denote the initial ($t = 0.0$ h) and final ($t = 1.0$ h) state of this closed-loop system trajectories, respectively

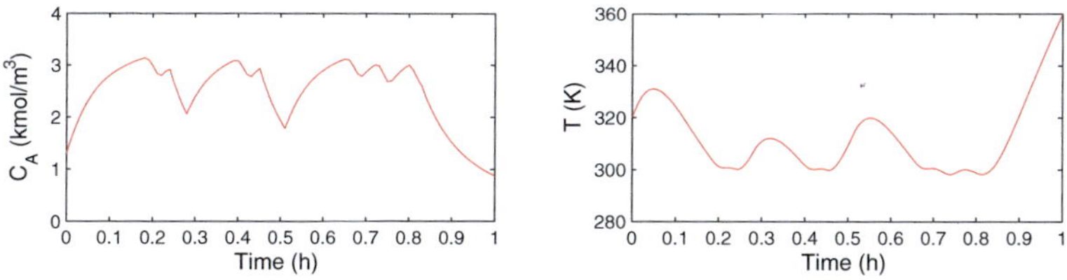

Fig. 5.5 State trajectories of the process under the state estimation-based LEMPC and initial state $(C_A(0), T(0)) = (1.3\,\text{kmol}\,\text{m}^{-3}, 320\,\text{K})$ for one period of operation subject to the constraint of Eq. 5.36f

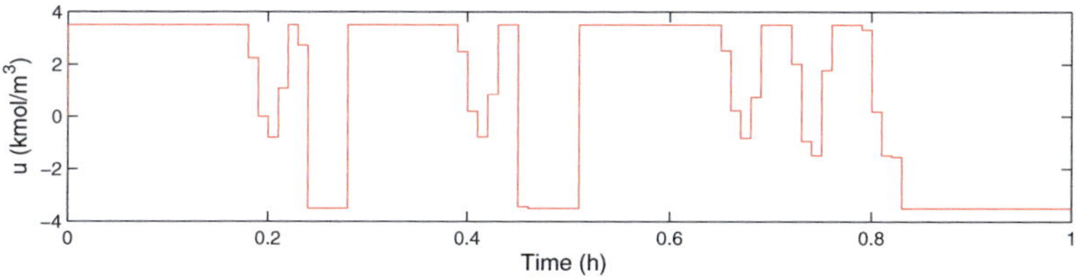

Fig. 5.6 Manipulated input trajectory under the state estimation-based LEMPC and initial state $(C_A(0), T(0)) = (1.3\,\text{kmol}\,\text{m}^{-3}, 320\,\text{K})$ for one period of operation subject to the constraint of Eq. 5.36f

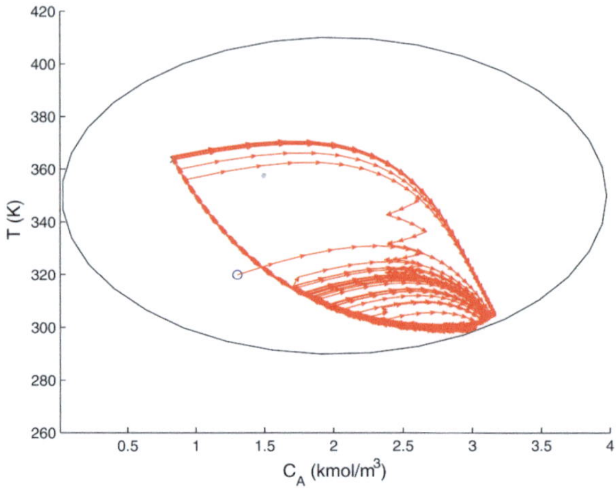

Fig. 5.7 The stability region Ω_ρ and the state trajectories of the process under the state estimation-based LEMPC and initial state $(C_A(0), T(0)) = (1.3\,\text{kmol m}^{-3}, 320\,\text{K})$ for 10 h operation in mode 1, followed by 10 h of operation in mode 2 and finally, 10 h of operation in mode 1. The symbols ∘ and × denote the initial ($t = 0.0$ h) and final ($t = 30.0$ h) state of this closed-loop system trajectories, respectively

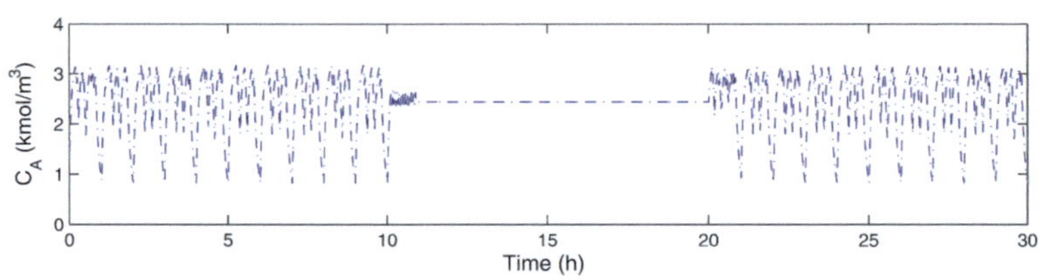

Fig. 5.8 Reactant concentration trajectory of the process under the state estimation-based LEMPC and initial state $(C_A(0), T(0)) = (1.3\,\text{kmol m}^{-3}, 320\,\text{K})$ for 10 h operation in mode 1, followed by 10 h of operation in mode 2 and finally, 10 h of operation in mode 1

tion over two consecutive 1 h periods, i.e., $t_f = 2.0$ h and $t_p = 1.0$ h. To be consistent in this comparison in the sense that both the LEMPC and the Lyapunov-based controller use the same, available amount of reactant material, we start the simulation in both cases from the same initial condition $(2.44\,\text{kmol m}^{-3}, 321.96\,\text{K})$, which corresponds to the steady-state of the process when the available reactant material is uniformly distributed over each period of operation. The objective of the Lyapunov-based controller is to keep the system state at this steady-state, while the output feedback LEMPC leads to time-varying operation that optimizes directly the economic cost. The corresponding economic costs for this 2-h operation are 26.50 for the LEMPC and 25.61 for the Lyapunov-based controller.

Furthermore, to demonstrate long-term reactor operation, i.e., $t_f = 30.0$ h and $t_p = 1.0$ h, we operate the process in a time-varying fashion to optimize the economic cost in mode 1 for the first 10 h, then switch to mode 2 to drive the closed-loop state

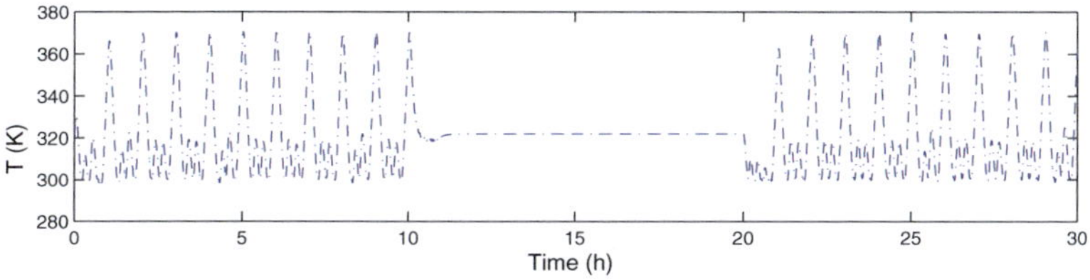

Fig. 5.9 Temperature trajectory of the process under the state estimation-based LEMPC and initial state $(C_A(0), T(0)) = (1.3\,\text{kmol}\,\text{m}^{-3}, 320\,\text{K})$ for 10 h operation in mode 1, followed by 10 h of operation in mode 2 and finally, 10 h of operation in mode 1

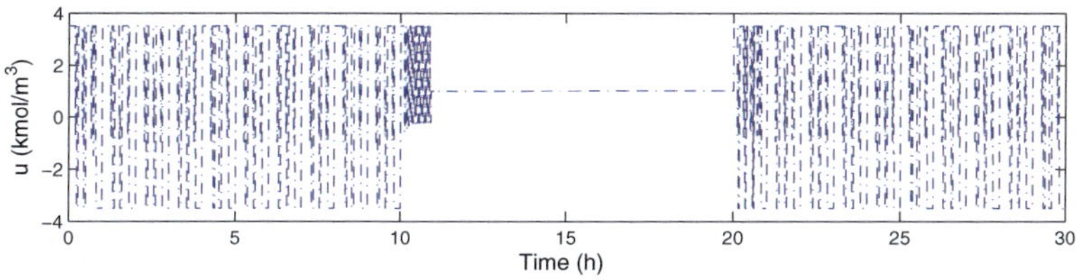

Fig. 5.10 Manipulated input trajectory under the state estimation-based LEMPC and initial state $(C_A(0), T(0)) = (1.3\,\text{kmol}\,\text{m}^{-3}, 320\,\text{K})$ for 10 h operation in mode 1, followed by 10 h of operation in mode 2 and finally, 10 h of operation in mode 1

to the steady-state corresponding to $u = 1.0$, i.e., equal distribution with time of the reactant material, for the next 10 h, and finally, operate the process in mode 1 for the last 10 h. Figures 5.7, 5.8, 5.9 and 5.10 display the results for this case, where the closed-loop system successfully alternates between the two different types (time-varying versus steady-state) of operation.

Finally, we performed a set of simulations to evaluate the effect of bounded measurement noise. Figures 5.11, 5.12 and 5.13 display the closed-loop system state and manipulated input of the state-estimation-based LEMPC subject to bounded output (temperature) measurement noise whose absolute value is bounded by 1.0 K. As it may be seen in Figs. 5.11, 5.12 and 5.13, the controller may tolerate the effect of measurement noise; in this case, Ω_{ρ_e} was reduced to 370 to improve the robustness margin of the controller to measurement noise. Economic closed-loop performance in this case is 12.95.

5.3 RMHE-Based EMPC Scheme

In the previous section, a high-gain observer is used in the design of the output feedback EMPC without explicitly considering process and measurement noise. In order to improve the robustness of the observer to plant-model mismatch and uncertainties while reducing its sensitivity to measurement noise significantly, a robust moving horizon estimation (RMHE) based output feedback LEMPC design is presented in this section. We consider systems that may be described by Eq. 5.1.

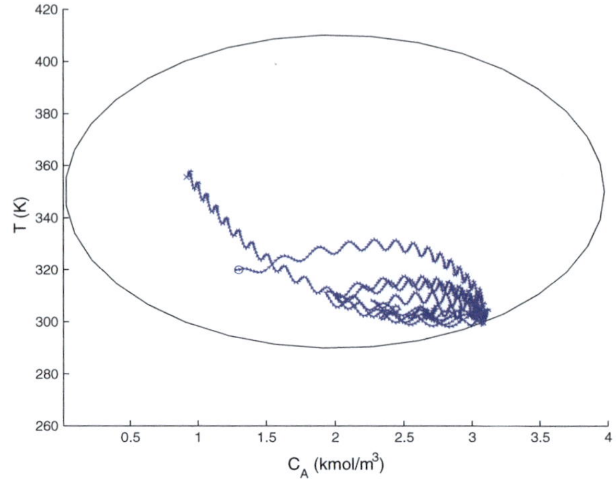

Fig. 5.11 The stability region Ω_ρ and the state trajectories of the process under the state estimation-based LEMPC and initial state $(C_A(0), T(0)) = (1.3\,\text{kmol m}^{-3}, 320\,\text{K})$ for one period of operation subject to the constraint of Eq. 5.36f and bounded measurement noise. The symbols $\circ$ and $\times$ denote the initial ($t = 0.0$ h) and final ($t = 1.0$ h) state of this closed-loop system trajectories, respectively

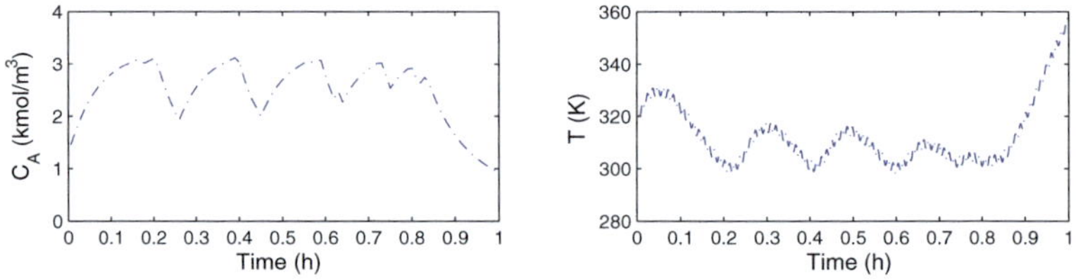

Fig. 5.12 State trajectories of the process under the state estimation-based LEMPC and initial state $(C_A(0), T(0)) = (1.3\,\text{kmol m}^{-3}, 320\,\text{K})$ for one period of operation subject to the constraint of Eq. 5.36f and bounded measurement noise

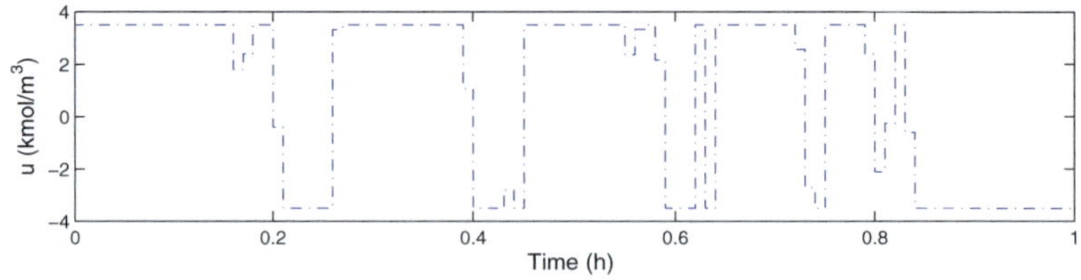

Fig. 5.13 Manipulated input trajectory under the state estimation-based LEMPC and initial state $(C_A(0), T(0)) = (1.3\,\text{kmol m}^{-3}, 320\,\text{K})$ for one period of operation subject to the constraint of Eq. 5.36f and bounded measurement noise

5.3.1 Observability Assumptions

It is assumed that there exists a deterministic observer that takes the following general form:

$$\dot{z} = F(\varepsilon, z, y) \tag{5.37}$$

where z is the observer state, y is the output measurement vector and ε is a positive parameter. This observer together with the state feedback controller $u = k(x)$ of Sect. 5.1.2 form an output feedback controller: $\dot{z} = F(\varepsilon, z, y)$, $u = k(z)$ which satisfies the following assumptions:

(1) there exist positive constants θ_w^*, θ_v^* such that for each pair $\{\theta_w, \theta_v\}$ with $\theta_w \leq \theta_w^*$, $\theta_v \leq \theta_v^*$, there exist $0 < \rho_1 < \rho$, $e_{m0} > 0$, $\varepsilon_L^* > 0$, $\varepsilon_U^* > 0$ such that if $x(t_0) \in \Omega_{\rho_1}$, $|z(t_0) - x(t_0)| \leq e_{m0}$ and $\varepsilon \in (\varepsilon_L^*, \varepsilon_U^*)$, the trajectories of the closed-loop system are bounded in Ω_ρ for all $t \geq t_0$;
(2) and there exists $e_m^* > 0$ such that for each $e_m \geq e_m^*$, there exists t_b such that $|z(t) - x(t)| \leq e_m$ for all $t \geq t_b(\varepsilon)$.

Note that a type of observer that satisfies the above assumptions is a high-gain observer like that presented in Sect. 5.2; see, also, [7] for results on high-gain observers subject to measurement noise. From an estimate error convergence speed point of view, it is desirable to pick the observer parameter ε as small as possible; however, when the parameter ε is too small, i.e., the observer gain is too large, it will make the observer very sensitive to measurement noise. In the observer assumptions, a key idea is to pick the gain ε in a way that balances the estimate error convergence speed to zero and the effect of the noise. In the remainder of this section, the estimate given by the observer F will be denoted as z.

Remark 5.5 It is important to point out the difference between the positive constants θ_w^* and θ_v^* and the bounds θ_w and θ_v. Specifically, the positive constants θ_w^* and θ_v^* correspond to theoretical bounds on the noise such that the closed-loop system under the output feedback controller: $\dot{z} = F(\varepsilon, z, y)$, $u = k(z)$ is maintained in Ω_ρ. The constants θ_w^* and θ_v^* depend on the stability properties of a given system under the output feedback controller. On the other hand, the bounds θ_w and θ_v correspond to the actual bound on the process and measurement noise for a given (open-loop) system.

5.3.2 Robust MHE

The idea of RMHE was initially developed in [8] integrating deterministic observer techniques and optimization-based estimation techniques in a unified framework. In the RMHE, an auxiliary deterministic nonlinear observer that is able to asymptotically track the nominal system state is used to calculate a confidence region. In the calculation of the confidence region, bounded process and measurement noise are taken into account. The RMHE problem is constrained to ensure that it computes a state estimate that is within the confidence region. By this approach, the RMHE gives bounded estimation error in the case of bounded process noise. Moreover, the RMHE could be used together with different arrival cost approximation techniques and was shown to compensate for the error in the arrival cost approximation [8]. The RMHE has been applied to the design of a robust output feedback Lyapunov-based MPC [9] and has also been extended to estimate the state of large-scale systems in a distributed manner [10]. The RMHE scheme in [8] will be adopted in this section to take advantage of the tunable convergence speed of the observer presented in the previous subsection while significantly reducing its sensitivity to measurement noise.

The RMHE is evaluated at discrete time instants denoted by the time sequence $\{t_k\}_{k\geq 0}$ with $t_k = t_0 + k\Delta$, $k = 0, 1, \ldots$ where t_0 is the initial time. In the RMHE scheme, the deterministic observer, which is denoted by F, will be used to calculate a reference state estimate at each sampling time from continuous output measurements. Based on the reference state estimate, the RMHE determines a confidence region for the actual system state. The RMHE computes a state estimate within the confidence region based on a sequence of previous output measurements, a system model, and bounds information of the process and measurement noise. The RMHE scheme at time instant t_k is formulated as follows:

$$\min_{\tilde{x}(t_{k-N_e}),\ldots,\tilde{x}(t_k)} \sum_{i=k-N_e}^{k-1} |w(t_i)|^2_{Q_m^{-1}} + \sum_{i=k-N_e}^{k} |v(t_i)|^2_{R_m^{-1}} + \hat{V}_T(t_{k-N_e}) \tag{5.38a}$$

$$\text{s.t.} \quad \dot{\tilde{x}}(t) = f(\tilde{x}(t)) + g(\tilde{x}(t))u(t) + l(\tilde{x}(t))w(t_i),\ t \in [t_i, t_{i+1}], \tag{5.38b}$$

$$v(t_i) = y(t_i) - h(\tilde{x}(t_i)),\ i = k - N_e, k - N_e + 1, \ldots, k \tag{5.38c}$$

$$w(t_i) \in \mathbb{W},\ i = k - N_e, k - N_e + 1, \ldots, k - 1 \tag{5.38d}$$

$$v(t_i) \in \mathbb{V},\ i = k - N_e, k - N_e + 1, \ldots, k \tag{5.38e}$$

$$\tilde{x}(t) \in \Omega_\rho,\ \forall\, t \in [t_{k-N_e}, t_k] \tag{5.38f}$$

$$|\tilde{x}(t_k) - z(t_k)| \leq \kappa |y(t_k) - h(z(t_k))| \tag{5.38g}$$

where N_e is the estimation horizon, Q_m and R_m are the estimated covariance matrices of w and v respectively, $\hat{V}_T(t_{k-N_e})$ denotes the arrival cost which summarizes past information up to t_{k-N_e}, $\tilde{x}$ is the predicted state x in the above optimization problem, $y(t_i)$ is the output measurement at t_i, $z(t_k)$ is an estimate given by the observer F based on continuous measurements of y, and κ is a positive constant which is a design parameter.

Once the optimization problem of Eq. 5.38 is solved, an optimal trajectory of the system state, $\tilde{x}^*(t_{k-N_e}), \ldots, \tilde{x}^*(t_k)$, is obtained. The optimal estimate of the current system state is denoted:

$$\hat{x}^*(t_k) = \tilde{x}^*(t_k). \tag{5.39}$$

Note that in the optimization problem of Eq. 5.38, w and v are assumed to be piece-wise constant variables with sampling time Δ to ensure that Eq. 5.38 is a finite dimensional optimization problem.

In the optimization problem of Eq. 5.38, $z(t_k)$ is a reference estimate calculated by the observer F. Based on the reference estimate and the current output measurement, $y(t_k)$, a confidence region that contains the actual system state is constructed, i.e., $\kappa|y(t_k) - h(z(t_k))|$. The estimate of the current state provided by the RMHE is only allowed to be optimized within the confidence region. This approach ensures that the RMHE inherits the robustness of the observer F and gives estimates with bounded errors.

Remark 5.6 In order to account for the effect of historical data outside the estimation window, an arrival cost which summarizes the information of those data is included in the cost function of an MHE optimization problem. The arrival cost plays an important role in the performance and stability of an MHE scheme. Different methods have been developed to approximate the arrival cost including Kalman filtering and smoothing techniques for linear systems [11], extended Kalman filtering for nonlinear systems [12], and particle filters for constrained systems [13].

5.3.3 RMHE-Based EMPC

Without loss of generality, it is assumed that the LEMPC is evaluated at time instants $\{t_k\}_{k\geq 0}$ with sampling time Δ as used in the RMHE. In the LEMPC design, we will take advantage of both the fast convergence rate of the observer F and the robustness of the RMHE to measurement noise.

5.3.3.1 Implementation Strategy

In the approach, the observer F is initially applied for a short period to drive the state estimate from the observer to a small neighborhood of the actual system state. Once the estimate has converged to a small neighborhood of the actual system state, the RMHE takes over the estimation task and provides smoother and optimal state estimates to the LEMPC. Without loss of generality, we assume that t_b is a multiple integer of the sampling time Δ in the sense that $t_b = b\Delta$ where b is a strictly positive integer. In the first b sampling periods, the observer F is applied with continuously output measurements, i.e., the observer is continuously evaluated and provides state estimates to the LEMPC at every sampling time. Starting from t_b, the RMHE is activated and provides an optimal estimates of the system state to the LEMPC at every subsequent sampling time. The LEMPC evaluates its optimal input trajectory based on either the estimates provided by the observer F or the estimates from the RMHE.

The two-mode operation scheme is adopted in the LEMPC design. From the initial time t_0 up to a time t_s, the LEMPC operates in the first operation mode to minimize the economic cost function while maintaining the closed-loop system state in the stability region Ω_ρ. In this operation mode, in order to account for the uncertainties in state estimates and process noise, a region Ω_{ρ_e} with $\rho_e < \rho$ is used. If the estimated state is in the region Ω_{ρ_e}, the LEMPC optimizes the cost function while constraining the predicted state trajectory be within the region Ω_{ρ_e}; if the estimated state is in the region $\Omega_\rho \backslash \Omega_{\rho_e}$, the LEMPC computes control actions such that the state is forced to the region Ω_{ρ_e}. After time t_s, the LEMPC operates in the second operation mode and calculates the inputs in a way that the state of the closed-loop system is driven to a neighborhood of the desired steady-state. The implementation strategy of the output feedback LEMPC described above may be summarized as follows:

Algorithm 5.2 RMHE-based LEMPC implementation algorithm

1. Initialize the observer F with $z(t_0)$ and continuously execute the observer F based on the output measurements y.
2. At a sampling time t_k, if $t_k < t_b$, go to Step 2.1; otherwise, go to Step 2.2.

   2.1. The LEMPC gets a sample of the estimated system state $z(t_k)$ at t_k from the observer F, and go to Step 3.

   2.2. Based on the estimate $z(t_k)$ provided by the observer F and output measurements at the current and previous N_e sampling instants, i.e., $y(t_i)$ with $i = k - N_e, \ldots, k$, the RMHE calculates the optimal state estimate $\hat{x}^*(t_k)$. The estimate $\hat{x}^*(t_k)$ is sent to the LEMPC. Go to Step 3.

3. If $t_k < t_s$ and if $z(t_k) \in \Omega_{\rho_e}$ (or if $\hat{x}^*(t_k) \in \Omega_{\rho_e}$), go to Step 3.1. Otherwise, go to Step 3.2.

   3.1. Based on $z(t_k)$ or $\hat{x}^*(t_k)$, the LEMPC calculates its input trajectory to minimize the economic cost function while ensuring that the predicted state trajectory over the prediction horizon lies within Ω_{ρ_e}. The first value of the optimal input trajectory is applied to the system. Go to Step 4.

   3.2. Based on $z(t_k)$ or $\hat{x}^*(t_k)$, the LEMPC calculates its input trajectory to drive the system state towards the origin. The first value of the input trajectory is applied to the system. Go to Step 4.

4. Go to Step 2 ($k \leftarrow k + 1$).

In the remainder, we will use $\hat{x}$ to denote the state estimate used in the LEMPC. Specifically, $\hat{x}$ at time t_k is defined as follows:

$$\hat{x}(t_k) = \begin{cases} z(t_k), & \text{if } t_k < t_b \\ \hat{x}^*(t_k), & \text{if } t_k \geq t_b \end{cases} \tag{5.40}$$

Remark 5.7 In the implementation Algorithm 5.2 as well as in the RMHE design of Eq. 5.38, the observer F provides state estimate to the RMHE at every sampling time and is independently evaluated from the RMHE. To improve the quality of the estimates provided by the observer F, the state of the observer F may be set to the estimate of the RMHE at every sampling time since the estimates obtained from the RMHE are expected to be more accurate. That is, at Step 2.2, the estimate $\hat{x}^*(t_k)$ is also sent to the observer F and the observer F resets its state to $z(t_k) = \hat{x}^*(t_k)$. The state estimate $z(t_{k+1})$ of the observer F at the next sampling time is computed with continuous output measurements received over the sampling period ($t \in [t_k, t_{k+1}]$) initialized with $z(t_k) = \hat{x}^*(t_k)$.

5.3.3.2 LEMPC Design

The LEMPC is evaluated every sampling time to obtain the optimal input trajectory based on estimated state $\hat{x}(t_k)$ provided by the observer F or the RMHE. Specifically, the optimization problem of the LEMPC is formulated as follows:

$$\min_{u \in S(\Delta)} \int_{t_k}^{t_{k+N}} l_e(\tilde{x}(\tau), u(\tau))\, d\tau \tag{5.41a}$$

$$\text{s.t. } \dot{\tilde{x}}(t) = f(\tilde{x}(t)) + g(\tilde{x}(t))u(t),\ t \in [t_k, t_{k+N}) \tag{5.41b}$$

$$\tilde{x}(t_k) = \hat{x}(t_k) \tag{5.41c}$$

$$u(\tau) \in \mathbb{U},\ \forall\, t \in [t_k, t_{k+N}) \tag{5.41d}$$

$$V(\tilde{x}(t)) \leq \rho_e,\ \forall\, t \in [t_k, t_{k+N}),\quad \text{if } t_k < t_s \text{ and } V(\hat{x}(t_k)) \leq \rho_e \tag{5.41e}$$

$$L_g V(\hat{x}(t_k))u(t_k) \leq L_g V(\hat{x}(t_k))k(\hat{x}(t_k)),\quad \text{if } t_k \geq t_s \text{ or } V(\hat{x}(t_k)) > \rho_e \tag{5.41f}$$

where N is the control prediction horizon and $\tilde{x}$ is the predicted trajectory of the system with control inputs calculated by this LEMPC. The constraint of Eq. 5.41b is the nominal system model used to predict the future evolution of the system initialized with the estimated state at t_k (Eq. 5.41c). The constraint of Eq. 5.41d accounts for the input constraint. The constraint of Eq. 5.41e is active only for mode 1 operation of the LEMPC which requires that the predicted state trajectory be within the region defined by Ω_{ρ_e}. The constraint of Eq. 5.41f is active for mode 2 operation of the LEMPC as well as mode 1 operation when the estimated system state is out of Ω_{ρ_e}. This constraint forces the LEMPC to generate control actions that drive the closed-loop system state towards the origin.

The optimal solution to this optimization problem is denoted by $u^*(t|t_k)$, which is defined for $t \in [t_k, t_{k+N})$. The manipulated input of the LEMPC is defined as follows:

$$u(t) = u^*(t|t_k),\ \forall t \in [t_k, t_{k+1}). \tag{5.42}$$

The control input applied to the closed-loop system from t_k to t_{k+1} is $u^*(t_k|t_k)$.

5.3.4 Stability Analysis

The stability of LEMPC of Eq. 5.41 based on state estimates obtained following Eq. 5.40 is analyzed in this subsection. A set of sufficient conditions is derived under which the closed-loop system state trajectory is ensured to be maintained in the region Ω_ρ and ultimately bounded in an invariant set.

In the remainder of this subsection, we first present two propositions and then summarize the main results in a theorem. Proposition 5.3 characterizes the continuity property of the Lyapunov function V. Proposition 5.4 characterizes the effects of bounded state estimation error and process noise.

Proposition 5.3 (Proposition 4.2) *Consider the Lyapunov function V of system of Eq. 5.1. There exists a quadratic function f_V such that*

$$V(x) \leq V(\hat{x}) + f_V(|x - \hat{x}|) \tag{5.43}$$

for all $x, \hat{x} \in \Omega_\rho$ with

$$f_V(s) = \alpha_4(\alpha_1^{-1}(\rho))s + M_v s^2 \tag{5.44}$$

where M_v is a positive constant.

Proposition 5.4 *Consider the systems*

$$\begin{aligned} \dot{x}_a(t) &= f(x_a) + g(x_a)u(t) + l(x_a)w(t) \\ \dot{x}_b(t) &= f(x_b) + g(x_b)u(t) \end{aligned} \tag{5.45}$$

with initial states $|x_a(t_0) - x_b(t_0)| \leq \delta_x$. If $x_a(t) \in \Omega_\rho$ and $x_b(t) \in \Omega_\rho$ for all $t \in [t_0, t']$, there exists a function $f_W(\cdot, \cdot)$ such that

$$|x_a(t) - x_b(t)| \leq f_W(\delta_x, t - t_0) \tag{5.46}$$

for all $x_a(t), x_b(t) \in \Omega_\rho$ and $u(t) \in \mathbb{U}$, $w(t) \in \mathbb{W}$ for all $t \in [t_0, t']$ with:

$$f_W(s, \tau) = \left(s + \frac{M_l \theta_w}{L_f + L_g u^{\max}}\right) e^{(L_f + L_g u^{\max})\tau} - \frac{M_l \theta_w}{L_f + L_g u^{\max}} \tag{5.47}$$

where L_f, L_g, M_l are positive constants associated with functions f, g, l.

Proof Define $e_x = x_a - x_b$. The time derivative of e_x is given by:

$$\dot{e}_x(t) = f(x_a) + g(x_a)u(t) + l(x_a)w(t) - f(x_b) - g(x_b)u(t). \tag{5.48}$$

By continuity and the smoothness property assumed for f, g, there exist positive constants L_f, L_g such that:

$$|\dot{e}_x(t)| \leq L_f|e_x(t)| + L_g u|e_x(t)| + |l(x_a)w(t)|, \tag{5.49}$$

for all $t \in [t_0, t']$ provided $x_a(t) \in \Omega_\rho$ and $x_b(t) \in \Omega_\rho$ for all $t \in [t_0, t']$. By the boundedness of x_a and the smoothness property assumed for l as well as the boundedness of u and w, there exist positive constants M_l such that:

$$|\dot{e}_x(t)| \leq (L_f + L_g u^{\max})|e_x(t)| + M_l \theta_w. \tag{5.50}$$

for all $t \in [t_0, t']$. Integrating the above inequality and taking into account that $|e_x(t_0)| \leq \delta_x$, the following inequality is obtained:

$$|e_x(t)| \leq \left(\delta_x + \frac{M_l \theta_w}{L_f + L_g u^{\max}}\right) e^{(L_f + L_g u^{\max})(t-t_0)} - \frac{M_l \theta_w}{L_f + L_g u^{\max}}. \tag{5.51}$$

This proves Proposition 5.4.

The following Theorem 5.2 summarizes the stability properties of the output feedback LEMPC. The stability of the closed-loop system is based on the observer F and controller k pair with F implemented continuously and k implemented in a sample-and-hold fashion.

Theorem 5.2 *Consider system of Eq. 5.1 in closed loop under LEMPC of Eq. 5.41 with state estimates determined following Eq. 5.40 based on an observer and controller pair satisfying the assumptions in Sect. 5.3.1. Let $\theta_w \leq \theta_w^*$, $\theta_v \leq \theta_v^*$, $\varepsilon \in (\varepsilon_L^*, \varepsilon_U^*)$ and $|z(t_0) - x(t_0)| \leq e_{m0}$. Also, let $\varepsilon_w > 0$, $\Delta > 0$ and $\rho > \rho_1 > \rho_e > \rho^* > \rho_s > 0$ and $\kappa \geq 0$ satisfy the following conditions:*

$$\rho_e \leq \rho - \max\{f_V(f_W(\delta_x, \Delta)) + f_V(\delta_x), M \max\{\Delta, t_b\}\alpha_4(\alpha_1^{-1}(\rho))\}, \tag{5.52}$$

$$-\alpha_3(\alpha_2^{-1}(\rho_s)) + \left(L_V^f + L_V^g u^{\max}\right)(M\Delta + \delta_x) + M_V^l \theta_w \leq -\varepsilon_w/\Delta \tag{5.53}$$

where $\delta_x = (\kappa L_h + 1)e_m + \kappa\theta_v$, L_V^f, L_V^g are Lipschitz constants associated with the Lie derivatives $L_f V$ and $L_g V$, respectively, M is a constant that bounds the time derivative of x, i.e., $|\dot{x}| \leq M$, and M_V^l is a constant that bounds $|L_l V|$ for $x \in \Omega_\rho$. If $x(t_0) \in \Omega_{\rho_e}$, then $x(t) \in \Omega_\rho$ for all $t \geq t_0$ and is ultimately bounded in an invariant set.

Proof In this proof, we consider $t \in [t_0, \max\{\Delta, t_b\})$ and $t \geq \max\{\Delta, t_b\}$ separately and prove that if the conditions stated in Proposition 5.2 are satisfied, the boundedness of the closed-loop state is ensured. The proof consists of three parts. In *Part I*, we prove that the closed-loop state trajectory is contained in Ω_ρ for $t \in [t_0, \max\{\Delta, t_b\})$; in *Part II*, we prove that the boundedness of the closed-loop state trajectory under the first operation mode of the LEMPC for $t \geq \max\{\Delta, t_b\}$ when the initial state is within Ω_{ρ_e}; and in *Part III*, we prove that the closed-loop state trajectory is bounded for the first operation mode when the initial state is within $\Omega_\rho \backslash \Omega_{\rho_e}$ and is ultimately bounded in an invariant set for the second operation mode for $t \geq \max\{\Delta, t_b\}$.

Part I: First, we consider the case that $t \in [t_0, \max\{\Delta, t_b\})$. The closed-loop system state may be described as follows:

$$\dot{x}(t) = f(x(t)) + g(x(t))u(t) + l(x(t))w(t) \tag{5.54}$$

with $u(t)$ determined by the LEMPC with $\hat{x} = z$. The Lyapunov function of the state trajectory may be evaluated as follows:

$$V(x(t)) = V(x(t_0)) + \int_{t_0}^{t} \dot{V}(x(t))d\tau = V(x(t_0)) + \int_{t_0}^{t} \frac{\partial V(x(\tau))}{\partial x}\dot{x}(\tau)d\tau \tag{5.55}$$

Using condition of Eq. 5.2 and the boundedness of $\dot{x}$ in the region of interest, if $x(t_0) \in \Omega_{\rho_e} \subset \Omega_{\rho_1} \subset \Omega_{\rho}$, it may be written for all $t \in [t_0, \max\{\Delta, t_b\})$ that:

$$V(x(t)) \leq \rho_e + M \max\{\Delta, t_b\}\alpha_4(\alpha_1^{-1}(\rho)) \tag{5.56}$$

with M a positive constant which bounds $\dot{x}$ in Ω_{ρ}, i.e., $|\dot{x}| \leq M$. If ρ_e is defined as in Proposition 5.2, then

$$V(x(t)) < \rho, \ \forall\, t \in [t_0, \max\{\Delta, t_b\}). \tag{5.57}$$

Part II: In this part, we consider the case that $t \geq \max\{\Delta, t_b\}$. In this case, we have that $|x(t) - z(t)| \leq e_m$. We consider that the LEMPC is operated in the first operation mode and focus on the evolution of the state trajectory from t_k to t_{k+1}. Moreover, we consider $\tilde{x}(t_k) = \hat{x}(t_k) \in \Omega_{\rho_e}$. In this case, the LEMPC will optimize the economic cost while keeping $\tilde{x}(t)$ within Ω_{ρ_e}. We prove that if $\tilde{x}(t_k) \in \Omega_{\rho_e}$, then $x(t_{k+1}) \in \Omega_{\rho}$ and $\hat{x}(t_{k+1}) \in \Omega_{\rho}$.

From t_k to t_{k+1}, the worst case scenario is as shown in Fig. 5.14. At time t_k, the estimate of the state $\hat{x}(t_k) = \tilde{x}(t_k)$ is on the boundary of Ω_{ρ_e} while the actual system state is outside of Ω_{ρ_e} and on the boundary of another set Ω_{ρ_2} due to uncertainty in

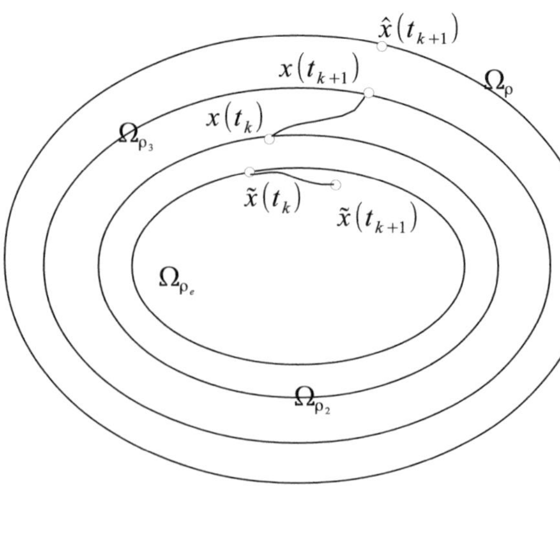

Fig. 5.14 Worst case scenario of the evolution of $\tilde{x}$ and x from t_k to t_{k+1} in the first operation mode

$\hat{x}$. The LEMPC will keep $\tilde{x}(t)$ inside Ω_{ρ_e} from t_k to t_{k+1}. However, due to the error in $\tilde{x}(t_k)$ and the presence of process noise, the actual system state $x(t_{k+1})$ may evolve to a point on the boundary of Ω_{ρ_3} in Fig. 5.14 that is further away of Ω_{ρ_e}. The distance between $\tilde{x}(t_{k+1})$ and $x(t_{k+1})$, is bounded. From Proposition 5.4, it may be obtained that:

$$|\tilde{x}(t_{k+1}) - x(t_{k+1})| \leq f_W(|\hat{x}(t_k) - x(t_k)|, \Delta). \tag{5.58}$$

Recall that when $t \geq t_b$, all the estimates are provided by the RMHE. From the design of the RMHE, it may be written that:

$$|\hat{x}(t_k) - z(t_k)| \leq \kappa |y(t_k) - h(z(t_k))| . \tag{5.59}$$

Using the relation that $|\hat{x} - x| \leq |\hat{x} - z| + |z - x|$, it may be obtained that:

$$|\hat{x}(t_k) - x(t_k)| \leq \kappa |y(t_k) - h(z(t_k))| + |z(t_k) - x(t_k)| . \tag{5.60}$$

Noticing that $|z(t_k) - x(t_k)| \leq e_m$ and $|y(t_k) - h(z(t_k))| = |h(x(t_k)) + v(t_k) - h(z(t_k))|$, and using the Lipschitz property of h, the boundedness of v, the following inequality may be written:

$$|\hat{x}(t_k) - x(t_k)| \leq (\kappa L_h + 1)e_m + \kappa\theta_v. \tag{5.61}$$

From Eqs. 5.58 and 5.61, it may be obtained that:

$$|\tilde{x}(t_{k+1}) - x(t_{k+1})| \leq f_W((\kappa L_h + 1)e_m + \kappa\theta_v, \Delta). \tag{5.62}$$

This implies that if $\tilde{x}$ is maintained in Ω_{ρ_e}, the actual system state x is ensured to be within the set Ω_{ρ_2} with $\rho_2 = \rho_e + f_V(f_W((\kappa L_h + 1)e_m + \kappa\theta_v, \Delta))$ which may be obtained from Proposition 5.3.

Taking into account Eq. 5.61 for $t = t_{k+1}$, the state estimate obtained at t_{k+1} could be outside the region Ω_{ρ_2} but the distance is bounded as follows:

$$|\hat{x}(t_{k+1}) - x(t_{k+1})| \leq (\kappa L_h + 1)e_m + \kappa\theta_v. \tag{5.63}$$

In order to ensure that $\hat{x}(t_{k+1})$ is within Ω_ρ which is required for the feasibility of LEMPC of Eq. 5.41, the following inequality should be satisfied:

$$\rho \geq \rho_e + f_V(f_W((\kappa L_h + 1)e_m + \kappa\theta_v, \Delta)) + f_V((\kappa L_h + 1)e_m + \kappa\theta_v) \tag{5.64}$$

which implies that ρ_e should be picked to satisfy the following condition:

$$\rho_e \leq \rho - f_V(f_W((\kappa L_h + 1)e_m + \kappa\theta_v, \Delta)) - f_V((\kappa L_h + 1)e_m + \kappa\theta_v). \tag{5.65}$$

If ρ_e is defined as in Proposition 5.2, the above condition is satisfied.

Part III: Next, we consider the case that $\hat{x}(t_k) = \tilde{x}(t_k) \in \Omega_\rho \backslash \Omega_{\rho_e}$ in the first operation mode or $t_k \geq t_s$ for the second operation mode. In either case, constraint of Eq. 5.41f will be active. The time derivative of the Lyapunov function may be evaluated as follows:

$$\dot{V}(x(t)) = \frac{\partial V(x(t))}{\partial x}(f(x(t) + g(x(t))u(t_k) + l(x(t))w(t)) \tag{5.66}$$

for $t \in [t_k, t_{k+1})$. Adding and subtracting the term:

$$\frac{\partial V(\hat{x}(t_k))}{\partial x}(f(\hat{x}(t_k) + g(\hat{x}(t_k))u(t_k))$$

to/from the above equation and considering constraint Eq. 5.41f as well as condition of Eq. 5.2, it is obtained that:

$$\begin{aligned}\dot{V}(x(t)) \leq -\alpha_3(|\hat{x}(t_k)|) + \frac{\partial V(x(t))}{\partial x}(f(x(t)) + g(x(t))u(t_k) + l(x(t))w(t)) \\ - \frac{\partial V(\hat{x}(t_k))}{\partial x}(f(\hat{x}(t_k) + g(\hat{x}(t_k))u(t_k))\end{aligned} \tag{5.67}$$

for all $t \in [t_k, t_{k+1})$. By the smooth properties of V, f, g and l, the boundedness of x, u and w, there exist positive constants L_V^f, L_V^g, M_V^l such that:

$$\dot{V}(x(t)) \leq -\alpha_3(|x(t_k)|) + \left(L_V^f + L_V^g u^{\max}\right)|x(t) - \hat{x}(t_k)| + M_V^l \theta_w \tag{5.68}$$

for all $x \in \Omega_\rho$. Noticing that $|x(t) - \hat{x}(t_k)| \leq |x(t) - x(t_k)| + |x(t_k) - \hat{x}(t_k)|$, it is obtained that:

$$|x(t) - \hat{x}(t_k)| \leq |x(t) - x(t_k)| + (\kappa L_h + 1)e_m + \kappa\theta_v. \tag{5.69}$$

By the continuity and smoothness properties of f, g, l and the boundedness of x, u and w, there exists a positive constant M such that $|\dot{x}| \leq M$. From the above inequalities, it may be obtained that:

$$\begin{aligned}\dot{V}(x(t)) \leq -\alpha_3(\alpha_2^{-1}(\rho_s)) + \left(L_V^f + L_V^g u^{\max}\right)(M\Delta + (\kappa L_h + 1)e_m + \kappa\theta_v) \\ + M_V^l \theta_w\end{aligned} \tag{5.70}$$

for all $x \in \Omega_\rho \backslash \Omega_{\rho_s}$. If condition of Eq. 5.53 is satisfied, it may be obtained from Eq. 5.70 that:

$$V(x(t_{k+1})) \leq V(x(t_k)) - \varepsilon_w. \tag{5.71}$$

This means that the function value $V(x)$ is decreasing in the first operation mode if $\tilde{x}(t_k) = \hat{x}(t_k)$ is outside of Ω_{ρ_e}. This implies that $\hat{x}(t_k)$ will eventually enter Ω_{ρ_e}.

This also implies that in the second operation mode, $V(x)$ decreases every sampling time and x will eventually enter Ω_{ρ_s}. Once $x \in \Omega_{\rho_s} \subset \Omega_{\rho^*}$, it will remain in Ω_{ρ^*} because of the definition of ρ^*. This proves Proposition 5.2.

Remark 5.8 Part I of Theorem 5.2 essentially treats the input as a perturbation to the system. Given that the input and the noise are bounded, a bound is derived for how large the Lyapunov function may increase over time t_b (which is small). This follows from the fact that the initial estimation error of the deterministic observer and actual closed-loop state are both bounded in a region containing the origin.

Remark 5.9 Parts II and III prove that if the current state $x(t_k) \in \Omega_\rho$ and if the current estimate $\hat{x}(t_k) \in \Omega_\rho$, the actual closed-loop state and the estimated state at the next sampling period are also within Ω_ρ. Since Part II considers mode 1 operation of the LEMPC, the worst case scenario is considered (Fig. 5.14). Part III considers mode 2 operation of the LEMPC. While the theoretical developments and corresponding bounding inequalities contained in this section are conservative, they do provide valuable insight and guidelines for selecting the parameters of the state feedback controller $k(x)$, the deterministic observer, the RMHE, and the output feedback LEMPC such that the closed-loop system of Eq. 5.1 with bounded process and measurement noise under the output feedback LEMPC of Eq. 5.41 is stable.

Remark 5.10 One could potentially apply the RMHE for t_0 to t_b instead of using the deterministic observer. However, it is difficult to prove closed-loop stability for this case owing to the fact that the estimation error may not have decayed to a small value over this time period with the RMHE, i.e., it is difficult to show that the RMHE satisfies the observability assumptions of Sect. 5.3.1.

5.3.5 Application to a Chemical Process Example

Consider the CSTR described in Sect. 5.2.4 with the same control objective and the same limitation on the available reactant material. To estimate the state from noisy temperature measurements, the RMHE scheme is used. The weighting matrices of the RMHE are given by $Q_e = \text{diag}([\sigma_{w_1}^2 \ \sigma_{w_2}^2])$ and $R_e = \sigma_v^2$ where σ denotes the standard deviation of the process or measurement noise. The design parameter of the RMHE is $\kappa = 0.4$, the sampling period is the same as the LEMPC, i.e., $\Delta_e = 0.01$ h), and the estimation horizon of the RMHE is $N_e = 15$. The robust constraint of the RMHE is based on a high-gain observer as in Sect. 5.2.4. For the first 15 sampling periods, the high-gain observer is used to provide the LEMPC with a state estimate. At each subsequent sampling periods, the LEMPC is initialized using the state estimate from the RMHE. To solve the optimization problems of the LEMPC and the RMHE at each sampling period, the open-source software Ipopt [6] is used. The process model is numerically simulated using an explicit Euler integration method with integration step $h_c = 1.0 \times 10^{-3}$ h. To simulate the process and measurement noise, new random numbers are generated and applied over each integration step. The process noise is

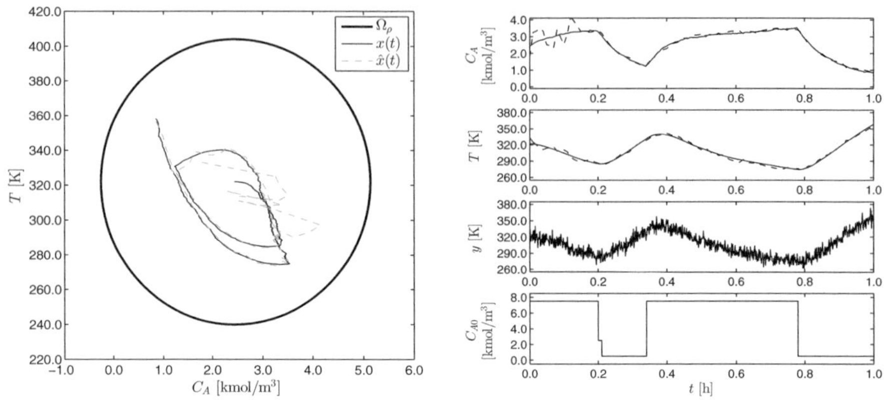

Fig. 5.15 The evolution of the closed-loop CSTR under the RMHE-based LEMPC scheme shown in state-space (*left*) and as a function of time (*right*). The *solid line* is the actual closed-loop state trajectory $x(t)$, while, the *dashed line* is the estimated state $\hat{x}(t)$

assumed to enter the system additively to the right-hand side of the process model ODEs. The random numbers are generated from a zero-mean, bounded Gaussian distribution.

Square bounds of $w_{\max} = [20.0\ 50.0]$ and $v_{\max} = 20.0$ are used to bound the process and measurement noise, respectively, and the standard deviation of the noise terms are $\sigma_w = [7.0\ 20.0]$ and $\sigma_v = 7.0$, respectively. The CSTR is initialized at $x_0^T = [2.44\,\text{kmol}\,\text{m}^{-3}\ 320.0\,\text{K}]$, which corresponds to the steady-state. The evolution of the closed-loop CSTR under the RMHE-based LEMPC is shown in Fig. 5.15. Initially, the estimated reactant concentration is significantly affected by the measurement noise which is expected since the state estimate is computed by the high-gain observer over this initial time period. After the RMHE is activated, the estimated state trajectories are nearly overlapping with the actual closed-loop state trajectories. Furthermore, the LEMPC computes a periodic-like input profile to optimize the process economics over the 1 h period of operation.

The average reaction rate over 1 h period of operation is 13.59 kmol m^{-3}. If, instead, the CSTR was maintained at the steady-state (x_0) without process and measurement noise (nominal operation), the average reaction rate over this 1 h operation would be 12.80 kmol m^{-3}. This is a 6.2 % improvement in the economic cost of the closed-loop system under the RMHE-based LEMPC with process and measurement noise over nominal steady-state operation. We note that the economic performance of the closed-loop system under LEMPC with full state feedback and nominal operation over 1 h operation is 13.60 kmol m^{-3} which is a 6.3 % economic performance improvement.

To assess the estimation performance of the RMHE, another simulation is performed with the same realization of the process and measurement noise and with the high-gain observer presented in Sect. 5.2.4. The evolution of the closed-loop CSTR under the high-gain observer state estimation-based LEMPC is shown in Fig. 5.16.

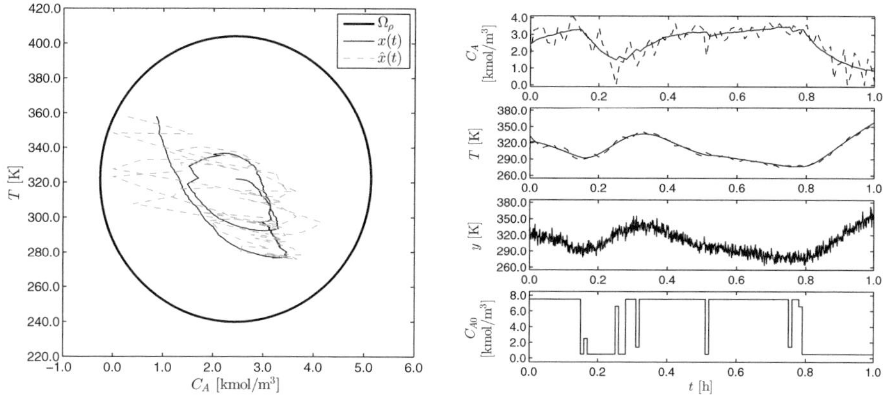

Fig. 5.16 The evolution of the closed-loop CSTR under the state estimation-based LEMPC scheme with the high-gain observer of Sect. 5.2.4 shown in state-space (*left*) and as a function of time (*right*). The *solid line* is the actual closed-loop state trajectory $x(t)$, while, the *dashed line* is the estimated state $\hat{x}(t)$

Not only does the noise impact the estimates provided by the high-gain observer in this case, but also, it impacts the computed input profile (Fig. 5.16). Comparing Figs. 5.15 and 5.16, the RMHE is able to provide estimates of the state within a small neighborhood of the actual process states, while the high-gain observer is not able to estimate the concentration as well as the RMHE. Furthermore, since the RMHE provides better (smoother) estimates of the states, the operation of the closed-loop system under the RMHE-based LEMPC is smoother which may be observed in the input trajectories.

Several additional closed-loop simulations with various bounds and standard deviations on the process and measurement noise and initial conditions are performed to further assess the estimation performance of RMHE compared to the one of the high-gain observer of Sect. 5.2.4. An estimation performance index which is defined as

$$J = \sum_{k=0}^{99} \left|\hat{x}(t_k) - x(t_k)\right|_S^2 \tag{5.72}$$

is used to assess the estimation performance where the matrix S is a positive definite weighting matrix given by $S = \text{diag}([50\ 1])$. The matrix S has been chosen to account for the different numerical ranges of the concentration and temperature. In addition to the assessment on the estimation performance, the total economic performance index over the length of the simulation is defined as

$$J_e = \frac{1}{100} \sum_{k=0}^{99} k_0 e^{-E/RT(t_k)} C_A^2(t_k) \tag{5.73}$$

Table 5.2 Estimation performance comparison of the closed-loop CSTR with various bounds and standard deviation of the disturbances and noise and initial conditions under the high-gain observer state estimation-based LEMPC and under the RMHE-based LEMPC (ordered below by increasing bounds and standard deviation)

	High-gain observer			RMHE		
	J	SSE of C_A	J_e	J	SSE of C_A	J_e
1	310.5	4.450	13.04	104.0	1.277	13.04
2	528.5	7.781	14.19	310.1	4.169	14.19
3	271.6	3.669	13.47	88.1	0.440	13.47
4	506.4	7.066	13.06	181.9	1.476	13.07
5	583.2	8.097	14.20	354.4	3.888	14.20
6	482.1	6.397	13.48	137.7	0.372	13.48
7	592.4	7.821	13.09	257.1	1.734	13.09
8	572.8	8.519	14.23	252.6	3.425	14.23
9	616.4	8.579	13.51	168.6	1.126	13.52
10	992.0	13.700	13.00	429.9	4.355	13.08
11	1079.8	14.871	14.14	888.7	12.076	14.21
12	1012.5	14.304	13.42	552.0	5.817	13.43
13	1643.6	22.606	13.02	665.3	3.523	12.99
14	1758.5	23.396	14.24	771.2	5.492	14.27
15	1591.0	21.740	13.51	561.5	1.717	13.55

The J column refers to the performance index of Eq. 5.72, the "SSE of C_A" column denotes the sum of squared errors of the concentration C_A estimation, and the J_e column refers to the economic performance index of Eq. 5.73

which is the time-averaged reaction rate over the simulation. From the results displayed in Table 5.2, the RMHE consistently provides a significantly better estimates of the state than the high-gain observer which demonstrates the robustness of the RMHE to process and measurement noise. However, the estimation performance does not translate into a significant closed-loop average economic performance improvement of the closed-loop system with the RMHE-based LEMPC over the closed-loop system with the high-gain observer and LEMPC. This relationship is due to the fact that the closed-loop average economic performance over one operation period is not strongly dependent on the initial condition of the LEMPC optimization problem, i.e., $\hat{x}(t_k)$, for this particular example. In other words, providing the LEMPC with an estimate of the actual state anywhere in a neighborhood around the actual state results in a computed input trajectory that leads to nearly the same economic cost for the closed-loop systems. For systems that are more sensitive to the estimate of the current state, it is expected that there would also be improved closed-loop economic performance with the RMHE-based LEMPC in addition to improved estimation performance.

5.4 Conclusions

In this chapter, two output feedback EMPC schemes were presented. In the first scheme, a high-gain observer-based EMPC for the class of full-state feedback linearizable nonlinear systems was introduced. A high-gain observer is used to estimate the nonlinear system state using output measurements and a Lyapunov-based approach is adopted to design the EMPC that uses the observer state estimates. It was proved, using singular perturbation arguments, that the closed-loop system is practically stable provided the observer gain is sufficiently large.

To achieve fast convergence of the state estimate to the actual system state as well as to improve the robustness of the estimator to measurement and process noise, a high-gain observer and a RMHE scheme were used to estimate the system states. In particular, the high-gain observer was first applied for a small time period with continuous output measurements to drive the estimation error to a small value. Once the estimation error had converged to a small neighborhood of the origin, the RMHE was activated to provide more accurate and smoother state estimates. In the design of the RMHE, the high-gain observer was used to provide reference estimates, which were subsequently used to calculate confidence regions. The RMHE was restricted to compute state estimates that are within these confidence regions. The output feedback EMPC was designed via Lyapunov techniques based on state estimates provided by the high-gain observer and the RMHE.

The application of the two schemes to a chemical reactor demonstrated the applicability and effectiveness of the schemes and the ability to deal with measurement noise.

References

1. Khalil H, Esfandiari F (1993) Semiglobal stabilization of a class of nonlinear systems using output feedback. IEEE Trans Autom Control 38:1412–1415
2. El-Farra NH, Christofides PD (2003) Bounded robust control of constrained multivariable nonlinear processes. Chem Eng Sci 58:3025–3047
3. Mahmoud NA, Khalil HK (1996) Asymptotic regulation of minimum phase nonlinear systems using output feedback. IEEE Trans Autom Control 41:1402–1412
4. Muñoz de la Peña D, Christofides PD (2008) Output feedback control of nonlinear systems subject to sensor data losses. Syst Control Lett 57:631–642
5. Christofides PD (2000) Robust output feedback control of nonlinear singularly perturbed systems. Automatica 36:45–52
6. Wächter A, Biegler LT (2006) On the implementation of an interior-point filter line-search algorithm for large-scale nonlinear programming. Math Program 106:25–57
7. Ahrens JH, Khalil HK (2009) High-gain observers in the presence of measurement noise: a switched-gain approach. Automatica 45:936–943
8. Liu J (2013) Moving horizon state estimation for nonlinear systems with bounded uncertainties. Chem Eng Sci 93:376–386
9. Zhang J, Liu J (2013) Lyapunov-based MPC with robust moving horizon estimation and its triggered implementation. AIChE J 59:4273–4286
10. Zhang J, Liu J (2013) Distributed moving horizon state estimation for nonlinear systems with bounded uncertainties. J Process Control 23:1281–1295

11. Rao CV, Rawlings JB, Lee JH (2001) Constrained linear state estimation-a moving horizon approach. Automatica 37:1619–1628
12. Rao CV, Rawlings JB (2002) Constrained process monitoring: moving-horizon approach. AIChE J 48:97–109
13. López-Negrete R, Patwardhan SC, Biegler LT (2011) Constrained particle filter approach to approximate the arrival cost in moving horizon estimation. J Process Control 21:909–919

Chapter 6
Two-Layer EMPC Systems

6.1 Introduction

As discussed in Chap. 1, in the traditional paradigm to optimization and control, a hierarchical strategy is employed using real-time optimization (RTO) to compute economically optimal steady-states that are subsequently sent down to a tracking MPC layer. The tracking MPC computes control actions that are applied to the closed-loop system to force the state to the optimal steady-state. RTO may periodically update the optimal steady-state to account for time-varying factors that may shift the optimal operating conditions and send the updated steady-state to the MPC layer. On the other hand, EMPC merges economic optimization and control and thus, employs a one-layer approach to optimization and control. While EMPC merges optimization and control, the extent that EMPC takes on all the responsibilities of RTO remains to be seen. For example, many EMPC methods are formulated using a steady-state, which potentially could be the economically optimal steady-state. RTO is also responsible for other tasks besides economic optimization. Therefore, one may envision that future optimization and control structures will maintain some aspects of the hierarchical approach within the context of industrial applications. Moreover, in some applications, maintaining a division between economic optimization and control is suitable, especially for applications where there is an explicit time-scale separation between the process/system dynamics and the update frequency or time-scale of evolution of economic factors and/or other factors that shift optimal operating conditions, e.g., disturbances.

In an industrial control architecture, which features a high degree of complexity, a hierarchical approach to dynamic economic optimization and control may be more applicable. Motivated by the aforementioned considerations, several two-layer approaches to dynamic economic optimization and control are discussed in this chapter. The upper layer, utilizing an EMPC, is used to compute economically optimal policies and potentially, also, control actions that are applied to the closed-loop system. The economically optimal policies are sent down to a lower layer MPC scheme which may be a tracking MPC or an EMPC. The lower layer MPC scheme

M. Ellis et al., *Economic Model Predictive Control*,
Advances in Industrial Control, DOI 10.1007/978-3-319-41108-8_6

forces the closed-loop state to closely follow the economically optimal policy computed in the upper layer EMPC.

The unifying themes of the two-layer EMPC implementations described in this chapter are as follows. First, the upper layer EMPC may employ a long prediction horizon. The long prediction horizon ideally prevents the EMPC from dictating an operating policy based on myopic decision-making, which may lead to poor closed-loop economic performance. Considering a one-layer EMPC approach with a long horizon, the computational time and complexity of the resulting optimization problem (thousands of decisions variables for large-scale systems) may make it unsuitable for real-time application. Second, the upper layer dynamic economic optimization problem, i.e., the EMPC problem, is formulated with explicit control-oriented constraints which allow for guaranteed closed-loop stability properties. This is a departure from other two-layer approaches to dynamic optimization and control such as those featuring dynamic-RTO, e.g. [1–10]. Third, the upper layer is solved infrequently in the sense that it is not solved every sampling time like a standard one-layer EMPC method with a receding horizon implementation. The rate that the upper layer EMPC is solved may be considered a tuning parameter of the optimization and control architectures. However, the upper layer does not need to wait until the system has reached steady-state owing to the fact that a dynamic model of the process is used in the optimization layer.

The lower layer MPC, which may be either a tracking MPC or an EMPC, may be formulated with a shorter prediction horizon and potentially, a smaller sampling period if state measurement feedback is available. It is used to force the closed-loop state to track the operating policy computed by the upper layer EMPC and to ensure closed-loop stability and robustness. Owing to the fact that the upper layer EMPC is solved infrequently and the lower layer MPC utilizes a shorter prediction horizon, one of the benefits of a two-layer EMPC implementation is improved computational efficiency compared to a one-layer EMPC method. The results of this chapter originally appeared in [11–13].

6.1.1 Notation

Given that this chapter deals with control elements arranged in a multi-layer configuration, an extended amount of notation is needed to describe the control system. To aid the reader, Table 6.1 summarizes the notation used in this chapter. Some of the notation will be made more precise in what follows. To clarify the difference between open-loop and closed-loop trajectories, consider a time sequence: $\{\bar{t}_i\}_{i=k}^{k+\bar{N}}$ where $\bar{t}_i = i\bar{\Delta}$, $\bar{\Delta} > 0$ is a constant and $\bar{N} \geq 1$ is a positive integer. Given a function $\bar{u} : [\bar{t}_k, \bar{t}_{k+\bar{N}}) \to \mathbb{U}$, which is right-continuous piecewise constant with constant hold period $\bar{\Delta}$, the open-loop predicted state trajectory under the open-loop input trajectory $\bar{u}$ is the solution to the differential equation:

Table 6.1 A summary of the notation used to describe the two-layer EMPC structure

Notation	Description
Δ_E	Zeroth-order hold period used for the upper layer control parameterization
N_E	Number of zeroth-order hold periods in the upper layer EMPC prediction horizon
K_E	Number of hold periods, Δ_E, that the upper layer EMPC is solved
t'	Operating period length with $t' = K_E \Delta_E$
$\{\hat{t}_k\}_{k\geq 0}$	Computation time sequence of upper layer with $\hat{t}_k = kt'$ ($k \in \mathbb{I}_+$)
$z(\cdot\|\hat{t}_k)$	Open-loop predicted state trajectory under an auxiliary controller computed at $\hat{t}_k$
$v(\cdot\|\hat{t}_k)$	Open-loop input trajectory computed by an auxiliary controller computed at $\hat{t}_k$
$x_E(\cdot\|\hat{t}_k)$	Open-loop predicted state trajectory under the upper layer MPC at $\hat{t}_k$
$u_E(\cdot\|\hat{t}_k)$	Open-loop input trajectory computed by the upper layer MPC at $\hat{t}_k$
Δ	Sampling period size of the lower layer
N	Number of sampling periods in the lower layer MPC prediction horizon
$\{t_j\}_{j\geq 0}$	Sampling time sequence of lower layer with $t_j = j\Delta$ ($j \in \mathbb{I}_+$)
$\tilde{x}(\cdot\|t_j)$	Open-loop predicted state trajectory under the lower layer MPC at t_j
$u(\cdot\|t_j)$	Open-loop input trajectory computed by the upper layer MPC at t_j
$x(\cdot)$	Closed-loop state trajectory under the two-layer control structure
$u^*(t_j\|t_j)$	Control action applied to the closed-loop system computed at t_j and applied from t_j to t_{j+1}
(x_s, u_s)	Steady-state and steady-state input pair

$$\dot{\bar{x}}(t) = f(\bar{x}(t), \bar{u}(t), 0) \tag{6.1}$$

for $t \in [\bar{t}_k, \bar{t}_{k+\bar{N}})$ with initial condition $\bar{x}(\bar{t}_k) = x(\bar{t}_k)$ where $x(\bar{t}_k)$ is a state measurement of the closed-loop system at time $\bar{t}_k$. The open-loop predicted state and input trajectories are denoted as $\bar{x}(\cdot|\bar{t}_k)$ and $\bar{u}(\cdot|\bar{t}_k)$ to make clear that both of these trajectories, which are functions of time, have been computed at $\bar{t}_k$ with a state measurement at $\bar{t}_k$.

The term closed-loop system refers to the resulting sampled-data system of Eq. 4.1 under an MPC scheme. The closed-loop state trajectory is the solution to:

$$\dot{x}(t) = f(x(t), k(x(t_j)), w(t)) \tag{6.2}$$

for $t \in [t_j, t_{j+1})$ with $t_j = j\Delta$ for some $\Delta > 0$ and $j = 0, 1, \ldots$. The initial time is taken to be zero. The mapping $k(\cdot)$ is a state feedback control law.

In the context of MPC, the state feedback control law is implicitly defined from the solution of an optimization problem and the receding horizon implementation. Specifically, the MPC receives a state measurement at a sampling time t_j, computes a control action, and applies it in a sample-and-hold fashion over the sample period, i.e., from t_j to t_{j+1}. The notation $u^*(t_j|t_j)$ is used to denote the computed control action by the MPC scheme at sampling time t_j with a state measurement $x(t_j)$. Under an MPC scheme, the closed-loop system is written similarly to Eq. 6.2 by replacing

$k(x(t_j))$ with $u^*(t_j|t_j)$. Finally, the notation $\cdot^*$, e.g., y^*, is used to denote that the quantity, which may be a vector with real elements or a function defined over an appropriate domain and range, is optimal with respect to a cost function (or cost functional) and some constraints.

6.2 Two-Layer Control and Optimization Framework

In this section, a two-layer dynamic economic optimization and control framework featuring EMPC in the upper layer and tracking MPC in the lower layer is discussed. The same nonlinear dynamic model is used in each layer to avoid modeling inconsistencies. Control-oriented constraints are employed in the dynamic optimization layer to ensure closed-loop stability. A rigorous theoretical treatment of the stability properties of the closed-loop system with the control architecture is provided. Variants and extensions of the two-layer optimization and control framework are discussed. The two-layer optimization and control framework is applied to a chemical process example.

6.2.1 Class of Systems

While a similar class of nonlinear systems is considered as that described by Eq. 4.1, the manipulated inputs are split into two groups; that is, the input vector is given by $u := [u_1\ u_2]^T$ where $u_1 \in \mathbb{R}^{m_1}$, $u_2 \in \mathbb{R}^{m_2}$, and $m_1 + m_2 = m$. Loosely speaking, the inputs are partitioned into two groups based on their main responsibility. The input u_1 is directly responsible for economic optimization and/or has the most significant impact on the closed-loop economic performance, while the input u_2 is responsible for maintaining closed-loop stability. In the chemical process example of Sect. 6.2.3, the inputs are partitioned using this rationale as a basis. Additional methods may be employed to help identify the inputs that have the most significant impact on the economic performance such as the methods presented in [14].

With the two sets of inputs, the following state-space model is written to emphasize the dependence of the vector field on each group of inputs:

$$\dot{x} = f(x, u_1, u_2, w) \tag{6.3}$$

where $x \in \mathbb{X} \subseteq \mathbb{R}^n$ denotes the state vector, $u_1 \in \mathbb{U}_1 \subset \mathbb{R}^{m_1}$ and $u_2 \in \mathbb{U}_2 \subset \mathbb{R}^{m_2}$ denote the two manipulated input vectors or the two sets of manipulated inputs, $w \in \mathbb{W} \subset \mathbb{R}^l$ denotes the disturbance vector and f is assumed to be a locally Lipschitz vector function on $\mathbb{X} \times \mathbb{U}_1 \times \mathbb{U}_2 \times \mathbb{W}$. The sets $\mathbb{U}_1$ and $\mathbb{U}_2$ are assumed to be nonempty compact sets. The disturbance is assumed to be bounded, i.e., $\mathbb{W} := \{w \in \mathbb{R}^l : |w| \leq \theta\}$ where $\theta > 0$. The origin of the nominal unforced system of Eq. 6.3 is assumed to be an equilibrium point ($f(0, 0, 0, 0) = 0$). The state of the system is sampled synchronously at the time instants indicated by the time sequence $\{t_j\}_{j\geq 0}$ where $t_j = j\Delta$, $j = 0, 1, \ldots$ and $\Delta > 0$ is the sampling period.

A stabilizability assumption is imposed on the system of Eq. 6.3 in the sense that the existence of a stabilizing feedback control law that renders the origin of the system of Eq. 6.3 asymptotically stable is assumed. The stabilizing feedback control law is given by the pair:

$$(h_1(x), h_2(x)) \in \mathbb{U}_1 \times \mathbb{U}_2 \tag{6.4}$$

for all $x \in \mathbb{X}$. While the domain of the stabilizing controller is taken to be $\mathbb{X}$, it renders the origin asymptotically stable with some region of attraction that may be a subset of $\mathbb{X}$. Applying converse Lyapunov theorems [15, 16], there exists a continuous differentiable Lyapunov function $V : D \to \mathbb{R}_+$ that satisfies the following inequalities:

$$\alpha_1(|x|) \leq V(x) \leq \alpha_2(|x|) \tag{6.5a}$$

$$\frac{\partial V(x)}{\partial x} f(x, h_1(x), h_2(x), 0) \leq -\alpha_3(|x|) \tag{6.5b}$$

$$\left| \frac{\partial V(x)}{\partial x} \right| \leq \alpha_4(|x|) \tag{6.5c}$$

for all $x \in D$ where $\alpha_i \in \mathcal{K}$ for $i = 1, 2, 3, 4$ and D is an open neighborhood of the origin. The region $\Omega_\rho \subseteq D$ such that $\Omega_\rho \subseteq \mathbb{X}$ is the (estimated) stability region of the closed-loop system under the stabilizing controller.

6.2.2 *Formulation and Implementation*

The dynamic economic optimization and control framework consists of EMPC in the upper layer and tracking MPC in the lower layer. A block diagram of the framework is given in Fig. 6.1. The prediction horizons of the EMPC and the MPC may be different. This allows for the EMPC to be formulated with a long prediction horizon. The number of sampling periods in the prediction horizon of the EMPC is denoted as $N_E \in \mathbb{I}_{\geq 1}$, and that of the MPC is denoted as $N \in \mathbb{I}_{\geq 1}$. For simplicity, the sampling periods of the upper layer EMPC and lower layer MPC are assumed to be the same ($\Delta_E = \Delta$) and Δ will be used to denote the sampling period. The two-layer framework may be extended to the case where $\Delta_E > \Delta$.

The upper layer EMPC problem is solved infrequently, i.e., not every sampling time. Let $K_E \geq \mathbb{I}_+$ be the number of sampling times that the upper layer is resolved. The time sequence $\{\hat{t}_k\}_{k\geq 0}$ denotes the time steps that the upper layer EMPC is solved. Owing to the implementation strategy, the time sequence is not necessarily a synchronous partitioning of time. For sake of simplicity, let $N \leq N_E - K_E$ to ensure that the upper layer EMPC problem is computed at a rate needed to ensure that the economically optimal trajectory is defined over the prediction horizon of

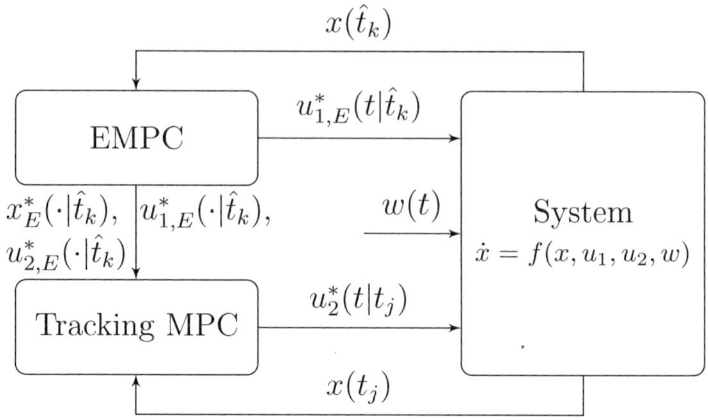

Fig. 6.1 A block diagram of the two-layer integrated framework for dynamic economic optimization and control with EMPC in the upper layer and tracking MPC in the lower layer. Both the upper and lower layers compute control actions that are applied to the system

the lower layer tracking MPC. If this is not satisfied, one could employ a shrinking horizon in the lower layer MPC. While the upper layer EMPC computes optimal input trajectories for both sets of manipulated inputs, it sends control actions for the manipulated input u_1 to the control actuators to be applied in an open-loop fashion. The implementation strategy is described below. The optimal operating trajectory over the prediction horizon of the EMPC is computed by the upper layer EMPC and sent to the lower layer tracking MPC to force the closed-loop state to track the optimal operating trajectory. In other words, the upper layer EMPC trajectory is used as the reference trajectory in the tracking MPC. The optimal operating trajectory is defined below.

Definition 6.1 Let $(u^*_{1,E}(t|\hat{t}_k), u^*_{2,E}(t|\hat{t}_k))$, which is defined for $t \in [\hat{t}_k, \hat{t}_k + N_E\Delta)$, be the optimal input pair computed by the upper layer EMPC and let $x(\hat{t}_k)$ be the state measurement at the sampling time $\hat{t}_k$. The economically optimal state trajectory $x^*_E(t|\hat{t}_k)$ for $t \in [\hat{t}_k, \hat{t}_k + N_E\Delta)$ of the system of Eq. 6.3 is the solution of

$$\dot{x}^*_E(t) = f(x^*_E(t), u^*_{1,E}(\tau_i|\hat{t}_k), u^*_{2,E}(\tau_i|\hat{t}_k), 0), \quad t \in [\tau_i, \tau_{i+1}) \tag{6.6}$$

for $i = 0, 1, \ldots, N_E - 1$ with $x_E(\hat{t}_k) = x(\hat{t}_k)$ where $\tau_i := \hat{t}_k + i\Delta$.

The lower layer MPC is implemented with a receding horizon implementation, i.e., is solved at every sampling time. The notation t_j will be reserved to denote a sampling time that the lower layer MPC problem is solved. To provide closed-loop stability guarantees on the resulting control framework, the upper layer EMPC is formulated as an LEMPC (Eq. 4.3) and the lower layer tracking MPC is formulated as an LMPC (Eq. 2.51). The advantage of the formulation of the upper and lower layer controllers is that the LEMPC computes a reference trajectory that the tracking LMPC layer may force the system to track and unreachable set-points are avoided.

An assumption is needed to ensure feasibility and stability with the resulting two-layer framework. Owing to the fact that the upper layer EMPC is solved infrequently and applies its computed trajectory for u_1 in an open-loop fashion, an assumption is needed to ensure that it is possible to maintain stability with the input u_2 in the sense that for any $u_1 \in \mathbb{U}_1$ it possible to find a $u_2 \in \mathbb{U}_2$ that ensures that the time-derivative of the Lyapunov function is negative. This is stated in the following assumption. Also, the assumption further clarifies how the manipulated inputs are divided into the two input groups.

Assumption 6.1 For any fixed $u_{1,E} \in U_1$, there exists $u_2 \in U_2$ such that:

$$\frac{\partial V(x)}{\partial x} f(x, u_{1,E}, u_2, 0) \leq \frac{\partial V(x)}{\partial x} f(x, h_1(x), h_2(x), 0) \tag{6.7}$$

for all $x \in \Omega_\rho$.

Variations of the assumption are discussed in Sect. 6.2.2.1. This assumption is not needed for the case that the upper layer LEMPC does not apply any control actions to the system.

The upper layer LEMPC has a similar formulation as that of Eq. 4.3 with one modification discussed below. The upper layer LEMPC problem is given by the following optimization problem:

$$\min_{u_{1,E}, u_{2,E} \in S(\Delta)} \int_{\hat{t}_k}^{\hat{t}_k + N_E \Delta} l_e(x_E(\tau), u_{1,E}(\tau), u_{2,E}(\tau)) \, d\tau \tag{6.8a}$$

$$\text{s.t.} \quad \dot{x}_E(t) = f(x_E(t), u_{1,E}(t), u_{2,E}(t), 0) \tag{6.8b}$$

$$x_E(\hat{t}_k) = x(\hat{t}_k) \tag{6.8c}$$

$$u_{1,E}(t) \in \mathbb{U}_1, \; u_{2,E}(t) \in \mathbb{U}_2, \; \forall \, t \in [\hat{t}_k, \hat{t}_k + N_E \Delta) \tag{6.8d}$$

$$V(x_E(t)) \leq \rho_e, \; \forall \, t \in [\hat{t}_k, \hat{t}_k + N_E \Delta), \quad \text{if } V(x(\hat{t}_k)) \leq \rho_e \text{ and } \hat{t}_k < t_s \tag{6.8e}$$

$$\frac{\partial V(x_E(\tau_i))}{\partial x} f(x_E(\tau_i), u_{1,E}(\tau_i), u_{2,E}(\tau_i), 0) \leq \frac{\partial V(x_E(\tau_i))}{\partial x} f(x_E(\tau_i), h_1(x_E(\tau_i)), h_2(x_E(\tau_i)), 0),$$
$$i = 0, 1, \ldots, N_E - 1, \; \text{if } V(x(\hat{t}_k)) > \rho_e \text{ or } \hat{t}_k \geq t_s \tag{6.8f}$$

where $\tau_i := \hat{t}_k + i\Delta$. The main difference between the upper layer LEMPC formulation and the LEMPC formulation of Eq. 4.3 is the mode 2 contractive constraint (Eq. 6.8f). In the upper layer LEMPC formulation, the mode 2 contractive constraint is imposed at each time instance of the prediction horizon. This ensures that the Lyapunov function value decays over the prediction horizon and thus, the lower layer LMPC attempts to force the closed-loop state along a reference trajectory that either converges to Ω_{ρ_e} if $\hat{t}_k < t_s$ or converges to a neighborhood of the origin if $\hat{t}_k \geq t_s$. The optimal trajectories computed by the upper layer LEMPC are denoted by $x_E^*(t|\hat{t}_k)$, $u_{E,1}^*(t|\hat{t}_k)$, and $u_{E,2}^*(t|\hat{t}_k)$ and are defined for $t \in [\hat{t}_k, \hat{t}_k + N_E \Delta)$.

The stage cost function used in the LMPC is formulated to penalize deviations of the state and inputs from the economically optimal trajectories. Additionally, the LMPC is equipped with dual-mode constraints similar to that imposed in LEMPC. The dual-mode LMPC problem is given by the following optimization problem:

$$\min_{u_2 \in S(\Delta)} \int_{t_j}^{t_{j+N}} l_T(\tilde{x}(\tau), x_E^*(\tau|\hat{t}_k), u_2(\tau), u_{2,E}^*(\tau|\hat{t}_k))\, d\tau \tag{6.9a}$$

$$\text{s.t. } \dot{\tilde{x}}(t) = f(\tilde{x}(t), u_{1,E}^*(t|\hat{t}_k), u_2(t), 0) \tag{6.9b}$$

$$\tilde{x}(t_j) = x(t_j) \tag{6.9c}$$

$$u_2(t) \in \mathbb{U}_2, \ \forall\, t \in [t_j, t_{j+N}) \tag{6.9d}$$

$$V(\tilde{x}(t)) \leq \rho_e, \ \forall\, t \in [t_j, t_{j+N}), \quad \text{if } V(x(t_j)) \leq \rho_e \text{ and } t_j < t_s \tag{6.9e}$$

$$\frac{\partial V(x(t_j))}{\partial x} f(x(t_j), u_{1,E}^*(t_j|\hat{t}_k), u_2(t_j), 0) \leq \frac{\partial V(x(t_j))}{\partial x} f(x(t_j), h_1(x(t_j)), h_2(x(t_j)), 0), \quad \text{if } V(x(t_j)) > \rho_e \text{ and } t_j \geq t_s \tag{6.9f}$$

where the stage cost of LMPC is given by:

$$l_T(\tilde{x}, x_E^*, u_2, u_{2,E}^*) = \left|\tilde{x} - x_E^*\right|_{Q_c}^2 + \left|u_2 - u_{2,E}^*\right|_{R_{c,2}}^2 \tag{6.10}$$

and Q_c and R_c are positive definite tuning matrices. The constraint of Eq. 6.9e defines mode 1 operation of the LMPC and serves a similar purpose as the mode 1 constraint of LEMPC (Eq. 4.3e). Under mode 2 operation of the LMPC, which is defined when the constraint of Eq. 6.9f is active, the LMPC computes control actions to ensure that the contractive Lyapunov-based constraint is satisfied. The optimal solution of Eq. 6.9 is denoted as $u_2^*(t|t_j)$ for $t \in [t_j, t_{j+N})$.

The two-layer optimization and control framework has a number of tunable parameters. Specifically, the tuning parameters include the weighting matrices Q_c and R_c, the prediction horizons N and N_E, the number of sampling times that the upper layer recomputes a solution K_E, the subset of the stability region that the control framework may operate the system in a time-varying fashion (Ω_{ρ_e}), the sampling period Δ and the triple (h_1, h_2, V), i.e., stabilizing controller design and Lyapunov function which are used in the Lyapunov-based constraints.

If the optimal state trajectory has been computed using mode 2 operation of the LEMPC and the current time is less than the switching time t_s, it is advantageous from a performance perspective, to recompute a new LEMPC solution using mode 1 once the state converges to the set Ω_{ρ_e}. This is captured in the implementation strategy of the two-layer optimization and control framework that is described by the following algorithm. Let the index $l \in \mathbb{I}_+$ be the number of sampling times since the last time that the upper layer LEMPC problem has been solved and m_k be the mode of operation of the LEMPC used to solve the LEMPC problem at $\hat{t}_k$. To initialize the algorithm, let $\hat{t}_0 = 0$, $k = 0$, and $j = 0$.

1. At $\hat{t}_k$, the upper layer LEMPC receives a state measurement $x(\hat{t}_k)$ and set $l = 0$. If $x(\hat{t}_k) \in \Omega_{\rho_e}$ and $\hat{t}_k < t_s$, go to Step 1.1. Else, go to Step 1.2.
   1.1 The mode 1 constraint (Eq. 6.8e) is active and the mode 2 constraint (Eq. 6.8f) is inactive. Set $m_k = 1$ and go to Step 1.3.
   1.2 The mode 2 constraint (Eq. 6.8f) is active and the mode 1 constraint (Eq. 6.8e) is inactive. Set $m_k = 2$ and go to Step 1.3.
   1.3 Solve the optimization problem of Eq. 6.8 to compute the optimal trajectories $x_E^*(t|\hat{t}_k)$, $u_{E,1}^*(t|\hat{t}_k)$, and $u_{E,2}^*(t|\hat{t}_k)$ defined for $t \in [\hat{t}_k, \hat{t}_k + N_E \Delta)$. Send these trajectories to the lower layer LMPC and go to Step 2.
2. At t_j, the lower layer LMPC receives a state measurement $x(t_j)$. If $x(t_j) \in \Omega_{\rho_e}$, $t_j < t_s$, and $m_k = 2$, set $\hat{t}_{k+1} = t_j$ and $k \leftarrow k + 1$, and go to Step 1. Else if $x(t_j) \in \Omega_{\rho_e}$ and $t_j < t_s$, go to Step 2.1. Else, go to Step 2.2.
   2.1 The mode 1 constraint (Eq. 6.9e) is active and the mode 2 constraint (Eq. 6.9f) is inactive. Go to Step 2.3.
   2.2 The mode 2 constraint (Eq. 6.9f) is active and the mode 1 constraint (Eq. 6.9e) is inactive. Go to Step 2.3.
   2.3 Solve the optimization problem of Eq. 6.9 to compute the optimal input trajectory $u^*(t|t_j)$ defined for $t \in [t_j, t_{j+N})$. Apply the input pair $(u_{1,E}^*(t_j|\hat{t}_k), u_2^*(t_j|t_j))$ to the system of Eq. 6.3 from t_j to t_{j+1}. Go to Step 3.
3. If $l + 1 = K_E$, set $\hat{t}_{k+1} = t_{j+1}$, $k \leftarrow k + 1$, and $j \leftarrow j + 1$, and go to Step 1. Else, go to Step 2 and set $l \leftarrow l + 1$ and $j \leftarrow j + 1$.

The two-layer implementation strategy allows for computational advantages over one-layer EMPC structures. When the LEMPC is operating in mode 1, the LEMPC problem is only computed once every K_E sampling times. The LMPC is less computationally expensive to solve than the LEMPC because the LMPC does not compute control actions for all of the manipulated inputs. Additionally, the LMPC may use a smaller prediction horizon than the LEMPC. Owing to these considerations, the two-layer framework is more computationally efficient compared to one-layer EMPC structures.

It is important to point out the two limiting cases of the optimization and control framework. If all the inputs are in the group u_2, then the control framework is reminiscent of current two-layer frameworks where economic optimization, which in this case is a dynamic optimization problem, and control are divided into separate layers. If, on the other hand, all inputs are placed in the group u_1 or the upper layer LEMPC is solved every sampling time ($K_E = 1$), then this would correspond to a one-layer implementation of LEMPC. For the case that all inputs are in the u_1 group, the LEMPC would need to be computed every sampling time to ensure stability and robustness of the closed-loop system.

6.2.2.1 Variants of the Two-Layer Optimization and Control Framework

While the stability results presented in Sect. 6.2.2.2 apply to the two-layer framework described above, one may consider many variations to the two-layer framework design. A few such variants are listed here.

1. The upper layer EMPC does not compute input trajectories for all inputs. For instance, some inputs in the u_2 set may have little impact on the economic performance. For these inputs, a constant input profile, for instance, could be assumed in the upper layer EMPC. This could further improve the computational efficiency of the two-layer framework. The inputs that are held constant in the EMPC problem could be used as additional degrees of freedom in the lower layer MPC to help force the closed-loop state to track the economically optimal state trajectory.
2. The lower layer MPC computes control actions for all the inputs, i.e., the upper layer EMPC does not apply any control actions directly to the system, but rather, is used to compute reference trajectories for the lower layer MPC. This approach is similar to current optimization and control structures but employs dynamic economic optimization with explicit control-oriented constraints imposed in the optimization layer.
3. Other assumptions to ensure feasibility and stability of the two-layer framework than Assumption 6.1 may be considered. For example, it may be possible to consider the input u_1 as a perturbation to the system and derive the explicit stabilizing controller on the basis of the inputs u_2. Specifically, if there exists an explicit controller $h_2 : \mathbb{X} \rightarrow \mathbb{U}_2$ and Lyapunov function that satisfies:

$$\frac{\partial V(x)}{\partial x} f(x, u_1, h_2(x), 0) \leq -\bar{\alpha}_3(|x|) \tag{6.11}$$

   for all $u_1 \in \mathbb{U}_1$ and $x \in \Omega_\rho \setminus B$ where $B \subset \Omega_\rho$ is some set containing the origin and $\bar{\alpha}(\cdot)$ is a class $\mathcal{K}$ function, then this assumption could be used to guarantee closed-loop stability and feasibility of the control problems. This assumption is essentially an input-to-state stability assumption of the closed-loop system of Eq. 6.3 under the controller h_2 with respect to the input u_1.
4. The two-layers could use a different sampling period size. In particular, the upper layer could use a larger sampling period than the lower layer.
5. Since the conditions that guarantee closed-loop stability presented below are independent of the objective function of the lower layer MPC, the stage cost used in the lower layer MPC may be readily modified. For example, one may include rate of change penalties on the inputs and soft constraints in the stage cost function or use the economic stage cost function. One such variant employing the latter concept, i.e., use an EMPC in the lower layer, is presented later in this chapter (Sect. 6.4).

6.2.2.2 Stability Analysis

In this section, sufficient conditions are presented that guarantee that the closed-loop system with the two-layer dynamic economic optimization and control framework is stable in the sense that the system state remains bounded in a compact set for all times. Two propositions are needed, which are straightforward extensions of Propositions 4.1 and 4.2, respectively. The propositions are restated here for convenience. The first proposition provides an upper bound on the deviation of the open-loop state trajectory, obtained using the nominal model (Eq. 6.3 with $w \equiv 0$), from the closed-loop state trajectory.

Proposition 6.1 (Proposition 4.1) *Consider the systems*

$$\begin{aligned} \dot{x}_a(t) &= f(x_a(t), u_1(t), u_2(t), w(t)) \\ \dot{x}_b(t) &= f(x_b(t), u_1(t), u_2(t), 0) \end{aligned} \tag{6.12}$$

with initial states $x_a(t_0) = x_b(t_0) \in \Omega_\rho$ and inputs $u_1(t) \in \mathbb{U}_1$ and $u_2(t) \in \mathbb{U}_2$ for $t \geq t_0$. If the states of the two systems are maintained in Ω_ρ for all $t \in [t_0, t_1]$ $(t_1 > t_0)$, there exists a class $\mathscr{K}$ function $\alpha_w(\cdot)$ such that

$$|x_a(t) - x_b(t)| \leq \alpha_w(t - t_0), \tag{6.13}$$

for all $w(t) \in W$ and $t \in [t_0, t_1]$.

The following proposition bounds the difference between the Lyapunov function of two different states in Ω_ρ.

Proposition 6.2 (Proposition 4.2) *Consider the Lyapunov function $V(\cdot)$ of the system of Eq. 6.3. There exists a quadratic function $\alpha_V(\cdot)$ such that:*

$$V(x) \leq V(\hat{x}) + \alpha_V(|x - \hat{x}|) \tag{6.14}$$

for all $x, \hat{x} \in \Omega_\rho$.

Theorem 6.1 provides sufficient conditions such that the two-layer dynamic economic optimization and control framework guarantees that the state of the closed-loop system is always bounded in Ω_ρ. The result is similar to that of the closed-loop stability properties under LEMPC.

Theorem 6.1 *Consider the system of Eq. 6.3 in closed-loop under the two-layer framework with the LEMPC of Eq. 6.8 in the upper layer and the LMPC of Eq. 6.9 in the lower layer both based on the explicit stabilizing controller that satisfies Eqs. 6.5a–6.5c. Let $\varepsilon_w > 0$, $\Delta > 0$, $N_E \geq 1$, $N \geq 1$ $(N \leq N_E - K_E)$, $\rho > \rho_e > \rho_{\min} > \rho_s > 0$, and L'_x, L'_w and M are positive constants (the existence of these constants follows from the assumptions on the system of Eq. 6.3) satisfy:*

$$\rho_e < \rho - \alpha_V(\alpha_w(\Delta)), \tag{6.15}$$

$$-\alpha_3(\alpha_2^{-1}(\rho_s)) + L'_x M\Delta + L'_w\theta \le -\varepsilon_w/\Delta \,, \tag{6.16}$$

and

$$\rho_{\min} = \max_{s\in[0,\Delta]} \{V(x(s)) : V(x(0)) \le \rho_s\} \,. \tag{6.17}$$

If $x(0) \in \Omega_\rho$ and Assumption 6.1 is satisfied, then the state $x(t)$ of the closed-loop system is always bounded in Ω_ρ for all $t \ge 0$. Moreover, if t_s is finite, the closed-loop state is ultimately bounded in $\Omega_{\rho_{\min}}$.

Proof The proof is organized into three parts. In part 1, feasibility of the optimization problems of Eqs. 6.8 and 6.9 is proved when the state measurement given to each problem is in Ω_ρ. In part 2, boundedness of the closed-loop state in Ω_ρ is established. Finally, (uniform) ultimate boundedness of the closed-loop state in a small state-space set containing the origin is proved when the switching time is finite.

Part 1: When the closed-loop state is maintained in Ω_ρ, which will be proved in Part 2, the sample-and-hold input trajectory obtained from the stabilizing feedback controller is a feasible solution to the upper layer LEMPC optimization problem of Eq. 6.8. Specifically, let $\hat{x}(t)$ denote the solution at time t to the system:

$$\dot{\hat{x}}(t) = f(\hat{x}(t), h_1(\hat{x}(\tau_i)), h_2(\hat{x}(\tau_i)), 0) \tag{6.18}$$

for $t \in [\tau_i, \tau_{i+1})$ $(\tau_i := \hat{t}_k + i\Delta), i = 0, 1, \ldots, N_E - 1$ with initial condition $\hat{x}(\hat{t}_k) = x(\hat{t}_k) \in \Omega_\rho$. Defining the pair $(\hat{u}_1(t), \hat{u}_2(t)) := (h_1(\hat{x}(\tau_i)), h_2(\hat{x}(\tau_i)))$ for $t \in [\tau_i, \tau_{i+1})$, $i = 0, 1, \ldots, N_E - 1$, the input trajectory pair $(\hat{u}_1, \hat{u}_2)$ is a feasible solution to the LEMPC problem. Specifically, for mode 2 operation of the LEMPC, the pair $(\hat{u}_1, \hat{u}_2)$ meets the input constraints since it is computed from the stabilizing controller, which satisfies the input constraints (Eq. 6.4). Also, the mode 2 contractive constraint of Eq. 6.8f is trivially satisfied with the input pair $(\hat{u}_1, \hat{u}_2)$. For mode 1 operation, the region Ω_{ρ_e} is forward invariant under the stabilizing controller applied in a sample-and-hold fashion when $\Omega_{\rho_{\min}} \subseteq \Omega_{\rho_e} \subset \Omega_\rho$ where $\Omega_{\rho_{\min}}$ will be explained further in Parts 2 and 3.

If Assumption 6.1 is satisfied, the feasibility of the lower layer LMPC problem of Eq. 6.9 follows because there exists an input trajectory $u_1(t)$ for $t \in [t_j, t_{j+N})$ that decreases the Lyapunov function by at least the rate given by the Lyapunov-based controller at each sampling time instance along the prediction horizon. Using similar arguments as that used for feasibility of the LEMPC, mode 2 operation of the LMPC is feasible. Assumption 6.1 further implies that there exists a sample-and-hold input trajectory such that Ω_{ρ_e} is forward invariant when $\Omega_{\rho_{\min}} \subseteq \Omega_{\rho_e} \subset \Omega_\rho$ which guarantees that mode 1 operation of the LMPC is feasible.

Part 2: To show that the state is maintained in Ω_ρ when $x(0) \in \Omega_\rho$, two cases must be considered. The first case occurs when the state $x(t_j) \in \Omega_{\rho_e}$ and $t_j < t_s$ and the second case occurs when $x(t_j) \in \Omega_\rho \setminus \Omega_{\rho_e}$ or $t_j \ge t_s$. It is sufficient to show that $x(t) \in \Omega_\rho$ for all $t \in [t_j, t_{j+1}]$. Through recursive application of this result, boundedness of the closed-loop state in Ω_ρ for all $t \geq 0$ follows if the initial state is in Ω_ρ.

Case 1: If $x(t_j) \in \Omega_{\rho_e}$ and $t_j < t_s$, the lower layer LMPC operates in mode 1 operation. Regardless if the upper layer LEMPC has been computed under mode 1 or mode 2, there exists a control action $\hat{u}_2(t_j)$ such that when applied to the model of Eq. 6.9b in a sample-and-hold fashion over one sampling period the state at the next sampling time will be predicted to be in Ω_{ρ_e} (this follows from Part 1). However, the closed-loop system of Eq. 6.3 does not evolve according to the model of Eq. 6.9b owing to the forcing of the disturbance w.

Let ρ_e satisfy Eq. 6.15. The proof proceeds by contradiction. Assume there exists a time $\tau^* \in [t_j, t_{j+1}]$ such that $V(x(\tau^*)) > \rho$. Define $\tau_1 := \inf\{\tau \in [t_j, t_{j+1}] : V(x(\tau)) > \rho\}$. A standard continuity argument in conjunction with the fact that $V(x(t_j)) \leq \rho_e < \rho$ shows that $\tau_1 \in (t_j, t_{j+1}]$, $V(x(t)) \leq \rho$ for all $t \in [t_j, \tau_1]$ with $V(x(\tau_1)) = \rho$, and $V(x(t)) > \rho$ for some $t \in (\tau_1, t_{j+1}]$. If ρ_e satisfies Eq. 6.15, then

$$\begin{aligned}\rho = V(x(\tau_1)) &\leq V(\tilde{x}(\tau_1)) + \alpha_V(\alpha_w(\tau_1)) \\ &\leq \rho_e + \alpha_V(\alpha_w(\Delta)) < \rho\end{aligned} \tag{6.19}$$

where the first inequality follows from Propositions 6.1, 6.2 and the second inequality follows from the fact that $\alpha_V \circ \alpha_w \in \mathcal{K}$ and $\tau_1 \leq \Delta$. Eq. 6.19 leads to a contradiction. Thus, $x(t_{j+1}) \in \Omega_\rho$ if Eq. 6.15 is satisfied.

Case 2: When $x(t_j) \in \Omega_\rho \setminus \Omega_{\rho_e}$ or $t_j \geq t_s$, the lower layer LMPC operates in mode 2. To cover both possibilities, consider any $x(t_j) \in \Omega_\rho$ and that mode 2 operation of the LEMPC is active. From the constraint of Eq. 6.9f and the condition of Eq. 6.5b, the computed control action at t_j satisfies:

$$\begin{aligned}&\frac{\partial V(x(t_j))}{\partial x} f(x(t_j), u^*_{1,E}(t_j|\hat{t}_k), u^*_2(t_j|t_j), 0) \\ &\leq \frac{\partial V(x(t_j))}{\partial x} f(x(t_j), h_1(x(t_j)), h_2(x(t_j)), 0) \leq -\alpha_3(|x(t_j)|)\end{aligned} \tag{6.20}$$

where $x(t_j)$ denotes the closed-loop state at sampling time t_j. Over the sampling period ($\tau \in [t_j, t_{j+1})$), the time derivative of the Lyapunov function of the closed-loop system is given by:

$$\dot{V}(x(\tau)) = \frac{\partial V(x(\tau))}{\partial x} f(x(\tau), u^*_{1,E}(t_j|\hat{t}_k), u^*_2(t_j|t_j), w(\tau)) \tag{6.21}$$

for $\tau \in [t_j, t_{j+1})$. Adding and subtracting the first term of Eq. 6.20 to/from Eq. 6.21 and accounting for the bound of Eq. 6.20, the time-derivative of the Lyapunov function over the sampling period is bounded by:

$$\dot{V}(x(\tau)) \leq -\alpha_3(|x(t_j)|) + \frac{\partial V(x(\tau))}{\partial x} f(x(\tau), u_{1,E}^*(t_j|\hat{t}_k), u_2^*(t_j|t_j), w(\tau)) - \frac{\partial V(x(t_j))}{\partial x} f(x(t_j), u_{1,E}^*(t_j|\hat{t}_k), u_2^*(t_j|t_j), 0) \tag{6.22}$$

for all $\tau \in [t_j, t_{j+1})$.

Since the sets Ω_ρ, $\mathbb{U}_1$, $\mathbb{U}_2$, and $\mathbb{W}$ are compact, the vector field f is locally Lipschitz, and the Lyapunov function is continuously differentiable, there exist $L'_x > 0$ and $L'_w > 0$ such that:

$$\left| \frac{\partial V(x)}{\partial x} f(x, u_1, u_2, w) - \frac{\partial V(x')}{\partial x} f(x', u_1, u_2, 0) \right| \leq L'_x |x - x'| + L'_w |w| \tag{6.23}$$

for all x, $x' \in \Omega_\rho$, $u_1 \in \mathbb{U}_1$, $u_2 \in \mathbb{U}_2$, and $w \in \mathbb{W}$. From Eqs. 6.22 to 6.23 and the fact that the disturbance is bounded in $\mathbb{W} = \{w \in \mathbb{R}^l : |w| \leq \theta\}$, the time-derivative of the Lyapunov function over the sampling period may be bounded as follows:

$$\dot{V}(x(\tau)) \leq -\alpha_3(|x(t_j)|) + L'_x \left| x(\tau) - x(t_j) \right| + L'_w \theta \tag{6.24}$$

for all $\tau \in [t_j, t_{j+1})$. Again, by the fact that the sets Ω_ρ, $\mathbb{U}_1$, $\mathbb{U}_2$, and $\mathbb{W}$ are compact and the vector field f is locally Lipschitz, there exists $M > 0$ such that

$$|f(x, u_1, u_2, w)| \leq M \tag{6.25}$$

for all $x \in \Omega_\rho$, $u_1 \in \mathbb{U}_1$, $u_2 \in \mathbb{U}_2$, and $w \in \mathbb{W}$. From Eq. 6.25 and continuity of $x(\tau)$ for $\tau \in [t_j, t_{j+1})$, the difference between the state at τ and t_j is bounded by:

$$\left| x(\tau) - x(t_j) \right| \leq M\Delta \tag{6.26}$$

for all $\tau \in [t_j, t_{j+1})$. From Eqs. 6.24 to 6.26 and for any $x(t_j) \in \Omega_\rho \setminus \Omega_{\rho_s}$, the inequality below follows:

$$\dot{V}(x(\tau)) \leq -\alpha_3(\alpha_2^{-1}(\rho_s)) + L'_x M\Delta + L'_w \theta \tag{6.27}$$

for all $\tau \in [t_j, t_{j+1})$ where the fact that $|x| \geq \alpha_2^{-1}(\rho_s)$ for all $x \in \Omega_\rho \setminus \Omega_{\rho_s}$ follows from Eq. 6.5a.

If the condition of Eq. 6.16 is satisfied, there exists $\varepsilon_w > 0$ such that the following inequality holds for $x(t_j) \in \Omega_\rho \setminus \Omega_{\rho_s}$

$$\dot{V}(x(\tau)) \leq -\varepsilon_w/\Delta$$

for all $\tau \in [t_j, t_{j+1})$. Integrating the bound for $\tau \in [t_j, t_{j+1})$, the following two bounds on the Lyapunov function value are obtained:

$$V(x(t_{j+1})) \le V(x(t_j)) - \varepsilon_w \tag{6.28}$$
$$V(x(\tau)) \le V(x(t_j)),\ \forall\, \tau \in [t_j, t_{j+1}) \tag{6.29}$$

for all $x(t_j) \in \Omega_\rho \setminus \Omega_{\rho_s}$ and when mode 2 operation of the lower layer LMPC is active.

If $x(t_j) \in \Omega_\rho \setminus \Omega_{\rho_e}$, the closed-loop state converges to Ω_{ρ_e} in a finite number of sampling times without leaving the stability region Ω_ρ which follows by applying Eq. 6.28 recursively. If $t_j \ge t_s$ and $x(t_j) \in \Omega_\rho \setminus \Omega_{\rho_s}$, the closed-lop state converges to Ω_{ρ_s} in a finite number of sampling times without leaving the stability region Ω_ρ (again, by recursive application of Eq. 6.28). Moreover, once the state converges to Ω_{ρ_s}, it remains inside $\Omega_{\rho_{\min}}$ for all times. This statement holds by the definition of $\rho_{\min}$. Therefore, from Case 1 and Case 2, the closed-loop state is bounded in Ω_ρ for all $t \ge 0$ when $x(0) \in \Omega_\rho$.

Part 3: If t_s is finite, the lower layer LMPC will switch to mode 2 operation only and the closed-loop state will be ultimately bounded in $\Omega_{\rho_{\min}}$, which follows from Part 2.

Remark 6.1 The closed-loop stability result presented in Theorem 6.1 is boundedness of the closed-loop state inside of Ω_ρ. Additional elements in the lower layer MPC are usually needed to guarantee that the closed-loop state will track the reference trajectories computed in the upper layer LEMPC. Nevertheless, acceptable tracking performance may usually be achieved in practice through a careful tuning of the weighting matrices Q_c and R_c and a sufficiently long prediction horizon in the lower layer LMPC, which is the case in the example below.

6.2.3 Application to a Chemical Process

The two-layer framework for dynamic economic optimization and process control is implemented on the benchmark chemical reactor example presented in Sect. 1.3.1. Recall, the nonlinear dynamic model that describes the evolution of the reactor (Eqs. 1.7–1.10) has four states: the vapor density in the reactor (x_1), the ethylene concentration in the reactor (x_2), the ethylene oxide concentration in the reactor (x_3), and the reactor temperature (x_4) and three inputs: the volumetric flow rate of the reactor feed, the ethylene concentration in the reactor feed, and the reactant coolant temperature. With abuse of notation, the notation u_1, u_2, and u_3 is used to denote the three inputs, respectively. The reactor has an asymptotically stable steady-state:

$$x_s^T = [0.998\ 0.424\ 0.032\ 1.002] \tag{6.30}$$

which corresponds to the steady-state input:

$$u_{1,s} = 0.35,\ u_{2,s} = 0.5,\ u_{3,s} = 1.0\,. \tag{6.31}$$

The control objective considered here is to optimize the time-averaged yield of ethylene oxide by operating the reactor in a time-varying fashion around the stable steady-state. Owing to the fact that closed-loop stability is not an issue for this application, the optimization and control framework operates with mode 1 operation only. The time-averaged yield of ethylene oxide over an operating length of t_f is given by

$$Y = \frac{\displaystyle\int_0^{t_f} x_3(t)x_4(t)u_1(t)\,dt}{\displaystyle\int_0^{t_f} u_1(t)u_2(t)\,dt} . \tag{6.32}$$

Owing to practical considerations, the average amount of ethylene that may be fed into the process over the length of operation is fixed, which is given by the following integral constraint:

$$\frac{1}{t_f}\int_0^{t_f} u_1(t)u_2(t)dt = u_{1,s}u_{2,s} = 0.175 \tag{6.33}$$

where $u_{1,s}$ and $u_{2,s}$ are the steady-state inlet volumetric flow rate and ethylene concentration, respectively. Since the average ethylene fed to the reactor is fixed, which fixes the denominator of the yield (Eq. 6.32), the economic stage cost used in the formulation of the upper layer LEMPC is

$$l_e(x, u) = -x_3x_4u_1 . \tag{6.34}$$

In the implementation of the two-layer dynamic optimization and control framework, the manipulated inputs are partitioned into two sets. The first set of manipulated inputs consists of the inlet flow rate and ethylene feed concentration inputs. As pointed out in Sect. 3.2, periodic switching of these two inputs may improve economic performance. Additionally, these two inputs are constrained by the integral constraint of Eq. 6.33. The first set of inputs is controlled by the upper layer LEMPC, i.e., the upper layer LEMPC computes control actions for these manipulated inputs that are applied to the reactor. The second set of manipulated inputs consists of the coolant temperature input that the lower layer LMPC (Eq. 6.9) controls.

To characterize the region Ω_{ρ_e}, which is used in the two-layer framework design, an explicit stabilizing controller is designed and a Lyapunov function is constructed. Specifically, the explicit controller is designed as a proportional controller for the input u_3: $h_2(x) = K(x_{s,3} - x_3) + u_{s,3}$ with $K = 0.1$. A quadratic Lyapunov function is found for the closed-loop reactor under the proportional controller, which is given by:

$$V(x) = (x - x_s)^T P(x - x_s)$$

where $P = \text{diag}([10\ 0.01\ 10\ 10])$. The closed-loop stability region of the reactor under the explicit controller with the inputs u_1 and u_2 fixed at their steady-state values is taken to be a level set of the Lyapunov function where the time-derivative of the Lyapunov function is negative definite for all points contained in the level set. The constructed level set is subsequently taken to be Ω_{ρ_e} with $\rho_e = 0.53$ and in this case, $\Omega_\rho = \Omega_{\rho_e}$.

The prediction horizon of the upper layer LEMPC and lower layer LMPC are $N_E = 47$ and $N = 3$, respectively, the sampling period is $\Delta = 1.0$, the number of sampling times that the upper layer LEMPC is recomputed is $K_E = 47$, which is the same as the prediction horizon in this case, and a shrinking horizon employed in the lower layer LMPC when the prediction horizon extends past the time that the upper layer optimal trajectory is defined. To ensure that the integral constraint of Eq. 6.33 is satisfied over the length of operation, the computed input trajectory of the upper layer LEMPC must satisfy the integral constraint, i.e., it is enforced over each operating windows of length 47 (dimensionless time). The weighting matrices of the lower layer LMPC are $Q_c = P$, and $R_c = 0.01$ which have been tuned to achieve close tracking of the optimal trajectory. The optimization problems of upper layer LEMPC and lower layer LMPC are solved using Ipopt [17].

In the first set of simulations, the two-layer framework is applied to the reactor without disturbances or plant-model mismatch. The reactor is initialized at a transient initial condition given by:

$$x_0^T = [0.997\ 1.264\ 0.209\ 1.004]\ .$$

The closed-loop state and input trajectories of the reactor under the two-layer optimization and control framework are shown in Figs. 6.2 and 6.3, respectively. From the state trajectories (Fig. 6.2), the lower layer LMPC is able to force the system to track the optimal state trajectory. Recall, the sampling periods of the upper and lower layer are the same, and the closed-loop system is not subjected to any uncertainties or disturbances. Therefore, perfect tracking of the optimal trajectory is expected.

As described above, a motivating factor for the design of a two-layer optimization and control architecture is to achieve a computation benefit relative to a one-layer EMPC approach. To compare the computational time of the two-layer framework with a one-layer EMPC approach, a one-layer LEMPC implementation is considered. The LEMPC is implemented with mode 1 operation only and with a shrinking prediction horizon. The shrinking horizon implementation is described as follows: the LEMPC is initialized with a prediction horizon of 47 (dimensionless time) and at every subsequent sampling time, the prediction horizon is decreased by one sampling period. Every 47 sampling times, the prediction horizon is reset to 47. It is important to point out that the closed-loop performance achieved under the two-layer LEMPC and that under the one-layer LEMPC are equal owing to the fact there is no plant-model mismatch. Also, a fixed-horizon one-layer LEMPC implementation strategy requires more computation time on average relative to the shrinking horizon implementation.

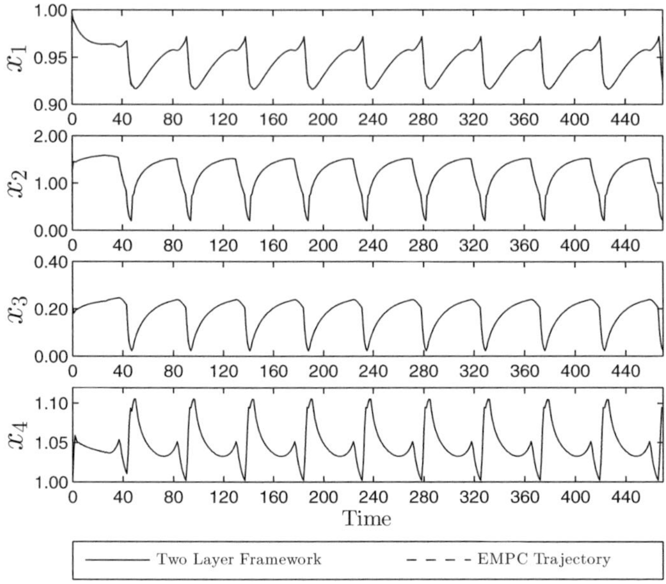

Fig. 6.2 The closed-loop state trajectories of the reactor under the two-layer dynamic economic optimization and control framework (the two trajectories are overlapping)

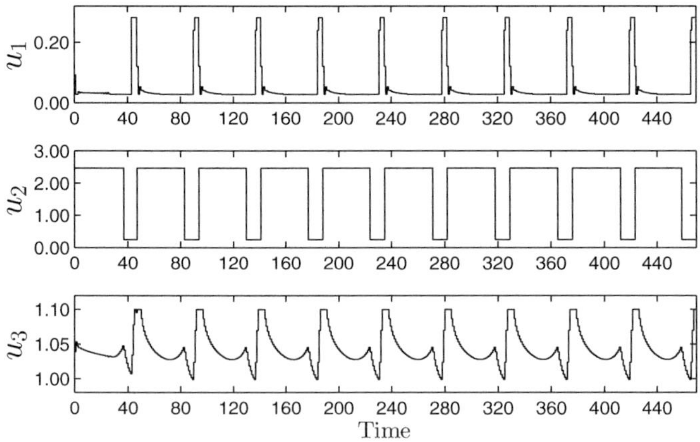

Fig. 6.3 The closed-loop input trajectories computed by two-layer dynamic economic optimization and control framework (the two trajectories are overlapping)

Figure 6.4 gives the computational time reduction achieved with the two-layer optimization and control framework relative to the one-layer LEMPC implementation. For this example, the lower layer LMPC computation time is insignificant compared to the computation time of the upper layer LEMPC. The two-layer framework only solves the LEMPC optimization problem once every 47 sampling times. Every

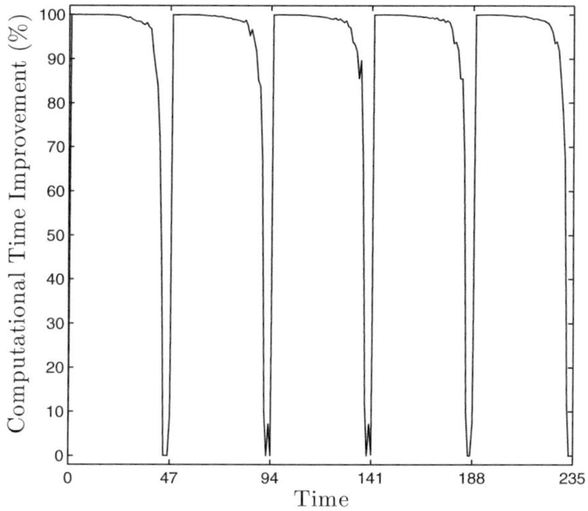

Fig. 6.4 The computational time reduction of the two-layer optimization and control framework relative to the one-layer implementation of LEMPC

47 sampling times when the upper layer LEMPC is solved and at the end of each operating interval of length 47 when the one-layer LEMPC horizon has decreased to a comparable length as the horizon of the lower layer LEMPC, the computational burden of the two-layer framework compared to that of the one-layer LEMPC is comparable, i.e., approximately a zero percent computational time improvement is achieved (Fig. 6.4). For the other sampling times, the computation of the LMPC which computes control actions for the set of manipulated inputs u_2 is much better than that compared to the one-layer LEMPC. For this case, an average of 89.4 % reduction of the computational time with the two-layer framework is achieved relative to that of the one-layer LEMPC implementation.

In the second set of simulations, significant process noise is added to the system states. The noise is assumed to be bounded Gaussian white noise with zero mean and standard deviation of 0.005, 0.03, 0.01, and 0.02 and bounds given by 0.02, 0.1, 0.03, and 0.08 for the four states, respectively. To simulate the process noise, a new random number is generated and applied to the process over each sampling period. The results of a closed-loop simulation are shown in Figs. 6.5 and 6.6. Because of the added process noise, the closed-loop trajectories do not perfectly track the reference trajectories. Also, the added process noise has an effect on the closed-loop economic performance. However, this effect was minimal in the sense that the time-averaged yield of the closed-loop system under the two-layer framework is 10.3 % with the added process disturbance and 10.4 % without the added process disturbance. Even with the process noise, the closed-loop reactor performance is better than that at the steady-state (the yield at steady-state is 6.4 %).

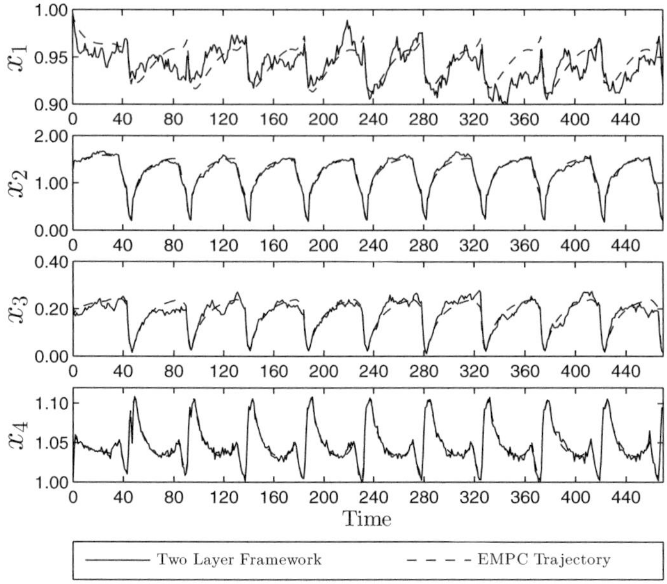

Fig. 6.5 The closed-loop state trajectories of the catalytic reactor under the two-layer dynamic economic optimization and control framework and with process noise added to the states

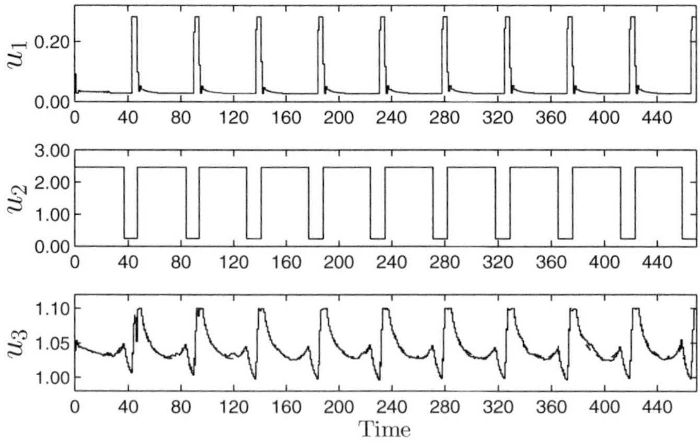

Fig. 6.6 The closed-loop input trajectories computed by two-layer dynamic economic optimization and control framework and with process noise added to the states (the two trajectories are nearly overlapping)

6.3 Unifying Dynamic Optimization with Time-Varying Economics and Control

In the previous section, a two-layer framework for dynamic optimization and control is presented. However, the framework treats the economic considerations, e.g., demand, energy pricing, variable feedstock quality, and product grade changes as time-invariant. This paradigm may be effective, especially for the applications where there is a sufficient time-scale separation between the time constants of the process/system dynamics and the update frequency of the economic parameters. However, including the time-variation of the economic considerations in the formulation of the economic stage cost may be needed to achieve good closed-loop performance when the time-scales are comparable. One class of examples where there may not be such a time-scale separation is energy systems with real-time pricing.

In this section, a two-layer framework for optimization and control of systems of the form of Eq. 4.1 is considered where the economic stage cost may be time-dependent in the sense that the system of Eq. 4.1 is equipped with a time-dependent economic cost $l_e : [0, \infty) \times \mathbb{X} \times \mathbb{U} \to \mathbb{R}$ ($(t, x, u) \mapsto l_e(t, x, u)$ where t is the time). The framework design is similar to that in the previous section. Specifically, the upper layer dynamic economic optimization problem (EMPC) is used to generate an economically optimal state trajectory defined over a finite-time horizon. In the lower layer, a tracking MPC is used to force the states to track the economically optimal trajectory. However, the main differences of the two-layer approach presented in this section compared to that of the previous section are the EMPC is formulated with an economic stage cost that may be explicitly time-dependent, the formulations of the layers are different, and the underlying theory and analysis are different.

Explicit constraints are used in the upper layer dynamic optimization problem to ensure that the lower layer tracking MPC may force the closed-loop state to track the trajectory computed in the optimization layer. In particular, the optimization layer is constrained to compute an optimal trajectory that is slowly time-varying. The resulting slowly time-varying trajectory vector is denoted as $x_E(t) \in \Gamma \subset \mathbb{R}^n$ for $t \geq 0$ where Γ is a compact set and the rate of change of the reference trajectory is bounded by

$$|\dot{x}_E(t)| \leq \gamma_E \tag{6.35}$$

for all $t \geq 0$. The deviation between the actual state trajectory and the slowly-varying reference trajectory is defined as $e := x - x_E$ with its dynamics described by

$$\begin{aligned} \dot{e} &= f(x, u, w) - \dot{x}_E \\ &= f(e + x_E, u, w) - \dot{x}_E \\ &=: g(e, x_E, \dot{x}_E, u, w) \, . \end{aligned} \tag{6.36}$$

The state e of the system of Eq. 6.36 will be referred to as the deviation state in the remainder of this section.

Assumption 6.2 The system of Eq. 6.36 has a continuously differentiable, isolated equilibrium for each fixed $x_E \in \Gamma$ in the sense that there exists a $\hat{u} \in \mathbb{U}$ for a fixed $x_E \in \Gamma$ to make $e = 0$ the equilibrium of Eq. 6.36 ($g(0, x_E, 0, \hat{u}, 0) = 0$).

In what follows, the upper layer EMPC computes a reference trajectory that evolves according to the nominal system dynamics $\dot{x}_E = f(x_E, u_E, 0)$ while maintaining the state trajectory to be in the set Γ where Γ is an equilibrium manifold in the sense that $\Gamma = \{x_E \in \mathbb{X} : \exists\, u_E \in \mathbb{U} \text{ s.t. } f(x_E, u_E, 0) = 0\}$. Nevertheless, the theory applies to a more general case where the following hold: $|\dot{x}_E| \leq \gamma_E$, Γ is compact and Assumption 6.2 is satisfied. One conceptually straightforward extension of the two-layer framework is to consider steady-state optimization in the upper layer instead of dynamic optimization. Specifically, x_E could be taken as a steady-state and varied slowly to account for the time-varying economic considerations.

6.3.1 Stabilizability Assumption

For each fixed $x_E \in \Gamma$, there exists a Lyapunov-based controller that makes the origin of the nonlinear system given by Eq. 6.36 without uncertainty ($w \equiv 0$) asymptotically stable under continuous implementation. This assumption is essentially equivalent to the assumption that the nominal system of Eq. 4.1 is stabilizable at each $x_E \in \Gamma$. More specifically, for each fixed $x_E \in \Gamma$, the existence of a mapping $h : D_s \times \Gamma \rightarrow \mathbb{U}$ and a continuously differentiable function $V : D_s \times \Gamma \rightarrow \mathbb{R}_+$ is assumed that satisfies:

$$\alpha_1(|e|) \leq V(e, x_E) \leq \alpha_2(|e|), \tag{6.37a}$$

$$\frac{\partial V(e, x_E)}{\partial e} g(e, x_E, 0, h(e, x_E), 0) \leq -\alpha_3(|e|), \tag{6.37b}$$

$$\left| \frac{\partial V(e, x_E)}{\partial e} \right| \leq \alpha_4(|e|), \tag{6.37c}$$

$$\left| \frac{\partial V(e, x_E)}{\partial x_E} \right| \leq \alpha_5(|e|), \tag{6.37d}$$

for all $e \in D_s$ where $\alpha_i \in \mathcal{K}$, $i = 1, 2, 3, 4, 5$, D_s is an open neighborhood of the origin, and h is the Lyapunov-based controller. In this sense, the function V is a Lyapunov function for each $x_E \in \Gamma$. While the inequalities of Eqs. 6.37a–6.37c are similar to the inequalities of standard Lyapunov functions, Eq. 6.37d is needed to account for the time-varying nature of x_E. More precisely, the special requirement that the inequalities hold uniformly in x_E is required to handle the perturbation, which results from the fact that x_E is not constant, but rather, a time-varying function.

For a fixed $x_E \in \Gamma \subset \mathbb{R}^n$, the symbol $\Omega_{\rho(x_E)}$ is a level set of the Lyapunov function, i.e., $\Omega_{\rho(x_E)} := \{e \in \mathbb{R}^n : V(e, x_E) \leq \rho(x_E)\}$ where $\rho(x_E) > 0$ depends on x_E.

The region $\Omega_{\rho*}$ is the intersection of stability regions $\Omega_{\rho(x_E)}$ of the closed-loop system under the Lyapunov-based controller for all $x_E \in \Gamma$.

For broad classes of nonlinear systems arising in the context of chemical process control applications, quadratic Lyapunov functions using state deviation variables, i.e., $V(x) = (x - x_s)^T P(x - x_s)$, where x_s is a steady-state, have been widely used and have been demonstrated to yield acceptable estimates of closed-loop stability regions (see [18] and the references therein). In the example of Sect. 6.3.3, a quadratic Lyapunov function is used where instead of a fixed equilibrium x_s a time-varying reference trajectory x_E is used, i.e., at time t, the Lyapunov function is given by: $V(e(t), x_E(t)) = e^T(t)Pe(t)$ where $e(t) = x(t) - x_E(t)$.

Remark 6.2 If the equilibrium point $e = 0$ of the frozen system forced by an explicit controller ($\dot{e} = g(e, x_E, 0, h(e, x_E), 0)$) is exponentially stable uniformly in x_E and under some additional mild smoothness requirements, then there exists a Lyapunov function satisfying Eqs. 6.37a–6.37d [15, Lemma 9.8].

Remark 6.3 The set $\Omega_{\rho*}$ is such that for any $e \in \Omega_{\rho*}$, the ability to drive the state with the Lyapunov-based controller asymptotically to any fixed $x_E \in \Gamma$ is guaranteed. This set may be estimated in the following way: first, the set Γ is chosen. Second, the regions $\Omega_{\rho(x_E)}$ for a sufficiently large number of x_E in the set Γ are estimated. The regions $\Omega_{\rho(x_E)}$ may be estimated as the level set (ideally the largest) of V for a fixed $x_E \in \Gamma$ where $\dot{V} < 0$ with the Lyapunov-based controller. Lastly, the stability region $\Omega_{\rho*}$ may be constructed from the intersection of these computed regions. It is important to point out that the design of the upper layer EMPC does not employ $\Omega_{\rho*}$. Therefore, for practical design purposes, an explicit construction of $\Omega_{\rho*}$ is not needed.

6.3.2 Two-Layer EMPC Scheme Addressing Time-Varying Economics

In this section, the two-layer framework for dynamic economic optimization and control for handling time-varying economics is described and the stability and robustness properties of the closed-loop system are given.

6.3.2.1 Formulation and Implementation

To address time-dependent economics, a two-layer framework is presented. The two-layer framework for optimization and control may be considered an intermediate approach between existing steady-state operation and one-layer EMPC schemes. A block diagram of the two-layer control framework is given in Fig. 6.7. In this framework, optimization and control are effectively divided into separate tasks. However, the upper optimization layer is formulated with specific control-oriented constraints

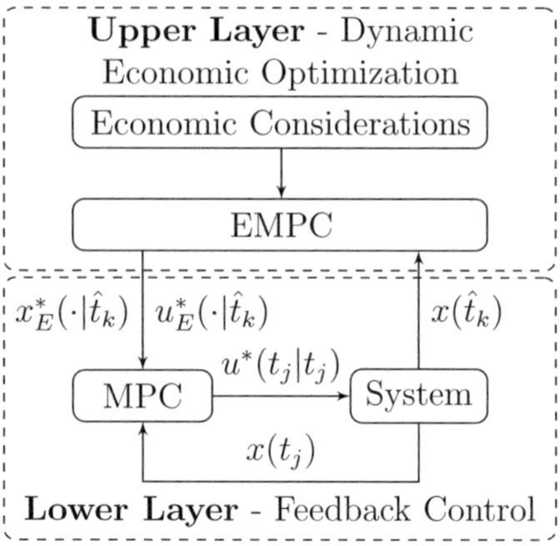

Fig. 6.7 A block diagram of the dynamic economic optimization and control framework for handling time-varying economics

to ensure stability. In the upper layer, an EMPC, formulated with a time-varying economic stage cost, computes economically optimal state and input trajectories over a finite-time horizon. The optimal trajectories are sent down to a lower layer tracking MPC to force the system to track the economically optimal state trajectory. For computational efficiency, the EMPC optimization problem is solved infrequently, i.e., it does not employ a standard receding horizon implementation strategy. Instead, the operating time is partitioned into finite-time intervals of length t' called operating periods. The operating period is chosen based on the time-scale of the process dynamics and update frequency of the economic parameters in the economic cost function, e.g., the update frequency of the energy price, product demand, or product transitions. The length of the operating period may be considered a tuning parameter of the control architecture. At the beginning of each operating period, the EMPC problem is initialized with a state measurement and is solved. The lower layer tracking MPC is solved every sampling time to maintain closed-loop stability and robustness and is formulated with a stage cost that penalizes deviations from the optimal trajectories. While in the lower layer any MPC tracking controller could be used, Lyapunov-based MPC (LMPC) is used here owing to its unique stability and robustness properties.

A summary of the implementation strategy and the notation is as follows (a detailed algorithm is given below after the formulations of the control problems are given). The operating time is divided into finite-time operating periods of length $t' = K_E \Delta_E$ where K_E is some integer greater than or equal to one. At the beginning of the operating period denoted by $\hat{t}_k = kt'$ where $k = 0, 1, \ldots$, the upper layer EMPC, with hold period $\Delta_E > 0$ (zeroth-order control parameterization is employed in the upper layer EMPC) and prediction horizon of $T_E = N_E \Delta_E$ where $N_E \in \mathbb{I}_+$, receives a state measurement and computes the economically optimal state and input trajectories. The prediction horizon of the EMPC is chosen to be sufficiently large to cover the operating period plus the transition to the next operating period, i.e., $T_E \geq t' + T$ where $T = N\Delta$ is the prediction horizon of the lower layer LMPC, $\Delta > 0$ denotes the sampling period of the lower layer LMPC that is less than or equal to Δ_E, and $N \in \mathbb{I}_+$ is the number of sampling periods in the prediction horizon of the LMPC. Between $\hat{t}_k$ and $\hat{t}_k + t'$, the lower layer LMPC computes control actions that work to force the closed-loop state to track the optimal trajectories.

The upper layer EMPC optimization problem is as follows:

$$\min_{u_E \in S(\Delta_E)} \int_{\hat{t}_k}^{\hat{t}_k+T_E} l_e(\tau, x_E(\tau), u_E(\tau))\, d\tau \tag{6.38a}$$

$$\text{s.t.} \quad \dot{x}_E(t) = f(x_E(t), u_E(t), 0) \tag{6.38b}$$

$$x_E(\hat{t}_k) = \underset{\Gamma}{\text{proj}}(x(\hat{t}_k)) \tag{6.38c}$$

$$u_E(t) \in \mathbb{U}, \ \forall\, t \in [\hat{t}_k, \hat{t}_k + T_E) \tag{6.38d}$$

$$|\dot{x}_E(t)| \leq \gamma_E, \ \forall\, t \in [\hat{t}_k, \hat{t}_k + T_E) \tag{6.38e}$$

$$x_E(t) \in \Gamma, \ \forall\, t \in [\hat{t}_k, \hat{t}_k + T_E) \tag{6.38f}$$

where $S(\Delta_E)$ is the family of piecewise constant functions with period Δ_E, l_e is the time-dependent economic measure which defines the cost function, the state x_E is the predicted trajectory of the system with the input trajectory u_E computed by the EMPC and $x(\hat{t}_k)$ is the state measurement obtained at time $\hat{t}_k$. The optimal state and input trajectory computed by the EMPC are denoted as $x_E^*(t|\hat{t}_k)$ and $u_E^*(t|\hat{t}_k)$ defined for $t \in [\hat{t}_k, \hat{t}_k + T_E)$, respectively.

In the optimization problem of Eq. 6.38, the constraint of Eq. 6.38b is the nominal dynamic model of the system ($w \equiv 0$) used to predict the future evolution under the sample-and-hold input trajectory. The constraint of Eq. 6.38c defines the initial condition of the optimization problem which is a projection of the state measurement at $\hat{t}_k$ onto the set Γ where the symbol $\text{proj}_\Gamma(x)$ denotes the projection of x onto the set Γ. The constraint of Eq. 6.38d ensures that the computed input trajectory takes values in the set of admissible inputs. The constraint of Eq. 6.38f limits the rate of change of the economically optimal state trajectory. Finally, the constraint of Eq. 6.38f ensures that the state evolution is maintained in the region Γ.

The constraint of Eq. 6.38c is used to ensure that the optimization problem is feasible. The projection operator may be any projection operator that projects the current state $x(\hat{t}_k)$ onto a near (ideally the nearest) point in the set Γ. In some cases, when the sampling periods of the upper and lower layers and the bounded disturbance are sufficiently small, it may also be sufficient to use the predicted state $x_E^*(\hat{t}_k|\hat{t}_{k-1})$ derived from the solution of the optimization problem of Eq. 6.38 that was solved at the beginning of the preceding operating period. Another potential option is to allow for the initial condition $x_E(\hat{t}_k)$ be a decision variable in the optimization problem by including another term in the objective function penalizing the deviation of the computed initial condition from the current state $x(\hat{t}_k)$. In this sense, the framework offers a degree of flexibility in the selection of the projection operator.

The last two constraints of the optimization problem of Eq. 6.38 are used to guarantee closed-loop stability under the integrated framework and to ensure that the lower layer may force the system to track the optimal state trajectory, i.e., they are control-oriented constraints. This is a departure from other types of two-layer dynamic economic optimization and control architectures featuring, for example, dynamic real-time optimization in the upper optimization layer. Also, the constraint imposed in the upper layer EMPC on the rate of change of the optimal trajectory (Eq. 6.38e) does pose a restriction on the feasible set of the optimization problem of Eq. 6.38 and could affect closed-loop economic performance achieved under the resulting two-layer framework. However, allowing the optimal state trajectory to have a large rate of change may be undesirable for many applications based on practical considerations like excessive strain on control actuators as well as the difficulty of forcing the system to track a rapidly changing reference trajectory in the presence of disturbances.

At the lower feedback control level, LMPC is employed to force the state to track the economically optimal state trajectory. The LMPC is implemented with a standard receding horizon implementation, i.e., the LMPC recomputes an updated input trajectory synchronously every sampling time. Let $\{t_j\}_{j\geq 0}$ where $t_j = j\Delta$, $j = 0, 1, \ldots$ denote the sampling time sequence of the LMPC. Also, the dynamic model used in the LMPC is that of Eq. 6.36, which is the deviation system. The LMPC optimization problem is given by:

$$\min_{u \in S(\Delta)} \int_{t_j}^{t_j+T} \left(|\tilde{e}(\tau)|_{Q_c}^2 + \left|u(\tau) - u_E^*(\tau|\hat{t}_k)\right|_{R_c}^2 \right) d\tau \tag{6.39a}$$

$$\text{s.t } \dot{\tilde{e}}(t) = g(\tilde{e}(t), x_E^*(t|\hat{t}_k), \dot{x}_E^*(t|\hat{t}_k), u(t), 0) \tag{6.39b}$$

$$\tilde{e}(t_j) = x(t_j) - x_E^*(t_j|\hat{t}_k) \tag{6.39c}$$

$$u(t) \in \mathbb{U},\ \forall\, t \in [t_j, t_j + T) \tag{6.39d}$$

$$\begin{aligned} &\frac{\partial V(\tilde{e}(t_j), x_E^*(t_j|\hat{t}_k))}{\partial e} g(\tilde{e}(t_j), x_E^*(t_j|\hat{t}_k), 0, u(t_j), 0) \\ &\leq \frac{\partial V(\tilde{e}(t_j), x_E^*(t_j|\hat{t}_k))}{\partial e} g(\tilde{e}(t_j), x_E^*(t_j|\hat{t}_k), 0, h(\tilde{e}(t_j), x_E(t_j|\hat{t}_k)), 0) \end{aligned} \tag{6.39e}$$

where $S(\Delta)$ is the family of piecewise constant functions with sampling period Δ, N is the prediction horizon of the LMPC, $\tilde{e}$ is the predicted deviation between the state trajectory predicted by the nominal model under the input trajectory computed by the LMPC and the economically optimal state trajectory $x_E^*(\cdot|\hat{t}_k)$. The optimal solution of the optimization problem of Eq. 6.39 is denoted by $u^*(t|t_j)$ defined for $t \in [t_j, t_{j+N})$.

In the optimization problem of Eq. 6.39, the constraint of Eq. 6.39b is the nominal model of the deviation system. The constraint of Eq. 6.39c is the initial condition to the dynamic optimization problem. The constraint of Eq. 6.39d defines the control energy available to all manipulated inputs. The constraint of Eq. 6.39e ensures that the Lyapunov function of the closed-loop system with the LMPC decreases by at least the rate achieved by the Lyapunov-based controller. The last constraint ensures that the closed-loop state trajectory converges to a neighborhood of the optimal state trajectory computed by the upper layer EMPC.

The implementation strategy of the dynamic economic optimization and control framework is summarized by the following algorithm.

1. At $\hat{t}_k$, the EMPC receives a state measurement $x(\hat{t}_k)$ and projects the current state $x(\hat{t}_k)$ onto the set Γ. Go to Step 2.
2. The EMPC computes the economically optimal state and input trajectories: $x_E^*(t|\hat{t}_k)$ and $u_E^*(t|\hat{t}_k)$ defined for $t \in [\hat{t}_k, \hat{t}_k + T_E)$. Go to Step 3.
3. For $\hat{t}_k$ to $\hat{t}_k + t'$ (one operating period), repeat:

   3.1 The LMPC receives a state measurement $x(t_j)$ and computes the deviation of the current state from the optimal state trajectory. The error $e(t_j)$ is used to initialize the dynamic model of the LMPC. Go to Step 3.2.
   3.2 The LMPC optimization problem is solved to compute an optimal input trajectory $u^*(t|t_j)$ defined for $t \in [t_j, t_j + T)$. Go to Step 3.3.
   3.3 The control action computed for the first sampling period of the prediction horizon is sent to the control actuators to be applied from t_j to t_{j+1}. If $t_{j+1} > \hat{t}_k + t'$, go to Step 4 and let $j \leftarrow j + 1$. Else, go to 3.1 and let $j \leftarrow j + 1$.

4. Go to Step 1, $k \leftarrow k + 1$.

6.3.2.2 Stability Analysis

In this subsection, the stability properties of the two-layer control framework with the EMPC of Eq. 6.38 in the upper layer and the LMPC of Eq. 6.39 in the lower layer when applied the system of Eq. 4.1. Before these properties may be presented, several properties are presented that are needed in the analysis. Owing to the fact that $\Omega_{\rho*}$, Γ, $\mathbb{U}$, and $\mathbb{W}$ are compact sets and f is locally Lipschitz, there exists $M_x > 0$ such that

$$|f(e + x_E, u, w)| \leq M_x \tag{6.40}$$

for all $e \in \Omega_{\rho*}$, $x_E \in \Gamma$, $u \in \mathbb{U}$, and $w \in \mathbb{W}$. From similar conditions and since the rate of change of x_E is bounded, there exists $M > 0$ such that

$$|g(e, x_E, \dot{x}_E, u, w)| \leq M \tag{6.41}$$

for all $e \in \Omega_{\rho*}$, $x_E \in \Gamma$, $u \in \mathbb{U}$, $w \in \mathbb{W}$ and $|\dot{x}_E| \leq \gamma_E$. In addition, since the Lyapunov function V is continuously differentiable (in both arguments) and the fact that f is locally Lipschitz, there exist positive constants L_e, L_w, L'_e, L'_E, L''_E, L'_w such that

$$\left|g(e, x_E, \dot{x}_E, u, w) - g(e', x_E, \dot{x}_E, u, 0)\right| \leq L_e|e - e'| + L_w|w|, \tag{6.42}$$

$$\left| \frac{\partial V(e, x_E)}{\partial e} g(e, x_E, \dot{x}_E, u, w) - \frac{\partial V(e', x'_E)}{\partial e} g(e', x'_E, \dot{x}'_E, u, 0) \right| \\ \leq L'_e |e - e'| + L'_E |x_E - x'_E| + L''_E |\dot{x}_E - \dot{x}'_E| + L'_w |w| \tag{6.43}$$

for all $e, e' \in \Omega_{\rho*}$, $x_E, x'_E \in \Gamma$, $u \in \mathbb{U}$, $w \in \mathbb{W}$, $|\dot{x}_E| \leq \gamma_E$, and $|\dot{x}'_E| \leq \gamma_E$.

The following Lemma gives the feasibility properties of the EMPC and therefore, by the constraint of Eq. 6.38f, the optimal state trajectory $x^*_E(t|\hat{t}_k)$ is embedded in the set Γ for $t \in [\hat{t}_k, \hat{t}_{k+1}]$.

Lemma 6.1 *Consider the system of Eq. 6.38b over the prediction horizon. If Assumption 6.2 is satisfied, the optimization problem of Eq. 6.38 is feasible and therefore, the optimal state trajectory $x^*_E(t|\hat{t}_k)$ for $t \in [\hat{t}_k, \hat{t}_k + T_E]$ computed by applying the optimal input trajectory $u^*_E(t|\hat{t}_k)$ defined for $t \in [\hat{t}_k, \hat{t}_k + T_E)$ takes values in the set Γ.*

Proof When the EMPC optimization problem of Eq. 6.38 is solved with an initial condition satisfying $x_E(\hat{t}_k) \in \Gamma$ (this is guaranteed through the projection procedure), the feasibility of the optimization problem follows if Assumption 6.2 is satisfied because maintaining the state at the initial condition along the predicted horizon is a feasible solution to the optimization problem as it satisfies all the constraints, i.e., there exists a constant input trajectory $u_E(t) = \bar{u}_E \in \mathbb{U}$ for $t \in [\hat{t}_k, \hat{t}_k + T_E)$ that maintains the state trajectory at its initial condition: $x_E(t) = \text{proj}_\Gamma(x(\hat{t}_k))$ for $t \in [\hat{t}_k, \hat{t}_k + T_E)$. Owing to the fact that the problem is feasible and imposing the constraint of Eq. 6.38f, the optimal state trajectory $x^*_E(t|\hat{t}_k)$ is bounded in the set Γ for $t \in [\hat{t}_k, \hat{t}_k + T_E]$.

Theorem 6.2 provides sufficient conditions such that the LMPC may track the economically optimal trajectory x^*_E. More specifically, the deviation state gets small over time until it is bounded in a small ball containing the origin.

Theorem 6.2 *Consider the system of Eq. 4.1 in closed-loop under the tracking LMPC of Eq. 6.39 based on the Lyapunov-based controller that satisfies the conditions of Eqs. 6.37a, 6.37d with the reference trajectory x^*_E computed by the upper layer EMPC of Eq. 6.38. Let $\varepsilon_{error} > 0$, $\mu > 0$, $\varepsilon_w > 0$, $\Delta > 0$, $\Delta_E > 0$, $N \geq 1$, $N_E \geq 1$, $N_E \Delta_E \geq t' + N\Delta$, and $\gamma_E > 0$ satisfy*

$$\mu > \alpha_3^{-1}\left(\frac{(L'_x M + L'_E \gamma_E)\Delta + (L''_E + \alpha_5(\alpha_1^{-1}(\rho*)))\gamma_E + L'_w \theta}{\hat{\theta}} \right) \tag{6.44}$$

for some $\hat{\theta} \in (0, 1)$,

$$\varepsilon_{error} = \max_{s \in [0, \Delta]} \{ |e(s)| : e(0) \in B_\mu \textit{ for all } x_E \in \Gamma \}, \tag{6.45}$$

and $B_\mu \subset B_{\varepsilon_{error}} \subset \Omega_{\rho*}$. If $(x(0) - x^*_E(0)) \in \Omega_{\rho*}$, then the deviation state of the system of Eq. 6.36 is always bounded in $\Omega_{\rho*}$ and therefore, also, the closed-loop *state trajectory x is always bounded. Furthermore, the deviation between the state trajectory of Eq. 4.1 and the economically optimal trajectory is ultimately bounded in* $B_{\varepsilon_{error}}$.

Proof The proof consists of two parts. First, the LMPC optimization problem of Eq. 6.39 is shown to be feasible for all deviation states in $\Omega_{\rho *}$. Subsequently, the deviation state is proved to be bounded in $\Omega_{\rho *}$ and to be ultimately bounded in $B_{\varepsilon_{\text{error}}}$.

Part 1: When the deviation state is maintained in $\Omega_{\rho *}$ (which will be proved in Part 2), the feasibility of the LMPC of Eq. 6.39 follows because the input trajectory obtained from the Lyapunov-based controller is a feasible solution. Specifically, define the trajectory v such that:

$$\begin{aligned}\dot{z}(t) &= g(z(t), x_E^*(t|\hat{t}_k), \dot{x}_E^*(t|\hat{t}_k), v(t), 0)\\ v(t) &= h(z(t_i), x_E^*(t_i|\hat{t}_k))\end{aligned}$$

for $t \in [t_i, t_{i+1})$, $i = j, j+1, \ldots, N-1$ where $z(t_j) = e(t_j)$. The trajectory v is a feasible solution to the optimization problem of Eq. 6.39 since the trajectory satisfies the input and the Lyapunov function constraints of Eq. 6.39. This is guaranteed by the closed-loop stability property of the Lyapunov-based controller.

Part 2: At $\hat{t}_k$, the EMPC computes an optimal trajectory $x_E^*(\cdot|\hat{t}_k)$ for the LMPC to track for one operating period. The computed trajectory is such that $x_E^*(t|\hat{t}_k)$ and $|\dot{x}_E^*(t|\hat{t}_k)| \leq \gamma_E$ for all $t \in [\hat{t}_k, \hat{t}_{k+1}]$ (Lemma 6.1). For simplicity of notation, let $x_E(\tau) = x_E^*(\tau|\hat{t}_k)$, $\dot{x}_E(\tau) = \dot{x}_E^*(\tau|\hat{t}_k)$,

$$\frac{\partial V(\tau)}{\partial e} := \frac{\partial V(e(\tau), x_E(\tau))}{\partial e}, \text{ and } \frac{\partial V(\tau)}{\partial x_E} := \frac{\partial V(e(\tau), x_E(\tau))}{\partial x_E} \tag{6.46}$$

for any $\tau \in [t_j, t_{j+1})$. At any sampling time $t_j \in [\hat{t}_k, \hat{t}_k + t')$ of the LMPC, consider $e(t_j) \in \Omega_{\rho *}$ (recursive arguments will be applied to show this is always the case when $e(0) \in \Omega_{\rho *}$). The computed control action at t_j satisfies:

$$\begin{aligned}\frac{\partial V(t_j)}{\partial e} g(e(t_j), x_E(t_j), 0, u^*(t_j|t_j), 0) &\leq \frac{\partial V(t_j)}{\partial e} g(e(t_j), x_E(t_j), 0, h(e(t_j), x_E(t_j)), 0)\\ &\leq -\alpha_3(|e(t_j)|)\end{aligned} \tag{6.47}$$

for all $e(t_j) \in \Omega_{\rho *}$. For all $\tau \in [t_j, t_{j+1})$, the time derivative of the Lyapunov function is given by:

$$\dot{V}(e(\tau), x_E(\tau)) = \frac{\partial V(\tau)}{\partial e}\dot{e}(\tau) + \frac{\partial V(\tau)}{\partial x_E}\dot{x}_E(\tau)\,. \tag{6.48}$$

Adding and subtracting the left-hand term of Eq. 6.47 and from the bound of Eq. 6.47, the time-derivative of the Lyapunov function may be upper bounded as follows:

$$\dot{V}(e(\tau), x_E(\tau)) \leq -\alpha_3(|e(t_j)|) + \frac{\partial V(\tau)}{\partial e} g(e(\tau), x_E(\tau), \dot{x}_E(\tau), u^*(t_j|t_j), w(\tau)) - \frac{\partial V(t_j)}{\partial e} g(e(t_j), x_E(t_j), 0, u^*(t_j|t_j), 0) + \frac{\partial V(\tau)}{\partial x_E} \dot{x}_E(\tau) \quad (6.49)$$

for all $\tau \in [t_j, t_{j+1})$. From Eq. 6.43, the time derivative of the Lyapunov function (Eq. 6.49) may be further upper bounded:

$$\begin{aligned} \dot{V}(e(\tau), x_E(\tau)) &\leq -\alpha_3(|e(t_j)|) + L'_x|e(\tau) - e(t_j)| + L'_E|x_E(\tau) - x_E(t_j)| \\ &\quad + L''_E|\dot{x}_E(\tau)| + L'_w|w(\tau)| + \alpha_5(|e(\tau)|)|\dot{x}_E(\tau)| \\ &\leq -\alpha_3(|e(t_j)|) + L'_x|e(\tau) - e(t_j)| + L'_E|x_E(\tau) - x_E(t_j)| \\ &\quad + (L''_E + \alpha_5(|e(\tau)|))\gamma_E + L'_w\theta \end{aligned} \quad (6.50)$$

for all $e(t_j) \in \Omega_{\rho*}$ and $\tau \in [t_j, t_{j+1})$ where the second inequality follows from the fact that $|\dot{x}_E(\tau)| \leq \gamma_E$ and $w(\tau) \in \mathbb{W}$.

Taking into account Eq. 6.41 and the fact that $|\dot{x}_E(\tau)| \leq \gamma_E$ and the continuity of e and x_E, the following bounds may be derived for all $\tau \in [t_j, t_{j+1})$:

$$\left|e(\tau) - e(t_j)\right| \leq M\Delta \;, \quad (6.51)$$

$$\left|x_E(\tau) - x_E(t_j)\right| \leq \gamma_E\Delta \;. \quad (6.52)$$

From Eqs. 6.50–6.52, the following inequality is obtained:

$$\begin{aligned} \dot{V}(e(\tau), x_E(\tau)) &\leq -\alpha_3(|e(t_j)|) + (L'_x M + L'_E\gamma_E)\Delta \\ &\quad + (L''_E + \alpha_5(|e(\tau)|))\gamma_E + L'_w\theta \end{aligned} \quad (6.53)$$

for all $\tau \in [t_j, t_{j+1})$.

If Δ, γ_E and θ are sufficiently small such that there exist $\hat{\theta} \in (0, 1)$ and $(\mu, \varepsilon_{\text{error}})$ satisfying Eqs. 6.44, 6.45 with $B_\mu \subset B_{\varepsilon_{\text{error}}} \subset \Omega_{\rho*}$, the following bound on the time-derivative of the Lyapunov function follows:

$$\dot{V}(e(\tau), x_E(\tau)) \leq -(1 - \hat{\theta})\alpha_3(|e(t_j)|) \quad (6.54)$$

for all $\tau \in [t_j, t_{j+1})$ and $e(t_j) \in \Omega_{\rho*} \setminus B_\mu$. Integrating this bound on $t \in [t_j, t_{j+1})$, one obtains that:

$$V(e(t_{j+1}), x_E(t_{j+1})) \leq V(e(t_j), x_E(t_j)) - (1 - \hat{\theta})\Delta\alpha_3(|e(t_j)|) \quad (6.55)$$

$$V(e(t), x_E(t)) \leq V(e(t_j), x_E(t_j)) \;\forall\, t \in [t_j, t_{j+1}) \quad (6.56)$$

for all $e(t_j) \in \Omega_{\rho*} \setminus B_\mu$. Using the above inequalities recursively, it may be proved that if $e(t_j) \in \Omega_{\rho*} \setminus B_\mu$, the deviation between the actual state trajectory and the economic optimal trajectory converges to B_μ in a finite number of sampling times without going outside the set $\Omega_{\rho*}$. Since the deviation state is always embedded in the set $\Omega_{\rho*}$ and from Lemma 6.1, the economically optimal state trajectory is always embedded in the set Γ, the boundedness of the closed-loop state trajectory of Eq. 4.1 under the lower layer LMPC follows because $\Omega_{\rho*}$ and Γ are compact sets.

To summarize, if $e(t_j) \in \Omega_{\rho*} \setminus B_\mu$, then

$$V(e(t_{j+1}), x_E(t_{j+1})) < V(e(t_j), x_E(t_j)) \ . \tag{6.57}$$

Furthermore, the deviation between the state trajectory and the economic optimal trajectory is ultimately bounded in $B_{\varepsilon_{\text{error}}}$ where satisfies Eq. 6.45 and $B_\mu \subset B_{\varepsilon_{\text{error}}} \subset \Omega_{\rho*}$. This statement holds because if the deviation state comes out of the ball B_μ, the deviation state is maintained within the ball $B_{\varepsilon_{\text{error}}}$ owing to Eq. 6.45. Once the deviation comes out of the ball B_μ, the Lyapunov function becomes decreasing.

Notes and remarks on results:

- Three factors influences the time-derivative of the Lyapunov function when $e(t_j) \in \Omega_{\rho*} \setminus B_\mu$ as observed in Eq. 6.53: the sampling period of the lower layer LMPC, the bound on the disturbance, and the bound on the rate of change of the economically optimal trajectory. While the bound on the disturbance is a property of the system, two of the other properties may be used to achieve a desired level of tracking: the sampling period of the lower level control loop and the rate of change of the economically optimal tracking trajectory.
- Theorem 6.2 clarifies how the parameter γ_E arises and why it is needed in the formulation of the EMPC of Eq. 6.38.
- No guarantee is made that the closed-loop economic performance with the two-layer framework is better compared to the performance using a steady-state model in the upper layer. In some cases, it may be the case that closed-loop performance is the same or possibly better using a steady-state model in the upper layer EMPC. In this case, the stability result presented here may be extended to the case where the optimal steady-state varies sufficiently slow.

6.3.3 Application to a Chemical Process Example

Consider a well-mixed, non-isothermal continuous stirred tank reactor (CSTR) where an elementary (first-order) reaction takes place of the form $A \rightarrow B$. The feed to the reactor consists of pure A at volumetric flow rate F, temperature $T_0 + \Delta T_0$ and molar concentration $C_{A0} + \Delta C_{A0}$ where ΔT_0 and ΔC_{A0} are disturbances. A jacket around the reactor is used to provide/remove heat to the reactor. The dynamic equations describing the behavior of the system, obtained through material and energy balances under standard modeling assumptions, are given below:

Table 6.2 Process parameters of the CSTR of Eq. 6.58

F	$5.0\,\mathrm{m^3 h^{-1}}$	ΔH	$-1.2 \times 10^4\,\mathrm{kJ\,kmol^{-1}}$
V_R	$1.0\,\mathrm{m^3}$	k_0	$3.0 \times 10^7\,\mathrm{h^{-1}}$
T_0	300 K	E	$5.0 \times 10^4\,\mathrm{kJ\,kmol^{-1}}$
R	$8.314\,\mathrm{kJ\,kmol^{-1}\,K^{-1}}$	ρ_L	$1000\,\mathrm{kg\,m^{-3}}$
C_p	$0.231\ \mathrm{kJ\,kg^{-1}\,K^{-1}}$		

$$\frac{dT}{dt} = \frac{F}{V_R}(T_0 + \Delta T_0 - T) - \frac{\Delta H k_0}{\rho_L C_p} e^{\frac{-E}{RT}} C_A + \frac{Q}{\rho_L C_p V_R} \tag{6.58a}$$

$$\frac{dC_A}{dt} = \frac{F}{V_R}(C_{A0} + \Delta C_{A0} - C_A) - k_0 e^{\frac{-E}{RT}} C_A \tag{6.58b}$$

where C_A is the concentration of the reactant A in the reactor, T is the reactor temperature, Q is the rate of heat input/removal, V_R is the reactor volume, ΔH is the heat of the reaction, k_0 and E are the pre-exponential constant and activation energy of the reaction, respectively, C_p and ρ_L denote the heat capacity and the density of the fluid in the reactor, respectively. The values of the process parameters are given in Table 6.2. The state vector is $x = [T\ C_A]^T$ and the manipulated inputs are the heat rate $u_1 = Q$ where $u_1 \in [-2, 2] \times 10^5\ \mathrm{kJ\,h^{-1}}$ and the inlet reactant concentration $u_2 = C_{A0}$ where $u_2 \in [0.5, 8.0]\,\mathrm{kmol\,m^{-3}}$. The feed disturbances are simulated as bounded Gaussian white noise with zero mean, variances $20\,\mathrm{K}^2$ and $0.1\,\mathrm{kmol^2\,m^{-6}}$, and bounds given by $|\Delta T_0| \leq 15\,\mathrm{K}$ and $|\Delta C_{A0}| \leq 1.0\,\mathrm{kmol\,m^{-3}}$.

The control objective is to force the system to track the economically optimal time-varying operating trajectories computed by the upper layer EMPC. The set Γ is defined as

$$\Gamma := \{x \in \mathbb{R}^2 : 340 \leq x_1 \leq 390\,\mathrm{K},\ 0.5 \leq x_2 \leq 3.0\,\mathrm{kmol\,m^{-3}}\}\,. \tag{6.59}$$

In this example, the time-varying economic stage cost penalizes energy consumption, credits conversion of the reactant to the product, and penalizes the deviation of temperature from 365.0 K and is given by:

$$l_e(t, x, u) = p_1(t)u_1^2 - p_2(t)\frac{(u_2 - x_2)}{u_2} + p_3(t)(x_1 - 365.0\,\mathrm{K})^2 \tag{6.60}$$

where p_1, p_2, and p_3 are the potentially time-varying weighting factors. The last term in the economic stage cost is used to prevent the system from operating on the boundary of Γ for long periods of time. The magnitudes of the economic weighting factors have been chosen so that all terms in the economic cost have the same order of magnitude. For this example, p_1 and p_3 are chosen to be time-varying and $p_2 = 10$ is constant. The time-varying weight $p_1(t)$, over four hours of operation, is given by

$$p_1(t) = \begin{cases} 1.0 \times 10^{-7}, & t < 1.0\,\text{h} \\ 5.0 \times 10^{-8}, & 1.0\,\text{h} \leq t < 2.0\,\text{h} \\ 1.0 \times 10^{-8}, & 2.0\,\text{h} \leq t < 3.0\,\text{h} \\ 5.0 \times 10^{-8}, & 3.0\,\text{h} \leq t \leq 4.0\,\text{h} \end{cases}$$

and is used to model the time-varying energy cost. The time-varying weight $p_3(t)$ is given by

$$p_3(t) = \begin{cases} 1.0 \times 10^{-2}, & t < 1.0\,\text{h} \\ 7.5 \times 10^{-3}, & 1.0\,\text{h} \leq t < 2.0\,\text{h} \\ 5.0 \times 10^{-3}, & 2.0\,\text{h} \leq t < 3.0\,\text{h} \\ 7.5 \times 10^{-3}, & 3.0\,\text{h} \leq t \leq 4.0\,\text{h} \end{cases}$$

The rationale for varying p_3 is to allow the CSTR be operated over a larger temperature range when the energy cost decreases and thus, take advantage of the decreased energy cost.

The upper layer EMPC is implemented with a sampling period of $\Delta_E = 36s$ and prediction horizon of $N_E = 60$ sampling periods. It is solved every 0.50 h, i.e., the operating period is chosen to be $t' = 0.50\,\text{h}$. The prediction horizon and operating period are chosen to account for the update frequency of the economic weighting parameters. It is found that defining and imposing a rate of change constraint in the upper layer EMPC, i.e., defining the parameter γ_E, is not needed for this particular example because the closed-loop system under the lower layer LMPC is able to achieve acceptable tracking performance without imposing a rate of change constraint in the upper layer EMPC. The projection operator is such that it projects the current state to the closest boundary of Γ if the current state is outside the set Γ, e.g., if $x = [400\,\text{K}\ 2.0\,\text{kmol m}^{-3}]^T$, then $\text{proj}_\Gamma(x) = [390\,\text{K}\ 2.0\,\text{kmol m}^{-3}]^T$.

To design the lower layer LMPC, a Lyapunov-based controller is designed for the CSTR, which is essentially two proportional controllers that account for the input constraints. Specifically, the two proportional controllers are given by:

$$\begin{aligned} &-K_1(x_1 - x^*_{E,1}) + u_{s,1}, \\ &-K_2(x_2 - x^*_{E,2}) + u_{s,2} \end{aligned} \tag{6.61}$$

where $K_1 = 8000$, $K_2 = 0.01$, and u_s is the steady-state input corresponding to the steady-state x^*_E, i.e., the input vector that makes the right-hand side of Eqs. 6.58a, 6.58b equal to zero with the state vector x^*_E. The resulting Lyapunov-based controller design for the CSTR is derived by accounting for the input constraints in the controller design of Eq. 6.61 as well as for the fact that u_s may be written as a function of x_E, i.e., the resulting Lyapunov-based controller is a mapping h that maps the pair (e, x_E) to $h(e, x_E) \in \mathbb{U}$. A quadratic Lyapunov function of the form $V(e, x_E) = e^T Pe$ is constructed for the closed-loop system under the Lyapunov-based controller with

$$P = \begin{bmatrix} 10 & 1 \\ 1 & 100 \end{bmatrix}. \tag{6.62}$$

The LMPC is implemented with a sampling time $\Delta = 36s$, prediction horizon $N = 5$, and weighting matrices of $Q_c = P$ and $R_c = \text{diag}\left[\, 10^{-7} \;\; 10 \,\right]$. The prediction horizon and weighting matrices of the lower layer LMPC are tuned to achieve a close tracking of the optimal state trajectory.

With the nonlinear system of Eqs. 6.58a–6.58b, the Lyapunov-based controller, and the Lyapunov function, the stability regions of the closed-system under the Lyapunov-based controller may be estimated for a sufficiently large number of points in Γ. This procedure is carried out as follows: fix $x_E \in \Gamma$ and compute a level set of the Lyapunov function where $\dot{V} < 0$ for all points contained in the level set. The intersection of all these level sets is taken to be an estimate of the closed-loop stability region $\Omega_{\rho*}$ of the CSTR under the Lyapunov-based controller. In this example, $\Omega_{\rho*}$ is estimated to be $\rho^* = 110$. Through the Lyapunov-based constraint on the LMPC of (Eq. 6.39e), the closed-loop system with the two-layer framework inherits the stability region Ω_{ρ^*}.

To simulate the closed-loop system, explicit Euler method with integration step $0.36s$ is used to integrate the ODEs and the open source interior point solver Ipopt [17] is used to solve the optimization problems. Three sets of closed-loop simulations are completed. In the first set of closed-loop simulations, the stability properties of the closed-loop system under the two-layer dynamic economic optimization and control framework are demonstrated. Second, time-varying operation with the two-layer dynamic economic optimization and control framework is analyzed. Third, the closed-loop economic performance of the CSTR under the two-layer framework is compared to the CSTR under a conventional approach to optimization and control.

To demonstrate the closed-loop stability properties of the proposed two-layer framework, the CSTR is initialized at $x_0 = [400\,\text{K},\;\; 0.1\,\text{kmol}\,\text{m}^{-3}]$ which is outside of Γ, but inside the stability region $\Omega_{\rho*}$. The projection operator of the upper layer EMPC projects the initial state onto the state $x_{E,0} = [390\,\text{K},\;\; 0.5\,\text{kmol}\,\text{m}^{-3}] \in \Gamma$ to use as an initial condition to the upper layer EMPC problem of Eq. 6.38. The evolution of the closed-loop system under the two-layer framework and with the inlet temperature and reactant concentration disturbance is shown in Figs. 6.8 and 6.9. From Fig. 6.9, the deviation of the actual closed-loop state and the economically optimal state is always maintained inside $\Omega_{\rho*}$. Moreover, the deviation becomes small over time until it is ultimately bounded in a small ball.

Two simulations of the closed-loop system without the feed disturbances added are shown in Figs. 6.10 and 6.11 with two different initial conditions to analyze the time-varying operation with the two-layer dynamic economic optimization and process control framework. The initial state in Fig. 6.10 is $x_0 = [400\,\text{K},\;\; 3.0\,\text{kmol}\,\text{m}^{-3}]^T$, while the initial state in Fig. 6.11 is $x_0 = [320\,\text{K},\;\; 3.0\,\text{kmol}\,\text{m}^{-3}]^T$. The closed-loop evolution of the two cases is initially different. For the CSTR starting at the larger temperature, heat should be removed from the reactor and the minimum amount of reactant material should be supplied to the reactor to decrease the temperature of the reactor. In contrast, when the CSTR is initialized at the smaller temperature, heat should be supplied to the reactor and reactant material should be fed to the reactor to increase the reactor temperature. After a sufficiently long length of operation, the effect of the initial condition diminishes and the closed-loop evolution of the two

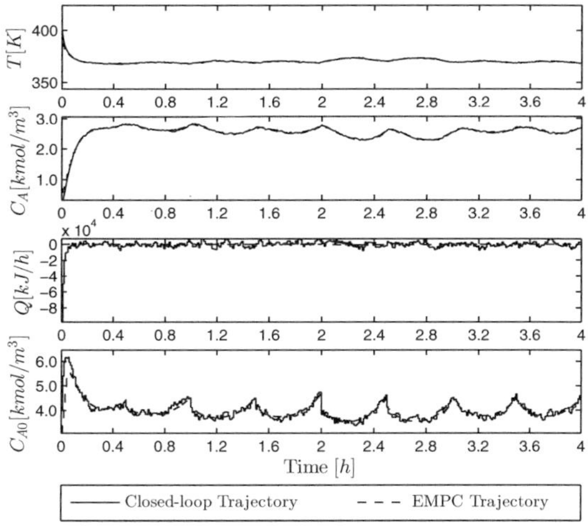

Fig. 6.8 The closed-loop state and input trajectories of Eqs. 6.58a, 6.58b under the two-layer optimization and control framework with the feed disturbances and starting from 400 K and 0.1 kmol m^{-3}

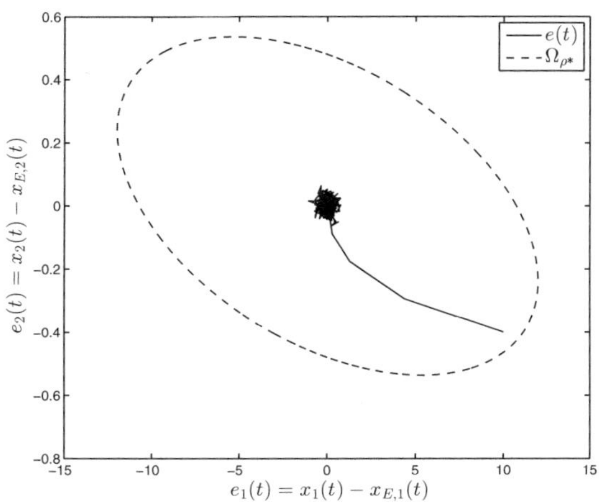

Fig. 6.9 The closed-loop state trajectory of Eqs. 6.58a, 6.58b under the two-layer optimization and control framework with the feed disturbances and starting from 400 K and 0.1 kmol m^{-3} shown in deviation state-space

cases becomes similar. For both of these cases, the reactor is operated in a time-varying fashion, i.e., never converges to a steady-state.

To compare the closed-loop economic performance under the dynamic economic optimization and control framework and under a conventional approach to optimization and control, the total economic cost over the length of operation is defined as

$$J_E = \sum_{j=0}^{M-1} \left(p_1(t_j) Q^2(t_j) + p_2 \frac{C_A(t_j)}{C_{A0}(t_j)} + p_3(t_j)(T(t_j) - 365)^2 \right) \tag{6.63}$$

where t_0 is the initial time of the simulation and $t_M = 4.0$ h is the end of the simulation. The conventional approach to optimization and control uses a steady-state economic

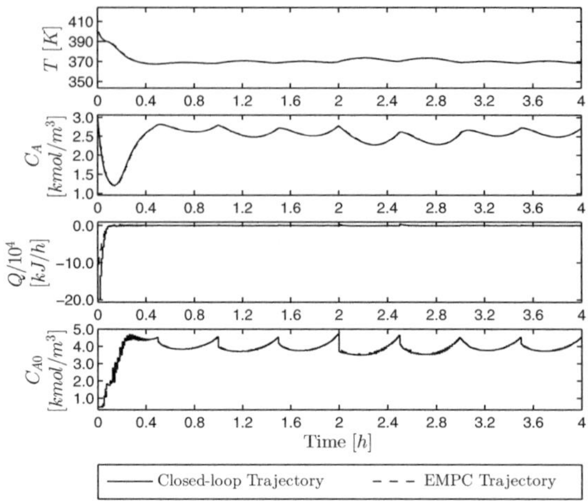

Fig. 6.10 The closed-loop system states and inputs of Eqs. 6.58a, 6.58b without the feed disturbances and starting from 400 K and 3.0kmol m^{-3}

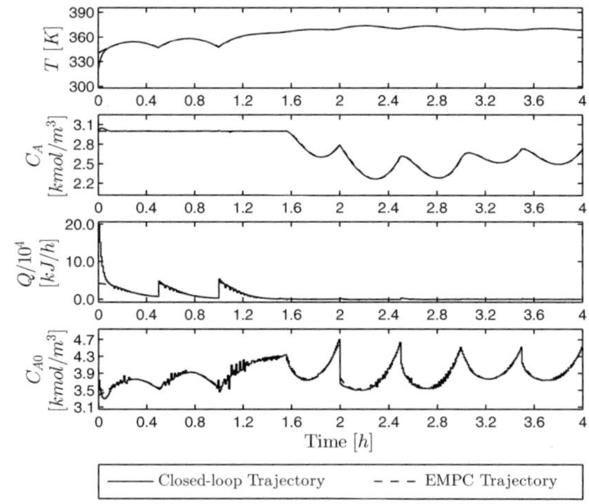

Fig. 6.11 The closed-loop system states and inputs of Eqs. 6.58a, 6.58b without the feed disturbances and starting from 320 K and 3.0 kmol m^{-3}

optimization problem to compute the optimal steady-states with respect to the time-varying economic cost weights. The optimal steady-states are used in a tracking MPC, which in this case is an LMPC, to force the CSTR states to the optimal steady-states. The optimal (time-varying) steady-state from steady-state economic optimization is

$$x_s^*(t) = \begin{cases} [370.0\,\text{K},\ 2.576\,\text{kmol}\,\text{m}^{-3}]^T, & t < 1.0\,\text{h} \\ [371.7\,\text{K},\ 2.447\,\text{kmol}\,\text{m}^{-3}]^T, & 1.0\,\text{h} \leq t < 2.0\,\text{h} \\ [375.2\,\text{K},\ 2.205\,\text{kmol}\,\text{m}^{-3}]^T, & 2.0\,\text{h} \leq t < 3.0\,\text{h} \\ [371.7\,\text{K},\ 2.447\,\text{kmol}\,\text{m}^{-3}]^T, & 3.0\,\text{h} \leq t \leq 4.0\,\text{h} \end{cases}$$

with the corresponding steady-state input of

$$u_s^*(t) = \begin{cases} [0.0\,\mathrm{kJ\,h^{-1}},\ 3.923\,\mathrm{kmol\,m^{-3}}]^T, & t < 1.0\,\mathrm{h} \\ [-0.5\,\mathrm{kJ\,h^{-1}},\ 3.827\,\mathrm{kmol\,m^{-3}}]^T, & 1.0\,\mathrm{h} \le t < 2.0\,\mathrm{h} \\ [0.0\,\mathrm{kJ\,h^{-1}},\ 3.653\,\mathrm{kmol\,m^{-3}}]^T, & 2.0\,\mathrm{h} \le t < 3.0\,\mathrm{h} \\ [-0.5\,\mathrm{kJ\,h^{-1}},\ 3.827\,\mathrm{kmol\,m^{-3}}]^T, & 3.0\,\mathrm{h} \le t \le 4.0\,\mathrm{h} \end{cases}$$

An LMPC is implemented to drive the system to the time-varying optimal steady-state, which is formulated as follows:

$$\begin{aligned} \min_{u \in S(\Delta)} \quad & \int_{t_j}^{t_{j+N}} \left(\left|\tilde{x}(\tau) - x_s^*(\tau)\right|_{Q_c} + \left|u(\tau) - u_s^*(\tau)\right|_{R_c} \right) d\tau \\ \text{s.t.} \quad & \dot{\tilde{x}}(t) = f(\tilde{x}(t), u(t), 0), \\ & \tilde{x}(t_j) = x(t_j), \\ & -2 \times 10^5 \le u_1(t) \le 2 \times 10^5,\ \forall\, t \in [t_j, t_{j+N}), \\ & 0.5 \le u_2(t) \le 8,\ \forall\, t \in [t_j, t_{j+N}), \\ & \frac{\partial V(x(t_j))}{\partial x} f(x(t_j), u(t_j), 0) \\ & \quad \le \frac{\partial V(x(t_j))}{\partial x} f(x(t_j), h(x(t_j), x_s^*(t_j)), 0) \end{aligned} \tag{6.64}$$

where the Lyapunov function, the Lyapunov-based controller, the weighting matrices R_c and Q_c, the sampling period Δ, and the prediction horizon N are all the same as the ones used in the tracking LMPC scheme.

To make a fair comparison, the same realization of the feed disturbances was applied to each closed-loop system simulation pair. The total economic cost values of several closed-loop simulations starting from different initial conditions and with and without the feed disturbances are given in Table 6.3. From the results of Table 6.3, substantial closed-loop economic performance is achieved under the two-layer framework than under the optimal steady-state tracking LMPC of Eq. 6.64. The largest economic cost improvement occurs when the CSTR is initialized at higher temperature. When the CSTR starts from a lower temperature, the amount of heat that needs to be supplied to the reactor initially is less than the amount of heat that needs to be initially removed when the CSTR starts at a higher temperature as explained above and demonstrated in Figs. 6.10 and 6.11. Thus, when the CSTR starts from a higher temperature, better closed-loop performance is achieved because less energy is required to be supplied/removed from the reactor.

Table 6.3 Comparison of the total economic cost, given by Eq. 6.63, of the closed-loop system with and without the feed disturbances for four hours of operation

Initial conditions		Total economic cost					
$T(0)$ K	$C_A(0)$ kmol m^{-3}	Steady-state optimization without disturbance	Two-layer Framework without Disturbance	Cost decrease (%)	Steady-state optimization with disturbance	Two-layer framework with disturbance	Cost decrease (%)
400.0	3.0	21970.5	14531.1	51.2	21642.4	14130.7	53.2
380.0	3.0	5235.4	3409.5	53.6	5060.1	3037.9	66.6
360.0	3.0	4261.8	3308.6	28.8	4083.2	2997.1	36.2
340.0	3.0	13732.2	10997.3	24.9	13554.9	10882.3	24.6
320.0	3.0	23719.4	19315.9	22.8	23729.1	19210.3	23.5
400.0	2.5	18546.8	10062.1	84.3	18283.4	9691.4	88.7
380.0	2.5	4558.7	3163.3	44.1	4387.2	2811.9	56.0
360.0	2.5	4496.4	3335.6	34.8	4322.7	3030.3	42.6
340.0	2.5	14078.3	11034.4	27.6	13910.2	10928.8	27.3
320.0	2.5	24052.2	19293.4	24.7	24002.2	19193.8	25.1
400.0	2.0	14831.5	6774.0	118.9	14682.4	6412.6	129.0
380.0	2.0	4073.2	3085.1	32.0	3905.0	2739.8	42.5
360.0	2.0	4765.4	3431.2	38.9	4596.4	3139.3	46.4
340.0	2.0	14395.5	11162.3	29.0	14236.8	11068.2	28.6
320.0	2.0	24202.7	19241.2	25.8	24223.5	19146.7	26.5
400.0	0.1	8146.1	4360.5	86.8	7999.4	4025.7	98.7

6.4 Addressing Closed-Loop Performance

An important theoretical consideration is the closed-loop performance of systems under EMPC because EMPC is formulated with a finite prediction horizon. The achievable closed-loop economic performance may strongly depend on the prediction horizon length. To address guaranteed closed-loop economic performance while formulating a computationally efficient control structure, a two-layer EMPC structure is presented in this section. In contrast to the two-layer EMPC methodologies presented in the previous sections, EMPC schemes are used in both layers of the two-layer EMPC structure to ensure economic performance improvement over a tracking controller, e.g., tracking MPC.

Each layer is formulated as an LEMPC scheme. The core idea of the two-layer LEMPC implementation is to solve the upper layer LEMPC infrequently (not every sampling period) while employing a long prediction horizon. Then, the solution generated by the upper layer LEMPC is subsequently used in the formulation of a lower layer LEMPC. The lower layer LEMPC is formulated with a shorter prediction horizon and smaller sampling time than the upper layer LEMPC and computes control actions that are applied to the closed-loop system. The control actions of the lower layer LEMPC are constrained to maintain the state near the economically optimal trajectories computed in the upper layer. For guaranteed performance improvement with the two-layer LEMPC implementation scheme, both layers are formulated with explicit performance-based constraints computed by taking advantage of the availability of an auxiliary stabilizing controller. The performance-based constraints, i.e., terminal constraints, are similar to that presented in Sect. 4.4, and guarantee that both the finite-time and infinite-time closed-loop economic performance under the two-layer LEMPC scheme are at least as good as that under the stabilizing controller. The use of the two-layer control implementation allows for the control architecture to be computationally efficient. The two-layer LEMPC structure is applied to a chemical process example to demonstrate the closed-loop performance, stability, and robustness properties of the two-layer LEMPC structure.

6.4.1 Class of Systems

In this section, nominally operated systems are considered, i.e., the system of Eq. 4.1 with $w \equiv 0$. Specifically, the class of continuous-time nonlinear systems considered is described by the following state-space form:

$$\dot{x} = f(x, u) \tag{6.65}$$

where the state vector is $x \in \mathbb{X} \subseteq \mathbb{R}^n$ and the input vector is $u \in \mathbb{U} \subset \mathbb{R}^m$. The vector function $f : \mathbb{X} \times \mathbb{U} \to \mathbb{X}$ is a locally Lipschitz vector function on $\mathbb{X} \times \mathbb{U}$. The set of admissible inputs $\mathbb{U}$ is assumed to be a compact set, and the state is synchronously sampled at time instances $j\Delta$ with $j = 0, 1, 2, \ldots$ where $\Delta > 0$ is the sampling period. As before, the initial time is taken to be zero, and the notation t will be used for the continuous-time, while the time sequence $\{t_j\}_{j \geq 0}$ is the discrete sampling time sequence which is a synchronous partitioning of $\mathbb{R}_+$ with $t_j = j\Delta$.

A time-invariant economic measure $l_e : \mathbb{X} \times \mathbb{U} \to \mathbb{R}$ is assumed for the system of Eq. 6.3 that describes the real-time system economics. The economic measure is assumed to be continuous on $\mathbb{X} \times \mathbb{U}$. The optimal steady-state x_s^* and steady-state input u_s^* pair with respect to the economic cost function is computed as follows:

$$(x_s^*, u_s^*) = \underset{x_s \in \mathbb{X}, u_s \in \mathbb{U}}{\arg\min} \; \{l_e(x_s, u_s) : f(x_s, u_s) = 0\} \; .$$

The existence of a minimizing pair where the minimum is attained. For the sake of simplicity, the optimal steady-state pair is assumed to be unique and to be $(x_s^*, u_s^*) = (0, 0)$.

6.4.2 Stabilizability Assumption

A stronger stabilizability-like assumption than the assumption imposed in previous sections and chapters is needed here (stated in Assumption 6.3). In this section, the existence of a stabilizing controller that renders the origin of the closed-loop system exponentially stable under continuous implementation is assumed whereas, previously, the existence of a stabilizing controller is assumed that renders the closed-loop system only asymptotically stable under continuous implementation. The stronger assumption is needed to ensure that the stabilizing controller renders the origin of the closed-loop system exponentially (and therefore, asymptotically) stable under sample-and-hold implementation. This will be required to consider infinite-time closed-loop economic performance. Specifically, asymptotic convergence to the origin and not just convergence to a neighborhood of the steady-state (practical stability of the origin) will be required.

Assumption 6.3 There exists a locally Lipschitz feedback controller $h : \mathbb{X} \to \mathbb{U}$ with $h(0) = 0$ for the system of Eq. 6.65 that renders the origin of the closed-loop system under continuous implementation of the controller locally exponentially stable. More specifically, there exist constants $\rho > 0$, $c_i > 0$, $i = 1, 2, 3, 4$ and a continuously differentiable Lyapunov function $V : \mathbb{X} \to \mathbb{R}_+$ such that the following inequalities hold:

$$c_1 |x|^2 \leq V(x) \leq c_2 |x|^2 , \tag{6.66a}$$

$$\frac{\partial V(x)}{\partial x} f(x, h(x)) \leq -c_3 |x|^2 , \tag{6.66b}$$

$$\left| \frac{\partial V(x)}{\partial x} \right| \leq c_4 |x| , \tag{6.66c}$$

for all $x \in \Omega_\rho \subseteq \mathbb{X}$.

Explicit feedback controllers that may be designed to satisfy Assumption 6.3 are, for example, feedback linearizing controller and some Lyapunov-based controllers, e.g., [15, 19]. The origin of the closed-loop system of Eq. 6.65 under the feedback controller, $h(x)$, implemented in a zeroth-order sample-and-hold fashion with a sufficiently small sampling period $\Delta > 0$, i.e., the controller is applied as an emulation controller may be shown to be exponentially stable (Corollary 2.2). Moreover, the proof of Corollary 2.2 shows that V is a Lyapunov function for the closed-loop sampled-data system in the sense that there exists a constant $\hat{c}_3 > 0$ such that

$$\frac{\partial V(x(t))}{\partial x} f(x(t), h(x(t_j))) \leq -\hat{c}_3 |x(t)|^2 \tag{6.67}$$

for all $t \in [t_j, t_{j+1})$ and integers $j \geq 0$ where $x(t)$ is the solution of Eq. 6.65 at time t starting from $x(t_j) \in \Omega_\rho$ and with the input $u(t) = h(x(t_j))$ for $t \in [t_j, t_{j+1})$. The stability region of the closed-loop system under the controller is defined as $\Omega_\rho \subseteq X$.

6.4.3 *Two-Layer EMPC Structure*

A detailed description of the two-layer LEMPC structure is provided which includes descriptions of the implementation strategy, the formulations of the upper and lower layer LEMPC schemes, and the provable stability and performance properties.

6.4.3.1 Implementation Strategy

The objective of the two-layer LEMPC design is to ensure that both the finite-time and infinite-time closed-loop economic performance of the resulting closed-loop system will be at least as good the closed-loop performance under a stabilizing controller. To address this objective, performance-based constraints are employed in the formulation of the upper and lower layer LEMPC that have been computed from the stabilizing controller. The stabilizing controller may be any controller that satisfies Assumption 6.3. For example, the stabilizing controller may be an explicit controller that satisfies Assumption 6.3 or an LMPC scheme, which is equipped with a contractive Lyapunov constraint designed using an explicit controller that satisfies Assumption 6.3. The formulation of such an LMPC scheme is provided below. However, it is important to point out that the amount of computation required to solve the LMPC is generally greater than that required for an explicit controller. The stabilizing controller will be referred as the auxiliary controller for the remainder.

A block diagram of the two-layer LEMPC is given in Fig. 6.12. In the upper layer, an LEMPC is used to optimize dynamic operation over a long horizon while accounting for the performance-based constraints generated from the auxiliary controller. Both the auxiliary controller and the upper layer LEMPC compute their input trajectories at the beginning of some operating window, and thus, the auxiliary controller and upper layer LEMPC are computed once each operating window for computational efficiency. In the lower layer, an LEMPC, using a shorter prediction horizon

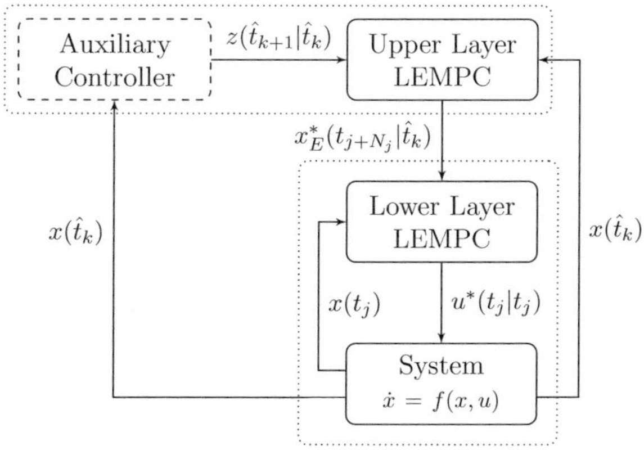

Fig. 6.12 Block diagram of the two-layer EMPC structure addressing closed-loop performance and computational efficiency

and a smaller sampling period than the upper layer LEMPC, computes control inputs that are applied to the process. Terminal constraints that have been generated from the upper layer LEMPC optimal solution are used to ensure that the lower layer LEMPC guides the system along the optimal solution computed in the upper layer since it uses a shorter prediction horizon and a smaller sampling period. In this manner, the lower layer LEMPC is used to ensure robustness of the closed-loop system (recomputes its optimal trajectory every sampling period to incorporate feedback). The lower layer LEMPC may also provide additional economic cost improvement over the upper layer LEMPC solution owing to the use of a smaller sampling time.

To maintain consistency of the notation, the operating window is denoted as t' and is equal to $N_E \Delta_E$ where $N_E \in \mathbb{I}_+$ is the number of hold periods in the prediction horizon of the upper layer LEMPC and $\Delta_E > 0$ is the hold period of the piecewise constant input trajectory computed in the upper layer (here, $K_E = N_E$). The time sequence $\{\hat{t}_k\}_{k \geq 0}$ denotes the discrete time steps that the upper layer computes a solution to its control problem where $\hat{t}_k = kt'$ and $k = 0, 1, \ldots$.

At the beginning of each operating window, the upper layer control problems are solved in a sequential manner: first, the auxiliary controller is solved to obtain its corresponding open-loop predicted state and input trajectories over the operating window and second, the upper layer LEMPC is solved to obtain its corresponding open-loop predicted state and input trajectories over the operation window. Specifically, the auxiliary controller computes the open-loop input trajectory that it would apply to the system over the time $\hat{t}_k$ to $\hat{t}_{k+1} = (k+1)t'$ along with the open-loop state trajectory If the auxiliary controller is an explicit controller, then the open-loop state trajectory is computed by recursively solving:

$$\dot{z}(t) = f(z(t), h(z(\tau_i))) \tag{6.68}$$

for $t \in [\tau_i, \tau_{i+1})$, $i = 0, 1, \ldots, N_E - 1$ where $\tau_i := \hat{t}_k + i\Delta_E$, $z(\hat{t}_k) = x(\hat{t}_k)$ is the initial condition, and $x(\hat{t}_k)$ is a state measurement obtained at $\hat{t}_k$. If, instead, the auxiliary controller is an LMPC, then the open-loop state trajectory may be obtained directly from the solution of the optimization problem. The open-loop state and input trajectories under the auxiliary controller are denoted as $z(t|\hat{t}_k)$ and $v(t|\hat{t}_k)$ for $t \in [\hat{t}_k, \hat{t}_{k+1}) = [kt', kt' + N_E\Delta_E)$, respectively. The terminal state of the open-loop state trajectory, $z(\hat{t}_{k+1}|\hat{t}_k)$, is then sent to the upper level LEMPC.

The upper layer LEMPC subsequently uses $z(\hat{t}_{k+1}|\hat{t}_k)$ as a terminal equality constraint in the optimization problem. In this framework, no restrictions are placed on the type of operation achieved under the two-layer framework, i.e., it could be steady-state operation or some more general time-varying operating behavior. Therefore, the upper level LEMPC is an LEMPC (Sect. 4.2) equipped with mode 1 operation only. If steady-state operation is desirable, one could formulate the upper level LEMPC with a mode 2 constraint similar to that of Eq. 6.8f to ensure that the optimal steady-state is asymptotically stable under the two-layer LEMPC. However, the mode 2 constraint of LEMPC is not discussed further. After receiving $z(\hat{t}_{k+1}|\hat{t}_k)$ from the auxiliary controller and a state measurement at $\hat{t}_k$, the upper layer LEMPC is solved to compute its optimal state and input trajectories over the operating window, which are denoted as $x_E^*(t|\hat{t}_k)$ and $u_E^*(t|\hat{t}_k)$ for $t \in [\hat{t}_k, \hat{t}_{k+1})$, respectively.

The upper layer hold period is divided into $\bar{N}$ subintervals of length Δ ($\Delta = \Delta_E/\bar{N}$ where $\bar{N}$ is a positive integer). The subintervals define the sampling period of the lower layer LEMPC and correspond to the sampling time sequence $\{t_j\}_{j\geq 0}$. The lower layer LEMPC recomputes its optimal input trajectory employing a shrinking horizon. Namely, at the beginning of each hold period of the upper layer, the lower layer is initialized with a prediction horizon $N_j = \bar{N}$. The lower layer LEMPC receives a state measurement, denoted as $x(t_j)$, as well as $x_E^*(t_{j+N_j}|\hat{t}_k)$ from the upper layer LEMPC. Using $x_E^*(t_{j+N_j}|\hat{t}_k)$ as a terminal equality constraint in the lower layer LEMPC, the lower layer LEMPC is solved. The optimal input trajectory computed by the lower layer LEMPC is denoted as $u^*(t|t_j)$, $t \in [t_j, t_{j+N_j})$. At the subsequent sampling period of the lower layer LEMPC, the prediction horizon decreases by one ($N_{j+1} = N_j - 1$). If decreasing the horizon results in the horizon being set to zero, the prediction horizon is reset to $\bar{N} = \Delta_E/\Delta$. This happens at the beginning of the next hold period of the upper layer LEMPC.

The implementation strategy is summarized below and an illustration of the closed-loop system is given in Fig. 6.13. The lower layer LMPC is initialized with a prediction horizon of $N_0 = \bar{N} = \Delta_E/\Delta$. To initialize the algorithm, let $k = 0$ and $j = 0$.

1. **Upper layer**: At $\hat{t}_k$, the auxiliary controller and the upper layer LEMPC are initialized with the state measurement $x(\hat{t}_k)$. Go to Step 1.1.

   1.1 The auxiliary controller computes its optimal input trajectory denoted as $v(t|\hat{t}_k)$ defined for $t \in [\hat{t}_k, \hat{t}_{k+1})$ and corresponding state trajectory denoted as $z(t|\hat{t}_k)$ defined for $t \in [\hat{t}_k, \hat{t}_{k+1})$. The terminal state $z(\hat{t}_{k+1}|\hat{t}_k)$ is sent to the upper layer LEMPC. Go to Step 1.2.

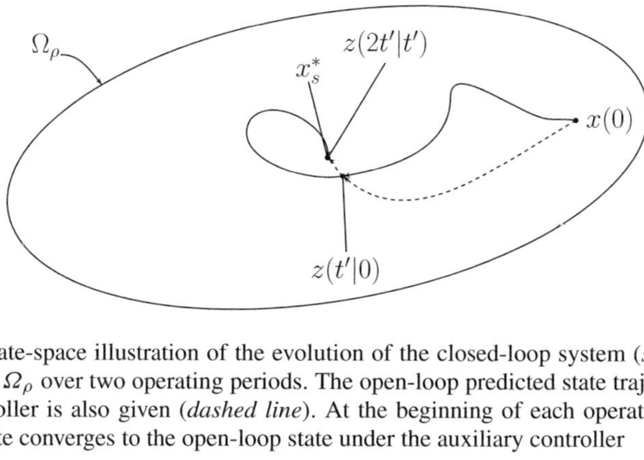

Fig. 6.13 A state-space illustration of the evolution of the closed-loop system (*solid line*) in the stability region Ω_ρ over two operating periods. The open-loop predicted state trajectory under the auxiliary controller is also given (*dashed line*). At the beginning of each operating window, the closed-loop state converges to the open-loop state under the auxiliary controller

1.2 The upper layer LEMPC receives $z(\hat{t}_{k+1}|\hat{t}_k)$ from the auxiliary controller and computes its optimal input trajectory $u_E^*(t|\hat{t}_k)$ defined for $t \in [\hat{t}_k, \hat{t}_{k+1})$ and state trajectory $x_E^*(t|\hat{t}_k)$ defined for $t \in [\hat{t}_k, \hat{t}_{k+1})$. Go to Step 2.

2. **Lower layer**: At t_j, the lower layer LEMPC receives a state measurement $x(t_j)$ and the terminal state $x_E^*(t_{j+N_j}|\hat{t}_k)$ from the upper layer LEMPC. Go to Step 2.1.

   2.1 The lower layer LEMPC computes its optimal input trajectory $u^*(t|t_j)$ defined for $t \in [t_j, t_{j+N_j})$. Go to Step 2.2.

   2.2 The control action $u^*(t_j|t_j)$, which is the computed input for the first sampling period of the lower layer LEMPC prediction horizon, is applied to the system from t_j to t_{j+1}. If $N_j - 1 = 0$, reset $N_{j+1} = \bar{N}$; else, let $N_{j+1} = N_j - 1$. If $t_{j+1} = \hat{t}_{k+1}$, set $j \leftarrow j + 1$ and $k \leftarrow k + 1$ and go to Step 1. Else, set $j \leftarrow j + 1$ and go to Step 2.

Remark 6.4 Even though the lower layer LEMPC uses a shrinking horizon and nominal operation is considered, recomputing the lower layer LEMPC input at every subsequent sampling time is necessary regardless if the solution to the lower level LEMPC is the same or not. The incorporation of feedback allows for stabilization of open-loop unstable systems that cannot be accomplished with an open-loop implementation and ensures the robustness of the control solution with respect to infinitesimally small disturbances/uncertainty. For further explanation on this point, see, for example, [20].

6.4.3.2 Formulation

The formulations of the two LEMPC schemes are given. For convenience, the specific formulation of the LMPC needed if the auxiliary controller is chosen to be an LMPC scheme is given first. Specifically, the LMPC is given by the following optimization problem:

$$\min_{v \in S(\Delta_E)} \int_{\hat{t}_k}^{\hat{t}_{k+1}} (|z(t)|_{Q_c} + |v(t)|_{R_c})\, dt \tag{6.69a}$$

$$\text{s.t.} \quad \dot{z}(t) = f(z(t), v(t)) \tag{6.69b}$$

$$z(\hat{t}_k) = x(\hat{t}_k) \tag{6.69c}$$

$$v(t) \in \mathbb{U}, \ \forall\, t \in [\hat{t}_k, \hat{t}_{k+1}) \tag{6.69d}$$

$$\frac{\partial V(z(\tau_i))}{\partial z} f(z(\tau_i), v(\tau_i)) \le \frac{\partial V(z(\tau_i))}{\partial z} f(z(\tau_i), h(x(\tau_i)))$$
$$\text{for } i = 0, 1, \ldots, N_E - 1 \tag{6.69e}$$

where $\tau_i := \hat{t}_k + i\Delta_E$ and z is the state trajectory of the system with input trajectory v calculated by the LMPC. The Lyapunov-based constraint of Eq. 6.69e differs from the Lyapunov-based constraint of Eq. 2.51e as it is imposed at each sampling period along the prediction horizon of the LMPC to ensure that the state trajectory with input computed by the LMPC converges to the steady-state. Through enforcement of the Lyapunov-based constraint, the LMPC inherits the same stability properties as that of the explicit controller. The optimal solution of the optimization problem of Eq. 6.69 is denoted as $v^*(t|\hat{t}_k)$ and is defined for $t \in [\hat{t}_k, \hat{t}_{k+1})$. From the optimal input trajectory, the optimal state trajectory $z^*(t|\hat{t}_k)$, $t \in [\hat{t}_k, \hat{t}_{k+1})$ may be computed for the operating window. When the LMPC is used as the auxiliary controller, the terminal state $z^*(\hat{t}_{k+1}|\hat{t}_k)$ is sent to the upper layer LEMPC.

The formulation of the upper layer LEMPC is similar to the mode 1 LEMPC formulation with a terminal equality constraint computed from the auxiliary controller:

$$\min_{u_E \in S(\Delta_E)} \int_{\hat{t}_k}^{\hat{t}_{k+1}} l_e(x_E(t), u_E(t))\, dt \tag{6.70a}$$

$$\text{s.t.}\quad \dot{x}_E(t) = f(x_E(t), u_E(t)) \tag{6.70b}$$

$$x_E(\hat{t}_k) = x(\hat{t}_k) \tag{6.70c}$$

$$u_E(t) \in \mathbb{U},\ \forall\, t \in [\hat{t}_k, \hat{t}_{k+1}) \tag{6.70d}$$

$$x_E(t) \in \Omega_\rho,\ \forall\, t \in [\hat{t}_k, \hat{t}_{k+1}) \tag{6.70e}$$

$$x_E(\hat{t}_{k+1}) = z(\hat{t}_{k+1}|\hat{t}_k) \tag{6.70f}$$

where x_E is the predicted state trajectory with the input trajectory u_E computed by the upper layer LEMPC. To ensure the existence of an input trajectory that has at least as good economic performance as the auxiliary LMPC input trajectory over the entire length of operation, the terminal constraint of Eq. 6.70f based on the auxiliary controller is used. The terminal constraint differs from traditional terminal equality constraints because $z(\hat{t}_{k+1}|\hat{t}_k)$ is not necessarily the steady-state. It does, however, asymptotically converge to the economically optimal steady-state. The optimal solution to the optimization problem of the upper layer LEMPC is denoted as $u_E^*(t|\hat{t}_k)$ and is defined for $t \in [\hat{t}_k, \hat{t}_{k+1})$. With the optimal solution, the optimal (open-loop) state trajectory may be computed and is denoted as $x_E^*(t|\hat{t}_k)$, for $t \in [\hat{t}_k, \hat{t}_{k+1})$.

The lower layer LEMPC formulation, which uses a terminal constraint computed from $x_E^*(\cdot|\hat{t}_k)$, is given by:

$$\min_{u \in S(\Delta)} \int_{t_j}^{t_{j+N_j}} l_e(\tilde{x}(t), u(t))\, dt \tag{6.71a}$$

$$\text{s.t.}\quad \dot{\tilde{x}}(t) = f(\tilde{x}(t), u(t)) \tag{6.71b}$$

$$\tilde{x}(t_j) = x(t_j) \tag{6.71c}$$

$$u(t) \in \mathbb{U},\ \forall\, t \in [t_j, t_{j+N_j}) \tag{6.71d}$$

$$\tilde{x}(t) \in \Omega_\rho,\ \forall\, t \in [t_j, t_{j+N_j}] \tag{6.71e}$$

$$\tilde{x}(t_{j+N_j}) = x_E^*(t_{j+N_j}|\hat{t}_k) \tag{6.71f}$$

where $\tilde{x}$ is the predicted state trajectory under the input trajectory u. The terminal constraint of Eq. 6.71f is computed from the upper layer LEMPC solution, and serves the same purpose as the terminal constraint of Eq. 6.70f. The optimal solution to the lower layer LEMPC is denoted as $u^*(t|t_j)$ which is defined for $t \in [t_j, t_{j+N_j})$. The control input $u^*(t_j|t_j)$ is sent to the control actuators to be applied to the system of Eq. 6.65 in a sample-and-hold fashion until the next sampling period.

Remark 6.5 When the economic stage cost does not penalize the use of control energy, one may consider formulating constraints in the LEMPC problems to prevent the LEMPC from computing an input trajectory that uses excessive control energy. In particular, one straightforward extension of the two-layer LEMPC structure is to compute the total control energy used by the auxiliary controller over the operating window, i.e., integral of the input trajectory v over $\hat{t}_k$ to $\hat{t}_{k+1}$. Then, enforce that the upper and lower layer LEMPCs compute an input trajectory that uses no more control energy than the auxiliary controller input profile over the operating window. This approach was employed in [21].

6.4.3.3 Closed-Loop Stability and Performance

The following proposition proves that the closed-loop system state under the two-layer EMPC structure is always bounded in the invariant set Ω_ρ and the economic performance is at least as good as the closed-loop state with the auxiliary LMPC over each operating period.

Proposition 6.3 *Consider the system of Eq. 6.65 in closed-loop under the lower layer LEMPC of Eq. 6.71. Let the terminal constraint of Eq. 6.71f computed from the upper layer LEMPC of Eq. 6.70, which has a terminal constraint formulated from the auxiliary controller that satisfies Assumption 6.3. Let* $\Delta_E \in (0, \ \Delta^*]$ *where* Δ^* *is defined according to Corollary 2.2,* $N_E \geq 1$, $\bar{N} \geq 1$, *and* $\Delta = \Delta_E/\bar{N}$. *If* $x(\hat{t}_k) \in \Omega_\rho$, *then the state remains bounded in* Ω_ρ *over the operating window with* $x(\hat{t}_{k+1}) = z(\hat{t}_{k+1}|\hat{t}_k) \in \Omega_\rho$, *the upper and lower LEMPCs remain feasible for all* $t \in [\hat{t}_k, \hat{t}_{k+1})$, *and the following inequality holds:*

$$\int_{\hat{t}_k}^{\hat{t}_{k+1}} l_e(x(t), u^*(t))\, dt \leq \int_{\hat{t}_k}^{\hat{t}_{k+1}} l_e(z(t|\hat{t}_k), v(t|\hat{t}_k))\, dt \tag{6.72}$$

where x *and* u^* *are the closed-loop state and input trajectories and* $z(\cdot|\hat{t}_k)$ *and* $v(\cdot|\hat{t}_k)$ *denote the open-loop predicted state and input trajectories under the auxiliary computed at* $\hat{t}_k$.

Proof **Stability**: If $\Delta_E \in (0, \ \Delta^*]$ and the auxiliary controller satisfies Assumption 6.3, Eq. 6.67 implies forward invariance of the set Ω_ρ under the auxiliary controller. The terminal constraint $z(\hat{t}_{k+1}|\hat{t}_k)$ computed by the auxiliary controller is therefore in Ω_ρ. If the optimization problems are feasible, boundedness of the closed-loop state in Ω_ρ over the operating window follows when $x(\hat{t}_k) \in \Omega_\rho$ owing to the fact that the constraint of Eq. 6.71e is imposed in the lower layer LEMPC, which is responsible for computing control action for the closed-loop system. Also, the terminal constraint of Eq. 6.71f imposed in the lower layer LEMPC is always in Ω_ρ as a result of the constraint of Eq. 6.70e imposed in the upper layer LEMPC.

Feasibility: Regarding feasibility of the upper layer LEMPC problem, the input trajectory $v(\cdot|\hat{t}_k)$ obtained from the auxiliary controller is a feasible solution to the upper layer LEMPC for any $x(\hat{t}_k) \in \Omega_\rho$ because it maintains the predicted state inside Ω_ρ and forces the predicted state to the terminal constraint of Eq. 6.70f. If the auxiliary controller is an explicit controller that satisfies Assumption 6.3, then the input trajectory v is obtained from recursively solving Eq. 6.68. On the other hand, if the LMPC of Eq. 6.69 is used as the auxiliary controller, then v is the solution to the optimization problem of Eq. 6.69.

Consider any sampling time $t_j \in [\hat{t}_k, \hat{t}_{k+1})$ such that $t_j = \hat{t}_k + i\Delta_E$ for some i in the set $\{0, \ldots, N_E - 1\}$, i.e., consider a sampling time of the lower layer LEMPC that corresponds to the beginning of a hold time of the upper layer. Let $\{\bar{t}_i\}_{i=0}^{N_E-1}$ denote the sequence of such times. The constant input trajectory $u(t) = u_E^*(t_j|\hat{t}_k)$ for all $t \in [t_j, t_{j+\bar{N}})$ where $t_{j+\bar{N}} = \bar{t}_{i+1} = \hat{t}_k + (i+1)\Delta_E$ is a feasible solution to the optimization problem of Eq. 6.71 because it maintains the state in Ω_ρ and it forces the state to the terminal constraint of Eq. 6.71f. Owing to the shrinking horizon implementation of the lower layer LEMPC, the computed input trajectory by the lower layer LEMPC at $t_j = \bar{t}_i$ is a feasible solution to the optimization problem at the next sampling time (t_{j+1}) in the sense that if $u^*(t|t_j)$ defined for $t \in [t_j, t_j + \bar{N}\Delta)$ is the optimal solution at t_j, then $u^*(t|t_j)$ for $t \in [t_{j+1}, t_{j+1} + (\bar{N} - 1)\Delta)$ is a feasible solution at t_{j+1}. Using this argument recursively until the sampling time $\bar{t}_{i+1} = \hat{t}_k + (i+1)\Delta_E$ when the horizon is reinitialized to $\bar{N}$ and then, repeating the arguments for $\bar{t}_{i+1}$, it follows that the lower layer LEMPC is feasible.

Performance: At $\bar{t}_i$, the lower layer LEMPC computes an optimal input trajectory that satisfies (by optimality):

$$\int_{\bar{t}_i}^{\bar{t}_{i+1}} l_e(x^*(t|\bar{t}_i), u^*(t|\bar{t}_i))\, dt \le \int_{\bar{t}_i}^{\bar{t}_{i+1}} l_e(x_E^*(t|\hat{t}_k), u_E^*(\bar{t}_i|\hat{t}_k))\, dt \tag{6.73}$$

for all $i \in \{0, \ldots, N_E - 1\}$ (recall, $\bar{t}_{i+1} = \bar{t}_i + \bar{N}\Delta$). Owing to the shrinking horizon and the principle of optimality, the closed-loop state and input trajectories are equal to the computed open-loop state and input trajectories computed at $\bar{t}_i$ and thus,

$$\int_{\bar{t}_i}^{\bar{t}_{i+1}} l_e(x^*(t|\bar{t}_i), u^*(t|\bar{t}_i))\, dt = \int_{\bar{t}_i}^{\bar{t}_{i+1}} l_e(x(t), u^*(t))\, dt \tag{6.74}$$

where $x^*(\cdot|\bar{t}_i)$ and $u^*(\cdot|\bar{t}_i)$ denote the optimal open-loop state and input trajectories computed at $\bar{t}_i$ and x and u^* are the closed-loop state and input trajectories. Therefore, from Eqs. 6.73–6.74, the closed-loop performance over one operating period is bounded by:

$$\begin{aligned}\int_{\hat{t}_k}^{\hat{t}_{k+1}} l_e(x(t), u^*(t))\, dt &= \sum_{i=0}^{N_E-1} \int_{\bar{t}_i}^{\bar{t}_{i+1}} l_e(x(t), u^*(t|\bar{t}_i))\, dt \\ &\leq \sum_{i=0}^{N_E-1} \int_{\bar{t}_i}^{\bar{t}_{i+1}} l_e(x_E^*(t|\hat{t}_k), u_E^*(t|\hat{t}_k))\, dt \\ &= \int_{\hat{t}_k}^{\hat{t}_{k+1}} l_e(x_E^*(t|\hat{t}_k), u_E^*(t|\hat{t}_k))\, dt\ . \end{aligned} \tag{6.75}$$

At $\hat{t}_k$, the upper layer LEMPC computes an optimal input trajectory. Owing to optimality, the computed (open-loop) state and input trajectories of the upper layer LEMPC satisfies:

$$\int_{\hat{t}_k}^{\hat{t}_{k+1}} l_e(x_E^*(t|\hat{t}_k), u_E^*(t|\hat{t}_k))\, dt \leq \int_{\hat{t}_k}^{\hat{t}_{k+1}} l_e(z(t|\hat{t}_k), v(t|\hat{t}_k))\, dt\ . \tag{6.76}$$

From Eqs. 6.75–6.76, the result of Eq. 6.72 follows.

The following theorem provides sufficient conditions such that the two-layer EMPC structure maintains the closed-loop state inside the region Ω_ρ and the closed-loop economic performance is at least as good as if the auxiliary LMPC was applied to the system of Eq. 6.65 over the entire length of operation which may be finite or infinite.

Theorem 6.3 *Consider the closed-loop system of Eq. 6.65 under the lower layer LEMPC of Eq. 6.71. Let the terminal constraint of Eq. 6.71f computed from the upper layer LEMPC of Eq. 6.70, which has a terminal constraint formulated from the auxiliary controller that satisfies Assumption 6.3, and let the assumptions of Proposition 6.3 hold. If $x(0) \in \Omega_\rho$, then $x(t) \in \Omega_\rho$ for all $t \geq 0$ and the following inequality holds for finite-time operation:*

$$\int_0^T l_e(x(t), u^*(t))\, dt \leq \int_0^T l_e(z(t), v(t))\, dt \tag{6.77}$$

where $T = K N_E \Delta_E$ and K is any strictly positive integer and x and u^* are the closed-loop state and input trajectory and z and v are the resulting state and input trajectory from the auxiliary controller defined over the interval $[0, T]$ with initial condition $z(0) = x(0) \in \Omega_\rho$. The following inequality holds for infinite-time operation:

$$\limsup_{K\to\infty} \frac{1}{K N_E \Delta_E} \int_0^{K N_E \Delta_E} l_e(x(t), u^*(t))\, dt \leq l_e(x_s^*, u_s^*)\ . \tag{6.78}$$

Proof Applying the results of Proposition 6.3 recursively over K operating periods, recursive feasibility of the optimization problems follows, and the closed-loop state is always bounded in Ω_ρ if $x(0) \in \Omega_\rho$, and $x(\hat{t}_k) = z(\hat{t}_k)$ for $k = 1, 2, \ldots, K$. To show the result of Eq. 6.77, the length of operation is divided into K operating periods and let $T = KN_E\Delta_E$:

$$\int_0^T l_e(x(t), u^*(t))\, dt = \int_0^{\hat{t}_1} l_e(x(t), u^*(t))\, dt + \cdots + \int_{\hat{t}_{K-1}}^{\hat{t}_K} l_e(x(t), u^*(t))\, dt \tag{6.79}$$

where $\hat{t}_K = T$. By Proposition 6.3, the inequality of Eq. 6.72 holds over each operating window when $x(\hat{t}_k) = z(\hat{t}_k)$ for $k = 1, 2, \ldots, K$ and thus, the inequality of Eq. 6.77 follows.

Owing to the result of Eq. 6.77, the average finite-time economic cost is given by:

$$\frac{1}{T}\int_0^T l_e(x(t), u^*(t))\, dt \le \frac{1}{T}\int_0^T l_e(z(t), v(t))\, dt \tag{6.80}$$

for $T = KN_E\Delta_E$ where K is any strictly positive integer. Recall, the economic cost function l_e is continuous on the compact set $\Omega_\rho \times \mathbb{U}$ and $x(t), z(t) \in \Omega_\rho$ and $u^*(t), v(t) \in \mathbb{U}$ for all $t \ge 0$. Thus, both integrals of Eq. 6.80 are bounded for any $T > 0$. Since the auxiliary controller satisfies Assumption 6.3 and $\Delta \in (0, \Delta^*]$, z and v asymptotically converge to the steady-state (x_s^*, u_s^*) (this follows from the inequality of Eq. 6.67).

Considering the limit of the right-hand side of Eq. 6.80 as T tends to infinity (or similarly, as K tends to infinity), the limit exists and is equal to $l_e(x_s^*, u_s^*)$ because z and v asymptotically converge to optimal steady-state (x_s^*, u_s^*) while remaining bounded for all $t \ge 0$. To prove this limit, the following result is shown: given $\varepsilon > 0$, there exists $T^* > 0$ such that for all $T > T^*$, the following holds:

$$\left| \frac{1}{T}\int_0^T l_e(z(t), v(t))\, dt - l_e(x_s^*, u_s^*) \right| < \varepsilon\,. \tag{6.81}$$

Define $I(0, T)$ as the following integral:

$$I(0, T) := \int_0^T l_e(z(t), v(t))\, dt \tag{6.82}$$

where the arguments of I represent the lower and upper limits of integration, respectively. The trajectories $z(t)$ and $v(t)$ converge to x_s^* and u_s^*, respectively, as t tends to infinity. Furthermore, $z(t) \in \Omega_\rho$ and $v(t) \in \mathbb{U}$ for all $t \ge 0$, so for every $\varepsilon > 0$, there exists a $\tilde{T} > 0$ such that

$$|l_e(z^*(t), v^*(t)) - l_e(x_s^*, u_s^*)| < \varepsilon/2 \tag{6.83}$$

for $t \geq \tilde{T}$. For any $T > \tilde{T}$:

$$\begin{aligned} \left|I(0,T) - Tl_e(x_s^*, u_s^*)\right| &= \left|I(0,\tilde{T}) + I(\tilde{T},T) - Tl_e(x_s^*, u_s^*)\right| \\ &\leq \int_0^{\tilde{T}} \left|l_e(z(t), v(t)) - l_e(x_s^*, u_s^*)\right| dt \\ &\quad + \int_{\tilde{T}}^{T} \left|l_e(z(t), v(t)) - l_e(x_s^*, u_s^*)\right| dt \\ &\leq \tilde{T}\tilde{M} + (T - \tilde{T})\varepsilon/2 \end{aligned} \tag{6.84}$$

where $\tilde{M} := \sup_{t\in[0,\tilde{T}]}\left\{|l_e(z(t), v(t)) - l_e(x_s^*, u_s^*)|\right\}$. For any $T > T^*$ where $T^* = 2\tilde{T}(\tilde{M} - \varepsilon/2)/\varepsilon$, the following inequality is satisfied:

$$\left|I(0,T)/T - l_e(x_s^*, u_s^*)\right| \leq (1 - \tilde{T}/T)\varepsilon/2 + \tilde{T}\tilde{M}/T < \varepsilon \tag{6.85}$$

which proves that the asymptotic average economic cost under the auxiliary controller is $l_e(x_s^*, u_s^*)$.

Considering the left hand side of Eq. 6.80, the limit as $K \to \infty$ may not exist owing to the possible time-varying system operation under the proposed two-layer LEMPC scheme. Therefore, an upper bound on the asymptotic average performance under the LEMPC scheme is considered. Since the limit superior is equal to the limit when the limit exists, the following is obtained:

$$\begin{aligned} &\limsup_{K\to\infty} \frac{1}{KN_E\Delta_E} \int_0^{KN_E\Delta_E} l_e(x(t), u^*(t))\, dt \\ &\leq \limsup_{K\to\infty} \frac{1}{KN_E\Delta_E} \int_0^{KN_E\Delta_E} l_e(z(t), v(t))\, dt = l_e(x_s^*, u_s^*) \end{aligned} \tag{6.86}$$

which is the desired result of Eq. 6.78.

Remark 6.6 The finite-time result of Theorem 6.3 may be extended to any $T > 0$ by, for instance, adjusting N_E and/or Δ_E in the last operating window.

6.4.4 Application to Chemical Process Example

Consider a three vessel chemical process network consisting of two continuously stirred tank reactors (CSTRs) in series followed by a flash tank separator. The process flow diagram of the process network is shown in Fig. 6.14. In each of the reactors, an irreversible second-order reaction of the form $A \to B$ takes place in an inert solvent D (A is the reactant and B is the desired product). The bottom stream of the flash tank is the product stream of the process network. Part of the overhead

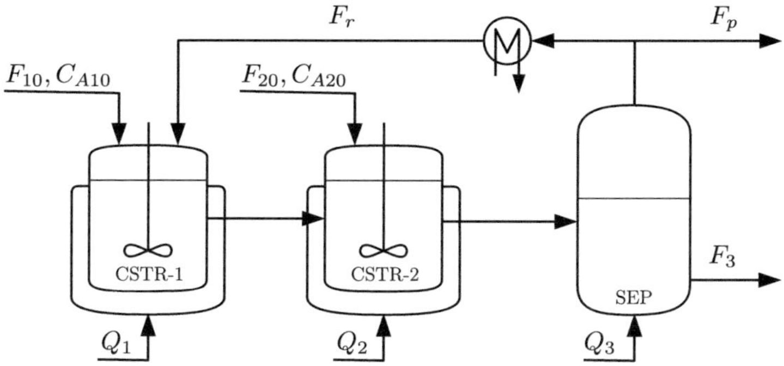

Fig. 6.14 Process flow diagram of the reactor and separator process network

vapor stream from the flash tank is purged from the process while the remainder is fully condensed and recycled back to the first reactor. Each of the vessels have a heating/cooling jacket to supply/remove heat from the liquid contents of the vessel. The following indexes are used to refer to each vessel: $i = 1$ denotes CSTR-1, $i = 2$ denotes CSTR-2, and $i = 3$ denotes SEP-1. The heat rate supplied/removed from the ith vessel is Q_i, $i = 1, 2, 3$. Furthermore, each reactor is fed with fresh feedstock containing A in the solvent D with concentration C_{Ai0}, volumetric flow rate F_{i0}, and constant temperature T_{i0} where $i = 1, 2$. Applying first principles and standard modeling assumptions, a dynamic model of the reactor-separator process network may be obtained (neglecting the dynamics of the condenser) and is given by the following ODEs (see Table 6.4 for variable definitions and values):

$$\begin{aligned}\frac{dT_1}{dt} &= \frac{F_{10}}{V_1}(T_{10} - T_1) + \frac{F_r - F_p}{V_1}(T_3 - T_1) \\ &\quad - \frac{\Delta H k_0}{\rho_L C_p} e^{-E/RT_1} C_{A1}^2 + \frac{Q_1}{\rho_L C_p V_1} \end{aligned} \tag{6.87a}$$

$$\frac{dC_{A1}}{dt} = \frac{F_{10}}{V_1}(C_{A10} - C_{A1}) + \frac{F_r - F_p}{V_1}(C_{Ar} - C_{A1}) - k_0 e^{-E/RT_1} C_{A1}^2 \tag{6.87b}$$

$$\frac{dC_{B1}}{dt} = -\frac{F_{10}}{V_1} C_{B1} + \frac{F_r - F_p}{V_1}(C_{Br} - C_{B1}) + k_0 e^{-E/RT_1} C_{A1}^2 \tag{6.87c}$$

$$\begin{aligned}\frac{dT_2}{dt} &= \frac{F_{20}}{V_2}(T_{20} - T_2) + \frac{F_1}{V_2}(T_1 - T_2) \\ &\quad - \frac{\Delta H k_0}{\rho_L C_p} e^{-E/RT_2} C_{A2}^2 + \frac{Q_2}{\rho_L C_p V_2} \end{aligned} \tag{6.87d}$$

$$\frac{dC_{A2}}{dt} = \frac{F_{20}}{V_2}(C_{A20} - C_{A2}) + \frac{F_1}{V_2}(C_{A1} - C_{A2}) - k_0 e^{-E/RT_2} C_{A2}^2 \tag{6.87e}$$

$$\frac{dC_{B2}}{dt} = -\frac{F_{20}}{V_2} C_{B2} + \frac{F_1}{V_2}(C_{B1} - C_{B2}) + k_0 e^{-E/RT_2} C_{A2}^2 \tag{6.87f}$$

$$\frac{dT_3}{dt} = \frac{F_2}{V_3}(T_2 - T_3) - \frac{\Delta H_{\text{vap}} F_r}{\rho_L C_p V_3} + \frac{Q_3}{\rho_L C_p V_3} \tag{6.87g}$$

Table 6.4 Process parameters of the reactor and separator process network

Symbol/Value	Description
$T_{10} = 300\,\text{K}$	Temp.: CSTR-1 inlet
$T_{20} = 300\,\text{K}$	Temp.: CSTR-2 inlet
$F_{10} = 5.0\,\text{m}^3\text{h}^{-1}$	Flow rate: CSTR-1 inlet
$F_{20} = 5.0\,\text{m}^3\text{h}^{-1}$	Flow rate: CSTR-2 inlet
$F_r = 3.0\,\text{m}^3\text{h}^{-1}$	Flow rate: SEP-1 vapor
$F_p = 0.5\,\text{m}^3\text{h}^{-1}$	Flow rate: purge stream
$V_1 = 1.5\,\text{m}^3$	Volume: CSTR-1
$V_2 = 1.0\,\text{m}^3$	Volume: CSTR-2
$V_3 = 1.0\,\text{m}^3$	Volume: SEP-1
$k_0 = 3.0 \times 10^6\,\text{m}^3\,\text{kmol}^{-1}\,\text{h}^{-1}$	Pre-exponential factor
$E = 3.0 \times 10^4\,\text{kJ}\,\text{kmol}^{-1}$	Activation energy
$\Delta H = -5.0 \times 10^3\,\text{kJ}\,\text{kmol}^{-1}$	Heat of reaction
$\Delta H_{\text{vap}} = 5.0\,\text{kJ}\,\text{kmol}^{-1}$	Heat of vaporization
$C_p = 0.231\,\text{kJ}\,\text{kg}^{-1}\,\text{K}^{-1}$	Heat capacity
$R = 8.314\,\text{kJ}\,\text{kmol}^{-1}\,\text{K}^{-1}$	Gas constant
$\rho_L = 1000\,\text{kg}\,\text{m}^{-3}$	Density
$\alpha_A = 5.0$	Relative volatility: A
$\alpha_B = 0.5$	Relative volatility: B
$\alpha_D = 1.0$	Relative volatility: D
$MW_A = 18.0\,\text{kg}\,\text{kmol}^{-1}$	Molecular weight: A
$MW_B = 18.0\,\text{kg}\,\text{kmol}^{-1}$	Molecular weight: B
$MW_D = 40.0\,\text{kg}\,\text{kmol}^{-1}$	Molecular weight: D

$$\frac{dC_{A3}}{dt} = \frac{F_2}{V_3}C_{A2} - \frac{F_r}{V_3}C_{Ar} - \frac{F_3}{V_3}C_{A3} \tag{6.87h}$$

$$\frac{dC_{B3}}{dt} = \frac{F_2}{V_3}C_{B2} - \frac{F_r}{V_3}C_{Br} - \frac{F_3}{V_3}C_{B3} \tag{6.87i}$$

and the following algebraic equations:

$$K = \frac{1}{\rho_L} \sum_{i \in \{A,B,D\}} \alpha_i C_{i3} MW_i, \tag{6.88a}$$

$$C_{ir} = \alpha_i C_{i3}/K, \quad i = A,\ B,\ D, \tag{6.88b}$$

$$F_1 = F_r - F_p + F_{10},\ \ F_2 = F_1 + F_{20}, \tag{6.88c}$$

$$F_3 = F_2 - F_r\ . \tag{6.88d}$$

where C_{ir} is the concentration of the ith component ($i = A, B, D$) in the flash separator overhead, purge, and recycle streams. The state variables of the process network include the temperatures and concentrations of A and B in each of the vessels:

$$x^T = \left[T_1\ C_{A1}\ C_{B1}\ T_2\ C_{A2}\ C_{B2}\ T_3\ C_{A3}\ C_{B3}\right] .$$

The manipulated inputs are the heat inputs to the three vessels, Q_1, Q_2, and Q_3, and the concentration of A in the inlet streams, C_{A10} and C_{A20}:

$$u^T = \left[Q_1\ Q_2\ Q_3\ C_{A10}\ C_{A20}\right] .$$

The control objective is to regulate the process in an economically optimal time-varying fashion to maximize the average amount of product B in the product stream F_3. Continuously feeding in the maximum concentration of A into each reactor maximizes the production of B owing to the second-order reaction. However, this may not be practical from an economic stand-point. Instead, the average amount of reactant material that may be fed to each reactor is fixed motivating the use of EMPC to control the process network. In addition, supplying/removing heat to/from the vessels is considered undesirable. To accomplish these economic considerations, the two-layer LEMPC structure is applied and the upper and lower layer LEMPCs are formulated with the following economic stage cost function and constraint, respectively:

$$l_e(x, u) = -F_3 C_{B3} + p_1 Q_1^2 + p_2 Q_2^2 + p_3 Q_3^2 \tag{6.89}$$

$$\frac{1}{t'} \int_{\hat{t}_k}^{\hat{t}_{k+1}} (C_{A10} + C_{A20})\ dt = 8.0\,\text{kmol}\,\text{m}^3 \tag{6.90}$$

where $t' = 1.0\,\text{h}$ is the operating period length and $p_i = 10^{-6}$, $i = 1, 2, 3$ are the penalty weights for using energy. The value for the heat rate penalty has been chosen to account for the different numerical range of the heat rate and the first term in the economic cost (molar flow rate of B in the product stream). The economically optimal steady-state with respect to the economic cost function of Eq. 6.89 is open-loop asymptotically stable and is the only steady-state in the operating region of interest. An explicit characterization of Ω_ρ is not needed for the LEMPC implementation.

The two-layer LEMPC structure, formulated with the cost function and reactant material constraint of Eqs. 6.89–6.90, respectively, is applied to the reactor-separator chemical process network. To numerically integrate the dynamic model of Eq. 6.87, explicit Euler method is used with an integration step of 1.0×10^{-3} h. The auxiliary controller is formulated as an auxiliary LMPC. The prediction horizon and sampling period of the auxiliary LMPC and upper layer LEMPC are $N_E = 10$ and $\Delta_E = 0.1$ h, respectively, while, the lower layer LEMPC is formulated with a prediction horizon of $\bar{N} = 2$ and sampling period $\Delta = 0.05$ h. Since the upper layer prediction horizon length is one hour, the reactant material constraint is enforced over each one hour operating period. However, the lower layer LEMPC prediction horizon does not cover the entire one hour operating window. Instead of using the material constraint of Eq. 6.90 directly in the lower layer LEMPC, a constraint is formulated on the basis of the upper layer LEMPC solution. Namely, over the prediction horizon of the lower layer LEMPC, the lower layer LEMPC solution must use the same amount of reactant material as that of the upper layer LEMPC solution over the same time so that the material constraint is satisfied over the operating window. To solve the optimization problems, Ipopt [17] is used and the simulations were completed on a desktop PC with an Intel® Core™ 2 Quad 2.66 GHz processor and a Linux operating system.

6.4.4.1 Effect of Horizon Length

In the first set of simulations, the length of the prediction horizon on closed-loop performance is considered. The closed-loop economic performance over 4.0 h is defined by the total economic cost given by:

$$J_E = \int_0^{4.0} \left(F_3 C_{B3} - p_1 Q_1^2 - p_2 Q_2^2 - p_3 Q_3^2 \right) \, dt \, . \tag{6.91}$$

In these simulations, only the upper layer LEMPC, formulated with a terminal constraint computed from the auxiliary LMPC, is considered. Figure 6.15 depicts the observed trend. As the prediction horizon increases, the closed-loop economic performance increases, which motivates the use of a long prediction horizon in EMPC.

6.4.4.2 Effect of the Terminal Constraint

Since for any optimization problem, the addition of constraints may restrict the feasible region of the optimization problem, a reasonable consideration is the effect of the terminal constraint on closed-loop performance. To address this issue, consider

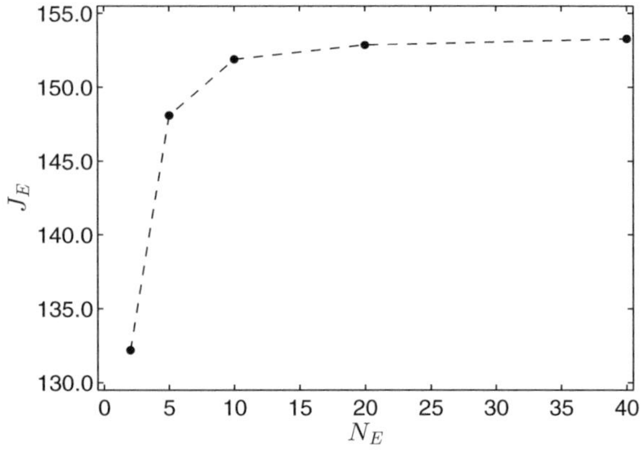

Fig. 6.15 The closed-loop economic performance (J_E) with the length of prediction horizon (N_E) for the reactor-separator process under the upper layer LEMPC with a terminal constraint computed from an auxiliary LMPC

the closed-loop system under the upper layer LEMPC formulated with a terminal equality constraint computed by the auxiliary LMPC and under an LEMPC (mode 1 operation only) formulated with the economic cost of Eq. 6.89 and the material constraint of Eq. 6.90, but without terminal constraints. Both use a prediction horizon of $N_E = 10$ and a sampling period of $\Delta = 0.01$ h. Figures 6.16 and 6.17 display the closed-loop state and input trajectories of the reactor-separator process network with the upper layer LEMPC; while, Figs. 6.18 and 6.19 display the closed-loop trajectories under LEMPC with no terminal constraints.

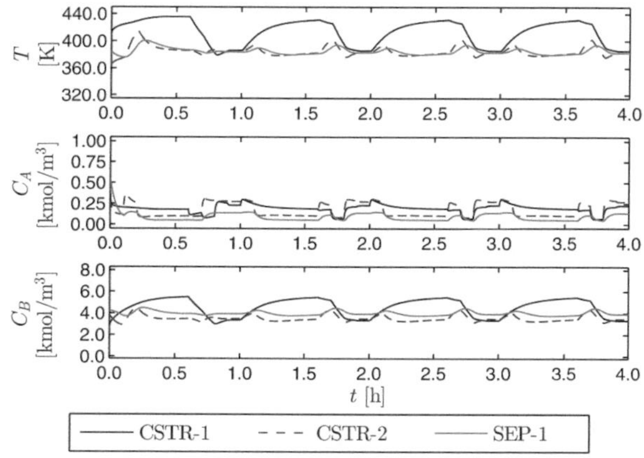

Fig. 6.16 Closed-loop state trajectories of the reactor-separator process network with the upper layer LEMPC formulated with a terminal constraint computed by the auxiliary LMPC

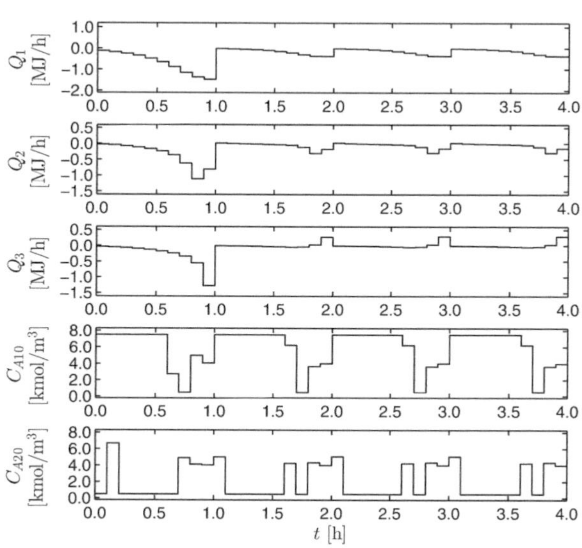

Fig. 6.17 Input trajectories of the reactor-separator process network computed by the upper layer LEMPC formulated with a terminal constraint computed by the auxiliary LMPC

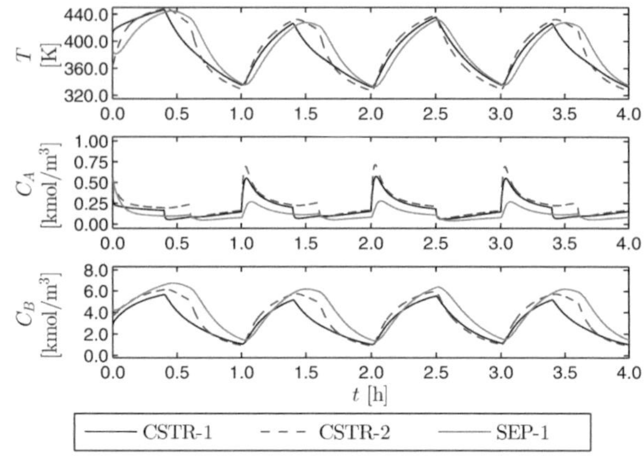

Fig. 6.18 Closed-loop state trajectories of the reactor-separator process network with an LEMPC formulated without terminal constraints

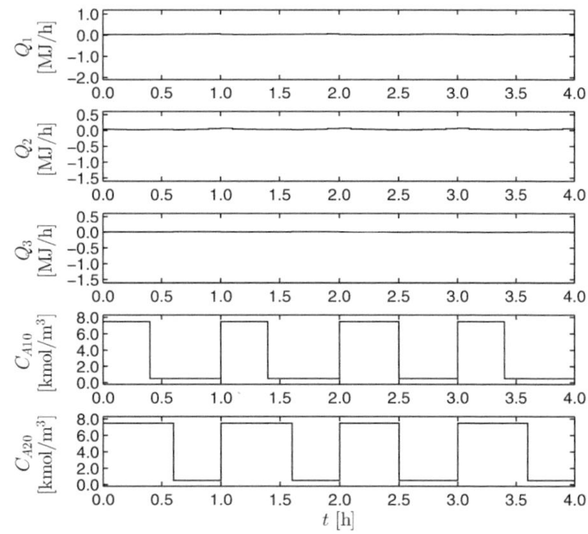

Fig. 6.19 Input trajectories of the reactor-separator process network computed by an LEMPC formulated without terminal constraints

The reactor-separator process network under the LEMPC with the terminal constraint evolves in a smaller operating range (370–430 K) than the evolution under the LEMPC without the terminal constraint (325– 440 K). The total economic cost with the upper layer LEMPC (based on the auxiliary LMPC) is 151.2, while the total economic cost with LEMPC formulated without terminal constraints is 159.3. The terminal constraint imposed in the LEMPC problem affects the achievable performance. However, the key advantage of the addition of this constraint is that for any system and any prediction horizon the closed-loop economic performance under the two-layer LEMPC structure is guaranteed to be at least as good as a stabilizing controller for both finite-time and infinite-time operating intervals.

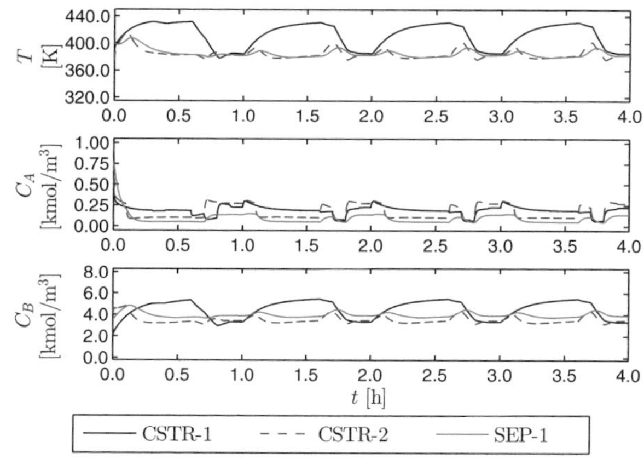

Fig. 6.20 Closed-loop state trajectories of the reactor-separator process network with the two-layer LEMPC structure

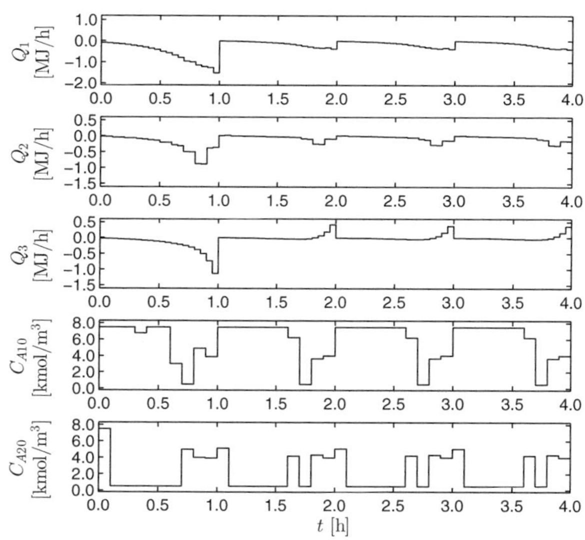

Fig. 6.21 Input trajectories of the reactor-separator process network computed by the two-layer LEMPC structure

6.4.4.3 Two-Layer LEMPC Structure

The two-layer LEMPC structure with a terminal constraint computed from an auxiliary LMPC is applied to the reactor-separator process network. Several closed-loop simulations over a 4.0 h length of operation are completed. The closed-loop state and input trajectories of one of the simulations are shown in Figs. 6.20 and 6.21, respectively and demonstrate time-varying operation of the process network. The economic performance (Eq. 6.91) is compared to the economic performance with the auxiliary LMPC (Table 6.5). From this comparison, an average of 10 % benefit with the two-layer LEMPC structure is realized over operation under the auxiliary LMPC, i.e., resulting in steady-state operation.

Table 6.5 Total economic cost and average computational time in seconds per sampling period for several 4.0 h simulations with: (a) the auxiliary LMPC, (b) the one-layer LEMPC and (c) the two-layer LEMPC structure

Sim.	LMPC	One-layer EMPC		Two-layer EMPC	
	Cost	CPU time	Cost	CPU time	Cost
1	140.1	5.68	151.5	1.10	151.1
2	150.3	4.24	153.9	1.05	153.4
3	142.0	4.65	152.4	0.98	152.0
4	130.7	6.45	152.3	1.24	151.9
5	126.0	4.67	151.9	1.11	151.5
6	140.2	4.63	151.6	1.33	151.2
7	144.6	4.60	150.6	1.08	150.2
8	138.1	5.01	152.5	1.06	152.1

Additionally, a comparison between the computational time required to solve the two-layer LEMPC system and that of a one-layer LEMPC system is completed. The one-layer LEMPC system consists of the upper layer LEMPC with a terminal constraint computed from the auxiliary LMPC. In the one-layer LEMPC system, the LEMPC applies its computed control actions directly to the process network, and there is no lower layer LEMPC. To make the comparison consistent, the one layer LEMPC is implemented with a prediction horizon of $N_E = 20$ and a sampling period of $\Delta_E = 0.05$ h, which are the same sampling period and horizon used in the lower layer LEMPC of the two-layer LEMPC system. Since the upper and lower layer controllers are sequentially computed, the computational time at the beginning of each operating window is measured as the sum of the computational time to solve the auxiliary LMPC, the upper layer LEMPC, and the lower layer LEMPC for the two-layer LEMPC system and as the sum of the time to solve the auxiliary LMPC and the LEMPC for the one-layer LEMPC system. From Table 6.5, the one-layer LEMPC achieves slightly better closed-loop economic performance because the one-layer LEMPC uses a smaller sampling period than the upper layer LEMPC in the two-layer LEMPC structure. However, the computational time required to solve the one-layer LEMPC structure is greater than the computational time of the two-layer LEMPC structure. The two-layer LEMPC structure is able to reduce the computational time by about 75 % on average.

6.4.4.4 Handling Disturbances

While the two-layer EMPC has been designed for nominal operation to guarantee finite-time and infinite-time closed-loop performance as is at least as good as that achieved under a stabilizing controller, it may be applied to the process model in the presence of disturbances, plant/model mismatch, and other uncertainties with some modifications to improve recursive feasibility of the optimization problems and to ensure greater robustness of the controller to uncertainties. For instance, if the disturbances are relatively small, it may be sufficient to relax the terminal constraints or treat them as soft constraints. If one were to simply relax the terminal constraints, e.g., use a terminal region instead of a point-wise terminal constraint, it is difficult to guarantee recursive feasibility of the optimization problem. Another potential methodology is to treat the terminal state constraints as a soft constraint instead of imposing them as hard constraints. For example, use a cost functional in the lower layer LEMPC of the form:

$$\left(\int_{t_j}^{t_{j+N}} l_e(\tilde{x}(t), u(t))\, dt\right) + \left|\tilde{x}(t_{j+N}) - x_E^*(t_j|t_k)\right|_Q \tag{6.92}$$

where Q is a positive definite weighting matrix. The cost functional works to optimize the economic performance while ensuring the predicted evolution is near the terminal state through the quadratic terminal cost. The resulting lower layer LEMPC has the same stability and robustness to bounded disturbances properties as the LEMPC (without terminal constraints), i.e., recursive feasibility and boundedness of the closed-loop state for all initial states starting in Ω_ρ. While no provable performance guarantees may be made on closed-loop performance in the presence of disturbances, the closed-loop performance benefit may be evaluated through simulations.

The two-layer LEMPC with the lower layer LEMPC designed with the cost described above in Eq. 6.92 and without terminal constraints is applied to the example with significant process noise added. The noise is modeled as bounded Gaussian white noise and is introduced additively to each model state. The closed-loop state and input trajectories are shown in Figs. 6.22 and 6.23, respectively. The closed-loop system performance under the two-layer LEMPC is compared to the system under auxiliary LMPC with the same realization of the process noise. The LMPC is formu-

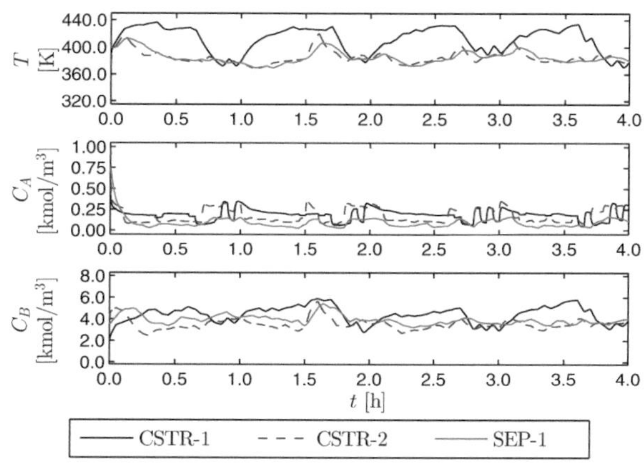

Fig. 6.22 Closed-loop state trajectories of the reactor-separator process network with process noise added with the two-layer LEMPC structure

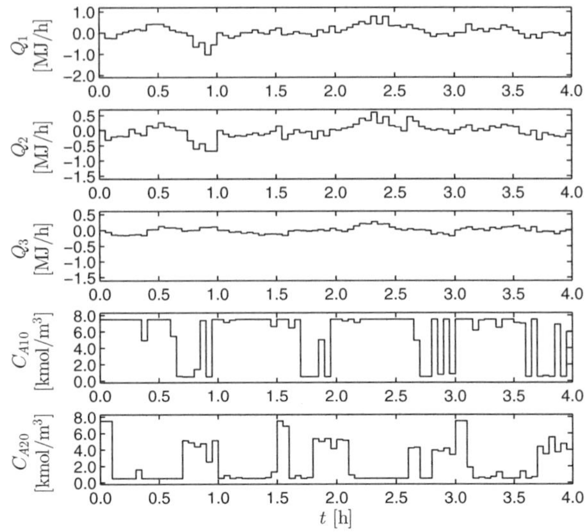

Fig. 6.23 Input trajectories of the reactor-separator process network with process noise added computed by the two-layer LEMPC structure

lated with a prediction horizon of $N = 2$ and sampling period $\Delta = 0.05$ h which is the same horizon and sampling period as the lower layer LEMPC. The closed-loop performance under the two-layer LEMPC is 2.6 % better than that under the LMPC for this particular realization of the process noise.

6.5 Conclusions

In this chapter, several computationally-efficient two-layer frameworks for integrating dynamic economic optimization and control of nonlinear systems were presented. In the upper layer, EMPC was employed to compute economically optimal time-varying operating trajectories. Explicit control-oriented constraints were employed in the upper layer EMPC. In the lower layer, an MPC scheme was used to force the system to track the optimal time-varying trajectory computed by the upper layer EMPC. The properties, i.e., stability, performance, and robustness, of closed-loop systems under the two-layer EMPC methods were rigorously analyzed. The two-layer EMPC methods were applied to chemical process examples to demonstrate the closed-loop properties. In all the examples considered, closed-loop stability was achieved, the closed-loop economic performance under the two-layer EMPC framework was better than that achieved under conventional approaches to optimization and control, and the total on-line computational time was better with the two-layer EMPC methods compared to that under one-layer EMPC methods.

References

1. Helbig A, Abel O, Marquardt W (2000) Structural concepts for optimization based control of transient processes. In: Allgöwer F, Zheng A (eds) Nonlinear model predictive control. Progress in systems and control theory, vol 26. Birkhäuser, Basel, pp 295–311
2. Kadam JV, Schlegel M, Marquardt W, Tousain RL, van Hessem DH, van den Berg J, Bosgra OH (2002) A two-level strategy of integrated dynamic optimization and control of industrial processes-a case study. In: Grievink J, van Schijndel J (eds) Proceedings of the European symposium on computer aided process engineering-12, computer aided chemical engineering, vol 10. The Netherlands, The Hague, pp 511–516
3. Marquardt W (2001) Nonlinear model reduction for optimization based control of transient chemical processes. In: Proceedings of the Sixth international conference on chemical process control, Tucson, AZ, , pp 12–42
4. Kadam JV, Marquardt W, Schlegel M, Backx T, Bosgra OH, Brouwer PJ, Dünnebier G, van Hessem D, Tiagounov A, de Wolf S (2003) Towards integrated dynamic real-time optimization and control of industrial processes. In: Proceedings of the 4th international conference of the foundations of computer-aided process operations. Coral Springs, Florida, pp 593–596
5. Tosukhowong T, Lee JM, Lee JH, Lu J (2004) An introduction to a dynamic plant-wide optimization strategy for an integrated plant. Comput Chem Eng 29:199–208
6. Zhu X, Hong W, Wang S (2004) Implementation of advanced control for a heat-integrated distillation column system. In: Proceedings of the 30th Conference of IEEE industrial electronics society, pp 2006–2011, Busan, Korea
7. Kadam JV, Marquardt W (2007) Integration of economical optimization and control for intentionally transient process operation. In: Findeisen R, Allgöwer F, Biegler LT (eds) Assessment and future directions of nonlinear model predictive control, vol 358., Lecture notes in control and information sciences, Springer, Berlin, pp 419–434
8. Würth L, Hannemann R, Marquardt W (2009) Neighboring-extremal updates for nonlinear model-predictive control and dynamic real-time optimization. J Process Control 19:1277–1288
9. Würth L, Rawlings JB, Marquardt W (2007) Economic dynamic real-time optimization and nonlinear model-predictive control on the infinite horizon. In: Proceedings of the 7th IFAC international symposium on advanced control of chemical processes, pp 219–224, Istanbul, Turkey
10. Würth L, Hannemann R, Marquardt W (2011) A two-layer architecture for economically optimal process control and operation. J Process Control 21:311–321
11. Ellis M, Christofides PD (2014) Optimal time-varying operation of nonlinear process systems with economic model predictive control. Ind Eng Chem Res 53:4991–5001
12. Ellis M, Christofides PD (2014) Integrating dynamic economic optimization and model predictive control for optimal operation of nonlinear process systems. Control Eng Pract 22:242–251
13. Ellis M, Christofides PD (2014) On finite-time and infinite-time cost improvement of economic model predictive control for nonlinear systems. Automatica 50:2561–2569
14. Ellis M, Christofides PD (2014) Selection of control configurations for economic model predictive control systems. AIChE J 60:3230–3242
15. Khalil HK (2002) Nonlinear systems, 3rd edn. Prentice Hall, Upper Saddle River
16. Massera JL (1956) Contributions to stability theory. Ann Math 64:182–206
17. Wächter A, Biegler LT (2006) On the implementation of an interior-point filter line-search algorithm for large-scale nonlinear programming. Math Progr 106:25–57
18. Christofides PD, El-Farra NH (2005) Control of nonlinear and hybrid process systems: designs for uncertainty. Constraints and time-delays. Springer, Berlin
19. Kokotović P, Arcak M (2001) Constructive nonlinear control: a historical perspective. Automatica 37:637–662
20. Sontag ED (1998) Mathematical control theory: deterministic finite dimensional systems, vol 6. Springer, Berlin

21. Heidarinejad M, Liu J, Christofides PD (2013) Algorithms for improved fixed-time performance of Lyapunov-based economic model predictive control of nonlinear systems. J Process Control 23:404–414

Chapter 7
EMPC Systems: Computational Efficiency and Real-Time Implementation

7.1 Introduction

While the two-layer EMPC structures of Chap. 6 were shown to successfully reduce the on-line computation time relative to that required for a centralized, one-layer EMPC scheme, EMPC optimization problems typically found within the context of chemical processes are nonlinear and non-convex because a nonlinear dynamic model is embedded in the optimization problem. Although many advances have been made in solving such problems and modern computers may efficiently perform complex calculations, it is possible that computation delay will occur that may approach or exceed the sampling time. If the computational delay is significant relative to the sampling period, closed-loop performance degradation and/or closed-loop instability may occur.

In this chapter, three EMPC design methodologies are presented to further address computational efficiency. In the first section, a composite control structure featuring EMPC is designed for systems with explicit two-time-scale dynamic behavior. Owing to the fact that the class of dynamic models describing such systems are typically stiff, a sufficiently small time step is required for forward numerical integration with explicit methods, which subsequently affects the computation time required to solve the EMPC problem. On the other hand, the composite control structure allows for larger time steps because it avoids the use of the stiff dynamic model embedded in the MPC problems of the composite control structure. In the second section, distributed EMPC (DEMPC), which computes the control actions by solving a series of distributed EMPC problems, is considered. Specifically, an application study whereby several DEMPC schemes are applied to a benchmark chemical process example is presented to evaluate the ability of the resulting DEMPC schemes to reduce the computation time relative to a centralized EMPC system. The closed-loop performance under DEMPC is compared with that achieved under a centralized EMPC approach. In the third section, a real-time implementation strategy for Lyapunov-based EMPC (LEMPC) is presented which addresses potentially unknown and time-varying

M. Ellis et al., *Economic Model Predictive Control*,
Advances in Industrial Control, DOI 10.1007/978-3-319-41108-8_7

computational time for control action calculation. Closed-loop stability under the real-time LEMPC strategy is rigorously analyzed.

7.2 Economic Model Predictive Control of Nonlinear Singularly Perturbed Systems

The development of optimal process control, automation, and management methodologies while addressing time-scale multiplicity due to the strong coupling of slow and fast phenomena occurring at different time-scales is an important issue in the context of chemical process control. For multiple-time-scale systems, closed-loop stability as well as controller design are usually addressed through explicit separation of fast and slow states in a standard singular perturbation setting [1–3] or by taking advantage of change of coordinates for two-time-scale systems in nonstandard singularly perturbed form [4]. In our previous work, we developed methods for slow time-scale (tracking) MPC as well as composite fast-slow (tracking) MPC for nonlinear singularly perturbed systems [5, 6]. In this section, these ideas are extended to EMPC of nonlinear singularly perturbed systems.

Specifically, in this section, an EMPC method for a broad class of nonlinear singularly perturbed systems is presented whereby a "fast" Lyapunov-based MPC (LMPC) using a standard tracking quadratic cost is employed to stabilize the fast closed-loop dynamics at their equilibrium slow manifold while a "slow" Lyapunov-based EMPC (LEMPC) is utilized for the slow dynamics to address economic performance considerations. Multi-rate sampling of the states is considered involving fast-sampling of the fast states and slow-sampling of the slow states. The states are subsequently used in the fast and slow MPC systems, respectively. Closed-loop stability of the control scheme is addressed through singular perturbation theory. The control method is demonstrated through a chemical process example which exhibits two-time-scale behavior.

7.2.1 *Class of Nonlinear Singularly Perturbed Systems*

Nonlinear singularly perturbed systems in standard form are considered in this section with the following state-space description:

$$\dot{x} = f(x, z, u_s, w, \varepsilon), \tag{7.1a}$$

$$\varepsilon\dot{z} = g(x, z, u_f, w, \varepsilon), \tag{7.1b}$$

where $x \in \mathbb{R}^{n_x}$ and $z \in \mathbb{R}^{n_z}$ denote the state vectors, $u_s \in \mathbb{U}_s \subset \mathbb{R}^{m_s}$ and $u_f \in \mathbb{U}_f \subset \mathbb{R}^{m_f}$ are the control (manipulated) inputs, $w \in \mathbb{R}^l$ denotes the vector of process disturbances, and $\varepsilon > 0$ is a small positive parameter. The sets $\mathbb{U}_s$ and $\mathbb{U}_f$ are assumed to be compact sets. The disturbance vector is assumed to be an absolutely continuous function of time and bounded in a sense that there exists a $\theta > 0$ such that $|w(t)| \leq \theta$ for all $t \geq 0$. To this end, let $\mathbb{W} = \{w \in \mathbb{R}^l : |w| \leq \theta\}$.

Owing to the multiplication of the small parameter ε with $\dot{z}$ in Eq. 7.1, there exists a time-scale separation in the two systems of differential equations of Eq. 7.1. Moreover, the system of Eq. 7.1 is said to be in singularly perturbed form. Through the rest of the section, x and z will be referred to as the slow and fast states, respectively. Furthermore, the vector functions f and g are assumed to be sufficiently smooth on $\mathbb{R}^{n_x} \times \mathbb{R}^{n_z} \times \mathbb{U}_s \times \mathbb{W} \times [0, \bar{\varepsilon})$ and $\mathbb{R}^{n_x} \times \mathbb{R}^{n_z} \times \mathbb{U}_f \times \mathbb{W} \times [0, \bar{\varepsilon})$, respectively, for some $\bar{\varepsilon} > 0$. The origin is assumed to be an equilibrium point of the unforced nominal system; that is, system of Eq. 7.1 with $u_s = 0$, $u_f = 0$, and $w \equiv 0$ possesses an equilibrium point at $(x, z) = (0, 0)$.

The fast states are sampled synchronously and are available at time instants indicated by the time sequence $\{t_{k_f}\}_{k_f \geq 0}$ with $t_{k_f} = t_0 + k_f \Delta_f$, $k_f = 0, 1, \ldots$ where t_0 is the initial time and $\Delta_f > 0$ is the measurement sampling period of the fast states. Similarly, the slow states are sampled synchronously and are available at time instants indicated by the time sequence $\{t_{k_s}\}_{k_s \geq 0}$ with $t_{k_s} = t_0 + k_s \Delta_s$, $k_s = 0, 1, \ldots$ where $\Delta_s > 0$ ($\Delta_s > \Delta_f$) is the measurement sampling period of the slow states. The initial time is taken to be zero, i.e., $t_0 = 0$. With respect to the control problem formulation, the controls u_f and u_s, which are responsible for the fast and slow dynamics, are computed every Δ_f and Δ_s, respectively. For the sake of simplicity, Δ_s / Δ_f is assumed to be a positive integer.

7.2.2 Two-Time-Scale Decomposition

The explicit separation of slow and fast states in the system of Eq. 7.1 allows for decomposing the system into two separate reduced-order systems evolving in different time-scales. To proceed with such a two-time-scale decomposition and to simplify the notation of the subsequent development, the issue of controlling the fast dynamics is addressed first. Similar to [5], a "fast" model predictive controller will be designed that renders the fast dynamics asymptotically stable in a sense to be made precise in Assumption 7.2. Moreover, u_f does not modify the open-loop equilibrium slow manifold for the fast dynamics. By setting $\varepsilon = 0$, the dimension of the state-space of the system of Eq. 7.1 reduces from $n_x + n_z$ to n_x because the differential equation of Eq. 7.1b degenerates into an algebraic equation. The following model is obtained that describes the slow dynamics with $u_f = 0$:

$$\dot{\bar{x}} = f(\bar{x}, \bar{z}, u_s, w, 0) \tag{7.2a}$$

$$0 = g(\bar{x}, \bar{z}, 0, w, 0) \tag{7.2b}$$

where the bar notation in $\bar{x}$ and $\bar{z}$ is used to indicate that these variables have been obtained by setting $\varepsilon = 0$.

Systems of the form of Eq. 7.1 satisfying the following assumption are said to be in *standard singularly perturbed form*, e.g. [1].

Assumption 7.1 The equation $g(\bar{x}, \bar{z}, 0, w, 0) = 0$ possesses a unique isolated root

$$\bar{z} = \tilde{g}(\bar{x}, w) \tag{7.3}$$

where $\tilde{g} : \mathbb{R}^{n_x} \times \mathbb{R}^l \to \mathbb{R}^{n_z}$ and its partial derivatives $\partial\tilde{g}/\partial\bar{x}$, $\partial\tilde{g}/\partial w$ are sufficiently smooth and $|\partial\tilde{g}/\partial w| \leq L_{\tilde{g}}$.

Assumption 7.1 ensures that the system of Eq. 7.1 has an isolated equilibrium manifold for the fast dynamics. While on this manifold, $\bar{z}$ may be expressed in terms of $\bar{x}$ and w using an algebraic expression. It should be emphasized that $g(\bar{x}, \bar{z}, 0, w, 0)$ is, in this case, independent of the expression of the "fast" input, u_f. Assumption 7.1 does not pose any significant limitations in practical applications, and it is a necessary one in the singular perturbation framework to construct a well-defined slow subsystem. Utilizing $\bar{z} = \tilde{g}(\bar{x}, w)$, the system of Eq. 7.2 may be re-written as follows:

$$\dot{\bar{x}} = f(\bar{x}, \tilde{g}(\bar{x}, w), u_s, w, 0) =: f_s(\bar{x}, u_s, w). \tag{7.4}$$

The system of Eq. 7.4 is referred to as the slow subsystem or the reduced system.

Introducing the fast time-scale $\tau = t/\varepsilon$ and the deviation variable $y = z - \tilde{g}(x, w)$, i.e., the deviation of the fast state from the equilibrium manifold, the nonlinear singularly perturbed system of Eq. 7.1 may be written in the (x, y) coordinates with respect to the fast time-scale as follows:

$$\begin{aligned} \frac{dx}{d\tau} &= \varepsilon f(x, y + \tilde{g}(x, w), u_s, w, \varepsilon) \\ \frac{dy}{d\tau} &= g(x, y + \tilde{g}(x, w), u_f, w, \varepsilon) \\ &\quad - \varepsilon\left(\frac{\partial\tilde{g}}{\partial x} f(x, y + \tilde{g}(x, w), u_f, w, \varepsilon) + \frac{\partial\tilde{g}}{\partial w}\dot{w}\right) \end{aligned} \tag{7.5}$$

Setting $\varepsilon = 0$, the following fast subsystem is obtained:

$$\frac{d\bar{y}}{d\tau} = g(x, \bar{y} + \tilde{g}(x, w), u_f, w, 0) \tag{7.6}$$

where the notation $\bar{y}$ is again used to indicate that its dynamics have been derived by setting $\varepsilon = 0$. In the system of Eq. 7.6, x and w are considered to be "frozen" to their initial values in the fast time-scale since their change in this time-scale is of order ε.

Remark 7.1 The difference between y and $\bar{y}$ is: y is the deviation between the singularly perturbed system state z (Eq. 7.1 with $\varepsilon > 0$) and the solution to the algebraic equation $g(x, \bar{z}, 0, w, 0) = 0$ for $x \in \mathbb{R}^{n_x}$ and $w \in \mathbb{W}$, which is denoted as $\bar{z} = \tilde{g}(x, w)$. The variable $\bar{y}$ is used to denote the solution to the fast subsystem obtained from Eq. 7.6 where x and w are frozen to their initial values and $\varepsilon = 0$; the initial condition of the ODE of Eq. 7.6 at some time $t_0 \geq 0$ is $\bar{y}(t_0) = y(t_0) = z(t_0) - \tilde{g}(x(t_0), w(t_0))$.

7.2.3 Stabilizability Assumption

A stabilizability assumption is imposed on the slow subsystem in the sense that the existence of a Lyapunov-based locally Lipschitz feedback controller $u_s = h_s(\bar{x})$ is assumed which, under continuous implementation, renders the origin of the nominal closed-loop slow subsystem of Eq. 7.4 asymptotically stable while satisfying the input constraints for all the states $\bar{x}$ inside a given stability region. Using converse Lyapunov theorems [7–9], this assumption implies that there exists a continuously differentiable Lyapunov function $V_s : \mathbb{D}_s \rightarrow \mathbb{R}_+$ for the nominal closed-loop slow subsystem that satisfy the following inequalities:

$$\alpha_{s_1}(|\bar{x}|) \leq V_s(\bar{x}) \leq \alpha_{s_2}(|\bar{x}|) \tag{7.7a}$$

$$\frac{\partial V_s(\bar{x})}{\partial \bar{x}} f_s(\bar{x}, h_s(\bar{x}), 0) \leq -\alpha_{s_3}(|\bar{x}|) \tag{7.7b}$$

$$\left| \frac{\partial V_s(\bar{x})}{\partial \bar{x}} \right| \leq \alpha_{s_4}(|\bar{x}|) \tag{7.7c}$$

$$h_s(\bar{x}) \in \mathbb{U}_s \tag{7.7d}$$

for all $\bar{x} \in \mathbb{D}_s \subseteq \mathbb{R}^{n_x}$ where D_s is an open neighborhood of the origin and the functions $\alpha_{s_i}(\cdot)$, $i = 1, 2, 3, 4$ are class $\mathcal{K}$ functions. The region $\Omega_{\rho^s} \subset \mathbb{D}_s$ denotes the stability region of the closed-loop slow subsystem under the Lyapunov-based controller $h_s(\bar{x})$.

Similarly, a stabilizability assumption is placed on the fast subsystem.

Assumption 7.2 There exists a feedback controller $u_f = p(x)\bar{y} \in \mathbb{U}_f$ where $p(x)$ is a sufficiently smooth vector function in x, such that the origin of the closed-loop fast subsystem:

$$\frac{d\bar{y}}{d\tau} = g(x, \bar{y} + \tilde{g}(x, w), p(x)\bar{y}, w, 0) \tag{7.8}$$

is globally asymptotically stable, uniformly in $x \in \mathbb{R}^{n_x}$ and $w \in \mathbb{W}$, in the sense that there exists a class $\mathcal{KL}$ function β_y such that for any $\bar{y}(0) \in \mathbb{R}^{n_z}$:

$$|\bar{y}(\tau)| \leq \beta_y(|\bar{y}(0)|, \tau) \tag{7.9}$$

for $\tau \geq 0$.

This assumption implies that there exist functions $\alpha_{f_i}(\cdot)$, $i = 1, 2, 3$ of class $\mathcal{K}$ and a continuously differentiable Lyapunov function $V_f(\bar{y})$ for the nominal closed-loop fast subsystem which satisfy the following inequalities:

$$\alpha_{f_1}(|\bar{y}|) \leq V_f(\bar{y}) \leq \alpha_{f_2}(|\bar{y}|)$$
$$\frac{\partial V_f(\bar{y})}{\partial \bar{y}} g(x, \bar{y} + \tilde{g}(x, 0), p(x)\bar{y}, 0, 0) \leq -\alpha_{f_3}(|\bar{y}|) \tag{7.10}$$
$$p(x)\bar{y} \in \mathbb{U}_f$$

for all $\bar{y} \in \mathbb{D}_f \subseteq \mathbb{R}^{n_z}$ and $x \in D_s$ where $\mathbb{D}_f$ is an open neighborhood of the origin. The region $\Omega_{\rho^f} \subseteq \mathbb{D}_f$ is used to denote the stability region of the closed-loop fast subsystem under the nonlinear controller $p(x)\bar{y}$.

Remark 7.2 The sets Ω_{ρ^s} and Ω_{ρ^f} denote the stability regions for the closed-loop slow and fast subsystems under the controllers $u_s = h_s(\bar{x})$ and $u_f = p(x)\bar{y}$, respectively, in the sense that the closed-loop states of the fast and slow subsystems, starting in Ω_{ρ^s} and Ω_{ρ^f}, remain in these sets thereafter. Regarding the construction of Ω_{ρ^s}, we have estimated it through the following procedure: $\dot{V}_s(\bar{x})$ is evaluated for different values of $\bar{x}$ while $h_s(\bar{x})$ is applied to the nominal system subject to the input constraint $h_s(\bar{x}) \in \mathbb{U}_s$. Then, we estimated Ω_{ρ^s} as the largest level set of the Lyapunov function $V_s(\bar{x})$ where $\dot{V}_s(\bar{x}) < 0$. The region Ω_{ρ^f} may be estimated in a similar fashion using the fast subsystem and the controller $p(x)\bar{y}$.

7.2.4 *LEMPC of Nonlinear Singularly Perturbed Systems*

In this section, the design of a composite control structure featuring an EMPC for nonlinear singularly perturbed systems is presented. In the control structure, a tracking Lyapunov-based MPC is used to stabilize the fast dynamics in a sense to be made precise below. To control the slow dynamics, a two-mode LEMPC of Sect. 4.2 is used to address economic considerations as well as address closed-loop stability of the slow subsystem.

Over the time interval $[0, t_s)$, the LEMPC operates in mode 1 operation to dictate a potentially time-varying operating policy of the slow subsystem. After t_s, the LEMPC operates in mode 2 operation to enforce convergence of the closed-loop subsystem state to the steady-state. In operation period $[0, t_s)$, the predicted slow subsystem state along the finite prediction horizon is constrained to be maintained in the region $\Omega_{\rho_e} \subset \Omega_{\rho^s}$. If the current slow subsystem state is in the region $\Omega_{\rho^s} \setminus \Omega_{\rho_e}$, the LEMPC drives the slow subsystem state to the region Ω_{ρ_e} and once the state converges to the set Ω_{ρ_e}, the LEMPC may dictate a time-varying operating policy. Under mode 2 operation of the LEMPC ($t \geq t_s$), LEMPC computes control actions such that the slow subsystem state is driven to a neighborhood of the steady-state.

7.2.4.1 Implementation Strategy

The above described implementation strategy corresponding to the fast subsystem under the fast LMPC may be described as follows:

1. The LMPC receives the fast subsystem state $\bar{y}(t_{k_f})$.
2. The LMPC obtains its manipulated input trajectory which ensures that the fast subsystem state is steered to a neighborhood of the equilibrium slow manifold of the fast state.
3. The LMPC sends the first step value of the computed input trajectory to the corresponding actuators.
4. Go to Step 1 ($k_f \leftarrow k_f + 1$).

Similarly, the implementation strategy corresponding to the slow subsystem under the slow LEMPC can be described as follows:

1. The LEMPC receives $x(t_{k_s})$ from the sensors.
2. If $t_{k_s} < t_s$, go to Step 3. Else, go to Step 4.
3. If $x(t_{k_s}) \in \Omega_{\rho_e}$, go to Step 3.1. Else, go to Step 3.2.
   3.1. The LEMPC computes an input trajectory over a finite-time prediction horizon that optimizes the economic cost function while maintains the predicted state trajectory to be within Ω_{ρ_e}. Go to Step 5.
   3.2. The LEMPC computes an input trajectory that forces the state closer to the region Ω_{ρ_e}. Go to Step 5.
4. The LEMPC computes an input trajectory that drives the slow subsystem state to a small neighborhood of the origin.
5. The LEMPC sends the first step value of the computed input trajectory to the corresponding actuators.
6. Go to Step 1 ($k_s \leftarrow k_s + 1$).

7.2.4.2 Fast LMPC Formulation

Referring to the fast subsystem of Eq. 7.6, the fast LMPC at sampling time t_{k_f} is formulated as follows

$$\min_{u_f \in S(\Delta_f)} \int_{t_{k_f}}^{t_{k_f}+N_f\Delta_f} [\tilde{y}^T(\hat{\tau})Q_f\tilde{y}(\hat{\tau}) + u_f^T(\hat{\tau})R_f u_f(\hat{\tau})]\, d\hat{\tau} \tag{7.11a}$$

$$\text{s.t.} \quad \frac{d\tilde{y}(\hat{\tau})}{d\hat{\tau}} = g(x(t_{k_s}), \tilde{y}(\hat{\tau}) + \tilde{g}(x(t_{k_s}), 0), u_f(\hat{\tau}), 0, 0) \tag{7.11b}$$

$$\tilde{y}(t_{k_f}) = \bar{y}(t_{k_f}) \tag{7.11c}$$

$$u_f(\hat{\tau}) \in \mathbb{U}_f, \ \forall\, \hat{\tau} \in [t_{k_f}, t_{k_f} + N_f\Delta_f) \tag{7.11d}$$

$$\begin{aligned} &\frac{\partial V_f(\bar{y}(t_{k_f}))}{\partial \bar{y}} g(x(t_{k_s}), \tilde{y}(t_{k_f}) + \tilde{g}(x(t_{k_s}), 0), u_f(t_{k_f}), 0, 0) \\ &\le \frac{\partial V_f(\bar{y}(t_{k_f}))}{\partial \bar{y}} g(x(t_{k_s}), \bar{y}(t_{k_f}) + \tilde{g}(x(t_{k_s}), 0), p(x(t_{k_s}))\bar{y}(t_{k_f}), 0, 0) \end{aligned} \tag{7.11e}$$

where $S(\Delta_f)$ is the family of piece-wise constant functions with sampling period Δ_f, N_f is the prediction horizon of LMPC, Q_f and R_f are positive definite weighting matrices that penalize the deviation of the fast subsystem state and manipulated input from their corresponding values at the equilibrium slow manifold, $\bar{y}(t_{k_f})$ is the fast subsystem state measurement obtained at t_{k_f}, $\tilde{y}$ denotes the predicted fast subsystem state trajectory of the nominal fast subsystem model of Eq. 7.11b over the prediction horizon. The input trajectory is subject to the manipulated input constraint of Eq. 7.11d. The constraint of Eq. 7.11e indicates that the amount of reduction in the value of the Lyapunov function when the manipulated input u_f computed by the LMPC of Eq. 7.11 is applied is at least the amount of reduction when the Lyapunov-based controller $p(x)y$ is applied in a sample-and-hold fashion. Since the LMPC of Eq. 7.11 obtains the manipulated input trajectory u_f every Δ_f, $x(t_{k_s})$ is the last available measurement of the slow process state, i.e., $t_{k_s} \leq t_{k_f}$. The optimal solution to this optimization problem is defined by $u_f^*(\hat{\tau}|t_{k_f})$ for all $\hat{\tau} \in [t_{k_f}, t_{k_f} + N_f \Delta_f)$ and the manipulated input of the closed-loop fast subsystem under the LMPC of Eq. 7.11 is defined as follows:

$$u_f(t) = u_f^*(t|t_{k_f}), \ \forall\, t \in [t_{k_f}, t_{k_f} + \Delta_f). \tag{7.12}$$

To analyze the closed-loop stability of the fast subsystem under the fast LMPC, a few properties are needed. By continuity, the smoothness property assumed for the vector function $g(\cdot)$ and taking into account that the manipulated input u_f and the disturbance w are bounded in compact sets, there exists a positive constant M_f such that

$$|g(x, \bar{y} + \tilde{g}(x, w), u_f, w, 0)| \leq M_f \tag{7.13}$$

for all $\bar{y} \in \Omega_{\rho^f}$, $u_f \in \mathbb{U}_f$, $x \in \Omega_{\rho^s}$ and $w \in \mathbb{W}$. Furthermore, by the continuous differentiable property of the Lyapunov function $V_f(\cdot)$ and the smoothness property assumed for the vector function $g(\cdot)$, there exist positive constants $L_{\bar{y}}$ and L_{w_f} such that

$$\begin{aligned} \left| \frac{\partial V_f(\bar{y})}{\partial \bar{y}} g(x, \bar{y}, u_f, w, 0) - \frac{\partial V_f(\bar{y}')}{\partial \bar{y}} g(x, \bar{y}', u_f, w, 0) \right| &\leq L_{\bar{y}} |\bar{y} - \bar{y}'| \\ \left| \frac{\partial V_f(\bar{y})}{\partial \bar{y}} g(x, \bar{y}, u_f, w, 0) - \frac{\partial V_f(\bar{y})}{\partial \bar{y}} g(x, \bar{y}, u_f, w', 0) \right| &\leq L_{w_f} |w - w'| \end{aligned} \tag{7.14}$$

for all $\bar{y}, \bar{y}' \in \Omega_{\rho^f}$, $u_f \in \mathbb{U}_f$, and $w, w' \in \mathbb{W}$.

Proposition 7.1 characterizes the closed-loop stability properties of the LMPC of Eq. 7.11. The analysis is similar to that of the LMPC presented in [10].

Proposition 7.1 *Consider the fast subsystem of Eq. 7.6 in closed-loop under the LMPC of Eq. 7.11 based on the feedback controller $p(x)y$ that satisfies the conditions of Eq. 7.10. Let $\varepsilon_{w_f} > 0$, $\Delta_f > 0$ and $\rho^f > \rho_s^f > 0$, $\theta > 0$ satisfy:*

$$-\alpha_{f_3}(\alpha_{f_2}^{-1}(\rho_s^f)) + L_{\bar{y}} M_f \Delta_f + (L_{\bar{y}} L_{\tilde{g}} + L_{w_f})\theta \leq -\varepsilon_{w_f}/\Delta_f. \tag{7.15}$$

Then, there exists a class $\mathcal{KL}$ function β_y and a class $\mathcal{K}$ function γ_y such that if $\bar{y}(0) \in \Omega_{\rho^f}$, then $\bar{y}(t) \in \Omega_{\rho^f}$ for all $t \geq 0$ and

$$|\bar{y}(t)| \leq \beta_y\left(|\bar{y}(0)|, \frac{t}{\varepsilon}\right) + \gamma_y(\rho_f^*) \tag{7.16}$$

with $\rho_f^* = \max_{s\in[0,\Delta_f]}\{V_f(\bar{y}(s)) : V_f(\bar{y}(0)) \leq \rho_s^f\}$, uniformly in $x \in \Omega_{\rho^s}$ and $w \in W$.

Proof The time derivative of the Lyapunov function along the state trajectory $\bar{y}(t)$ of fast subsystem of Eq. 7.6 for $t \in [t_{k_f}, t_{k_f} + \Delta_f)$ is written as follows:

$$\dot{V}_f(\bar{y}(t)) = \frac{\partial V_f(\bar{y}(t))}{\partial \bar{y}} g(x(t_{k_s}), \bar{y}(t) + \tilde{g}(x(t_{k_s}), w), u_f^*(t_{k_f}|t_{k_f}), w, 0). \tag{7.17}$$

Adding and subtracting the term:

$$\frac{\partial V_f(\bar{y}(t_{k_f}))}{\partial \bar{y}} g(x(t_{k_s}), \bar{y}(t_{k_f}) + \tilde{g}(x(t_{k_s}), 0), u_f^*(t_{k_f}|t_{k_f}), 0, 0)$$

to the right-hand-side of Eq. 7.17 and taking Eq. 7.10 into account, we obtain the following inequality:

$$\begin{aligned}\dot{V}_f(\bar{y}(t)) \leq -\alpha_{f_3}(|\bar{y}(t_{k_f})|) &+ \frac{\partial V_f(\bar{y}(t))}{\partial \bar{y}} g(x(t_{k_s}), \bar{y}(t) + \tilde{g}(x(t_{k_s}), w), u_f^*(t_{k_f}|t_{k_f}), w, 0)\\ &- \frac{\partial V_f(\bar{y}(t_{k_f}))}{\partial \bar{y}} g(x(t_{k_s}), \bar{y}(t_{k_f}) + \tilde{g}(x(t_{k_s}), 0), u_f^*(t_{k_f}|t_{k_f}), 0, 0)\end{aligned} \tag{7.18}$$

for $t \in [t_{k_f}, t_{k_f} + \Delta_f)$. From Eq. 7.14, Assumption 7.1 and the inequality of Eq. 7.18, the following inequality is obtained for all $\bar{y}(t_{k_f}) \in \Omega_{\rho^f} \backslash \Omega_{\rho_s^f}$:

$$\dot{V}_f(\bar{y}(t)) \leq -\alpha_{f_3}(\alpha_{f_2}^{-1}(\rho_s^f)) + L_{\bar{y}}|\bar{y}(t) - \bar{y}(t_{k_f})| + (L_{\bar{y}}L_{\tilde{g}} + L_{w_f})\theta. \tag{7.19}$$

for $t \in [t_{k_f}, t_{k_f} + \Delta_f)$. Taking into account Eq. 7.13 and the continuity of $\bar{y}(t)$, the following bound can be written for all $t \in [t_{k_f}, t_{k_f} + \Delta_f)$, $|\bar{y}(t) - \bar{y}(t_{k_f})| \leq M_f \Delta_f$. Using this expression, the following bound is obtained on the time derivative of the Lyapunov function for $t \in [t_{k_f}, t_{k_f} + \Delta_f)$:

$$\dot{V}_f(\bar{y}(t)) \leq -\alpha_{f_3}(\alpha_{f_2}^{-1}(\rho_s^f)) + L_{\bar{y}}M_f\Delta_f + (L_{\bar{y}}L_{\tilde{g}} + L_{w_f})\theta.$$

for all $\bar{y}(t_{k_f}) \in \Omega_{\rho^f} \backslash \Omega_{\rho_s^f}$. If the condition of Eq. 7.15 is satisfied, then

$$\dot{V}_f(\bar{y}(t)) \leq -\varepsilon_{w_f}/\Delta_f, \ \forall\, t \in [t_{k_f}, t_{k_f} + \Delta_f)$$

for $\bar{y}(t_{k_f}) \in \Omega_{\rho^f} \setminus \Omega_{\rho_s^f}$. Integrating this bound over $t \in [t_{k_f}, t_{k_f} + \Delta_f)$, the following is obtained:

$$V_f(\bar{y}(t_{k_f} + \Delta_f)) \leq V_f(\bar{y}(t_{k_f})) - \varepsilon_{w_f} \tag{7.20a}$$

$$V_f(\bar{y}(t)) \leq V_f(\bar{y}(t_{k_f})), \ \forall\, t \in [t_{k_f}, t_{k_f} + \Delta_f) \tag{7.20b}$$

for all $\bar{y}(t_{k_f}) \in \Omega_{\rho^f} \setminus \Omega_{\rho_s^f}$. Using Eq. 7.20 recursively, it can be proved that, if $x(t_{k_f}) \in \Omega_{\rho^f} \setminus \Omega_{\rho_s^f}$, the state converges to $\Omega_{\rho_s^f}$ in a finite number of sampling times without leaving the stability region. Once the state converges to $\Omega_{\rho_s^f} \subseteq \Omega_{\rho_f^*}$, it remains inside $\Omega_{\rho_f^*}$ for all times. This statement holds because of the definition of ρ_f^*. This proves that the closed-loop system under the fast LMPC design is ultimately bounded in $\Omega_{\rho_f^*}$. Thus, due to the continuity of the Lyapunov function $V_f(\cdot)$, it can be concluded that there exists a class $\mathscr{KL}$ function β_y and a class $\mathscr{K}$ function γ_y such that if $\bar{y}(0) \in \Omega_{\rho^f}$, then $\bar{y}(t) \in \Omega_{\rho^f}$ for all $t \geq 0$ and

$$|\bar{y}(t)| \leq \beta_y\left(|\bar{y}(0)|, \frac{t}{\varepsilon}\right) + \gamma_y(\rho_f^*). \tag{7.21}$$

Remark 7.3 The purpose of the fast LMPC scheme is to stabilize the fast subsystem dynamics while economic considerations are addressed through the slow LEMPC. Depending on the application and certain optimality specifications, the fast LMPC is desired in processes where the fast time-scale is large enough to warrant the use of MPC to achieve optimal performance compared to explicit fast feedback controllers that achieve fast stabilizability without necessarily achieving optimal performance. However, when the fast time-scale is short, an explicit feedback controller may be needed to ensure sufficiently fast computation of the "fast" control action; please see the example section. Additionally, since the closed-loop stability analysis does not require certain conditions on the stage cost of the LMPC, i.e., the stage cost does not need to be a quadratic stage cost, one can readily use an economic stage cost in the fast LMPC formulation.

7.2.4.3 Slow LEMPC Formulation

Referring to the slow subsystem of Eq. 7.4, the slow LEMPC at sampling time t_{k_s} is formulated as follows

$$\max_{u_s \in S(\Delta_s)} \int_{t_{k_s}}^{t_{k_s}+N_s\Delta_s} l_e(\tilde{x}(\tilde{\tau}), u_s(\tilde{\tau}))\, d\tilde{\tau} \tag{7.22a}$$

$$\text{s.t.} \quad \frac{d\tilde{x}(\tilde{\tau})}{d\tilde{\tau}} = f_s(\tilde{x}(\tilde{\tau}), u_s(\tilde{\tau}), 0) \tag{7.22b}$$

$$u_s(\tilde{\tau}) \in \mathbb{U}_s, \ \forall\, \tilde{\tau} \in [t_{k_s}, t_{k_s} + N_s\Delta_s) \tag{7.22c}$$

$$\tilde{x}(t_{k_s}) = \bar{x}(t_{k_s}) \tag{7.22d}$$

$$V_s(\tilde{x}(\tilde{\tau})) \le \rho_e,\ \forall\, \tilde{\tau} \in [t_{k_s}, t_{k_s} + N_s\Delta_s),\ \text{ if } t_{k_s} \le t_s \text{ and } V_s(\bar{x}(t_{k_s})) \le \rho_e \tag{7.22e}$$

$$\frac{\partial V_s(\bar{x}(t_{k_s}))}{\partial \bar{x}} f_s(\bar{x}(t_{k_s}), u_s(t_{k_s}), 0) \le \frac{\partial V_s(\bar{x}(t_{k_s}))}{\partial \bar{x}} f_s(\bar{x}(t_{k_s}), h_s(\bar{x}(t_{k_s})), 0),$$
$$\text{if } t_{k_s} > t_s \text{ or } V_s(x(t_{k_s})) > \rho_e \tag{7.22f}$$

where $S(\Delta_s)$ is the family of piece-wise constant functions with sampling period Δ_s, N_s is the prediction horizon of LEMPC and $l_e(\cdot)$ denotes an economic stage cost function. The symbol $\tilde{x}$ denotes the predicted slow subsystem state trajectory obtained by utilizing the nominal slow subsystem model of Eq. 7.22b initialized by the state feedback of Eq. 7.22d. For mode one operation which corresponds to sampling times $t_{k_s} \le t_s$, the constraint of Eq. 7.22e maintains the predicted slow state within Ω_{ρ_e} if $V_s(x(t_{k_s})) \le \rho_e$. If $V_s(x(t_{k_s})) > \rho_e$ or the LEMPC operates under mode two, the constraint of Eq. 7.22f ensures that the amount of reduction in the value of the Lyapunov function $V_s(\cdot)$ when u_s computed by the LEMPC of Eq. 7.22 is applied, is at least at the level when the feedback controller $h_s(\cdot)$ is applied in a sample-and-hold fashion. The optimal solution to the optimization problem of Eq. 7.22 is defined by $u_s^*(\tilde{\tau}|t_{k_s})$ for $\tilde{\tau} \in [t_{k_s}, t_{k_s} + N_s\Delta_s)$ and the manipulated input of the closed-loop slow subsystem under the LEMPC of Eq. 7.22 is defined as follows:

$$u_s(t) = u_s^*(t|t_{k_s}),\ \forall\, t \in [t_{k_s}, t_{k_s} + \Delta_s). \tag{7.23}$$

Next, the stability properties of the slow subsystem under the slow LEMPC is analyzed. By continuity, the smoothness property assumed for the vector field $f_s(\cdot)$ and taking into account that the manipulated input u_s and the disturbance w are bounded in compact sets, there exists a positive constant M_s such that

$$|f_s(\bar{x}, u_s, w)| \le M_s \tag{7.24}$$

for all $\bar{x} \in \Omega_{\rho^s}$, $u_s \in \mathbb{U}_s$, and $w \in \mathbb{W}$. In addition, by the continuous differentiable property of the Lyapunov function $V_s(\cdot)$ and the smoothness property assumed for the vector field $f_s(\cdot)$, there exist positive constants $L_{\bar{x}}$ and L_{w_s} such that

$$\left|\frac{\partial V_s(\bar{x})}{\partial \bar{x}} f_s(\bar{x}, u_s, w) - \frac{\partial V_s(\bar{x}')}{\partial \bar{x}} f_s(\bar{x}', u_s, w)\right| \le L_{\bar{x}}|\bar{x} - \bar{x}'| \tag{7.25a}$$

$$\left|\frac{\partial V_s(\bar{x})}{\partial \bar{x}} f_s(\bar{x}, u_s, w) - \frac{\partial V_s(\bar{x})}{\partial \bar{x}} f_s(\bar{x}, u_s, w')\right| \le L_{w_s}|w - w'| \tag{7.25b}$$

for all $\bar{x}, \bar{x}' \in \Omega_{\rho^s}$, $u_s \in \mathbb{U}_s$, and $w, w' \in \mathbb{W}$. Also, by the smoothness property assumed for the vector function $f(x, z, u_s, w, \varepsilon)$, there exist positive constants L_z, L_ε, L_x, $\bar{L}_z$, $\bar{L}_\varepsilon$, and $\bar{L}_w$ such that

$$\left|f(x, z, u_s, w, \varepsilon) - f(x, z', u_s, w, \varepsilon')\right| \le L_z|z - z'| + L_\varepsilon|\varepsilon - \varepsilon'| \tag{7.26}$$

$$\left|\frac{\partial V_s(x)}{\partial \bar{x}} f(x, z, u_s, w, \varepsilon) - \frac{\partial V_s(x')}{\partial \bar{x}} f(x', z', u_s, w', \varepsilon')\right| \leq L_x|x - x'| + \bar{L}_z|z - z'| + \bar{L}_\varepsilon|\varepsilon - \varepsilon'| + \bar{L}_w|w - w'| \tag{7.27}$$

for all $x, x' \in \Omega_{\rho^s}$, $z - \tilde{g}(x, w)$, $z' - \tilde{g}(x, w') \in \Omega_{\rho^f}$, $u_f \in \mathbb{U}_f$, and $w, w' \in \mathbb{W}$.

Proposition 7.2 characterizes the closed-loop stability properties of the LEMPC of Eq. 7.11.

Proposition 7.2 (Theorem 4.1) *Consider the slow subsystem of Eq. 7.4 under the LEMPC design of Eqs. 7.22 based on a controller $h_s(\cdot)$ that satisfies the conditions of Eq. 7.7. Let $\varepsilon_{w_s} > 0$, $\Delta_s > 0$, $\rho^s > \rho_e > 0$ and $\rho^s > \rho_s^s > 0$ satisfy*

$$\rho_e \leq \rho^s - f_V(f_W(\Delta_s)) \tag{7.28}$$

where

$$f_V(s) = \alpha_{s_4}(\alpha_{s_1}^{-1}(\rho^s))s + M_v s^2 \tag{7.29}$$

with M_v being a positive constant and

$$f_W(s) = \frac{L_{w_s}\theta}{L_{\bar{x}}}(e^{L_{\bar{x}}s} - 1) \tag{7.30}$$

and

$$-\alpha_{s_3}(\alpha_{s_2}^{-1}(\rho_s^s)) + L_{\bar{x}} M_s \Delta_s + L_{w_s}\theta \leq -\frac{\varepsilon_{w_s}}{\Delta_s}. \tag{7.31}$$

If $\bar{x}(0) \in \Omega_{\rho^s}$, $\rho_s^s < \rho_e$, $\rho_s^* < \rho^s$ and $N_s \geq 1$ then the state $\bar{x}(t)$ of the closed-loop slow subsystem is always bounded in Ω_{ρ^s}. Furthermore, there exists a class $\mathcal{KL}$ function β_x and a class $\mathcal{K}$ function γ_x such that

$$|\bar{x}(t)| \leq \beta_x(|\bar{x}(t^*)|, t - t^*) + \gamma_x(\rho_s^*) \tag{7.32}$$

with $\rho_s^* = \max_{s\in[0,\Delta_s]}\{V_s(\bar{x}(s)) : V_s(\bar{x}(0)) \leq \rho_s^s\}$, for all $\bar{x}(t^*) \in B_\delta \subset \Omega_{\rho^s}$ and for $t \geq t^* > t_s$ where t^* is chosen such that $\bar{x}(t^*) \in B_\delta$ and $B_\delta = \{x \in \mathbb{R}^{n_x} : |x| \leq \delta\}$.

The proof includes similar steps as the proof of Proposition 7.1. For the specific details, refer to Theorem 4.1.

7.2.4.4 Closed-Loop Stability

The closed-loop stability of the system of Eq. 7.1 under the LMPC of Eq. 7.11 and LEMPC of Eq. 7.22 is established in the following theorem under appropriate conditions.

Theorem 7.1 *Consider the system of Eq. 7.1 in closed-loop with u_f and u_s computed by the LMPC of Eq. 7.11 and LEMPC of Eq. 7.22 based on the Lyapunov-based controllers $p(x)y$ and $h_s(\cdot)$ that satisfy the conditions of Eqs. 7.7 and 7.10, respectively. Let Assumptions 7.1 and 7.2 and the conditions of Propositions 7.1 and 7.2 hold and*

$$\rho_e + (M_s + L_z M_y + L_\varepsilon \varepsilon)\Delta_s \alpha_{s_4}(\alpha_{s_1}^{-1}(\rho^s)) < \rho^s \tag{7.33}$$

and

$$-\alpha_{s_3}(\alpha_{s_1}^{-1}(\rho_s^s)) + d_1 < 0 \tag{7.34}$$

where L_z, M_y and d_1 are positive constants to be defined in the proof. Then there exist functions β_x of class $\mathcal{KL}$ and γ_x of class $\mathcal{K}$, a pair of positive real numbers (δ_x, d) and $\varepsilon^* > 0$ such that if

$$\max\{|x(0)|, |y(0)|, \|w\|, \|\dot{w}\|\} \le \delta_x$$

where $\|w\|$ denotes $\text{ess sup}_{t\ge0}|w(t)|$ and $\varepsilon \in (0, \varepsilon^*]$, then $x(t) \in \Omega_{\rho^s}$ and $y(t) \in \Omega_{\rho^f}$ for $t \ge 0$ and

$$|x(t)| \le \beta_x(|x(t^*)|, t - t^*) + \gamma_x(\rho_s^*) + d \tag{7.35}$$

for all $t \ge t^* > t_s$ where t^* has been defined in Proposition 7.2.

Proof When $u_f = u_f^*$ and $u_s = u_s^*$ are determined by the LMPC of Eq. 7.11 and LEMPC of Eq. 7.22, respectively, the closed-loop system takes the following form:

$$\begin{aligned} \dot{x} &= f(x, z, u_s^*, w, \varepsilon) \\ \varepsilon\dot{z} &= g(x, z, u_f^*, w, \varepsilon). \end{aligned} \tag{7.36}$$

We will first compute the slow and fast closed-loop subsystems.

Setting $\varepsilon = 0$ in Eq. 7.36 and taking advantage of the fact that $u_f^* = 0$ when $\varepsilon = 0$, one obtains that:

$$\begin{aligned} \frac{d\bar{x}}{dt} &= f(\bar{x}, \bar{z}, u_s^*, w, 0) \\ 0 &= g(\bar{x}, \bar{z}, 0, w, 0). \end{aligned} \tag{7.37}$$

Using Assumption 7.1, Eq. 7.37 can be written as follows:

$$\frac{d\bar{x}}{dt} = f(\bar{x}, \tilde{g}(\bar{x}, w), u_s^*, w, 0) = f_s(\bar{x}, u_s^*, w) \tag{7.38}$$

Let $\tau = t/\varepsilon$ and $y = z - \tilde{g}(x, w)$, and the closed-loop system of Eq. 7.36 can be written as:

$$\begin{aligned}\frac{dx}{d\tau} &= \varepsilon f(x, y + \tilde{g}(x, w), u_s^*, w, \varepsilon)\\ \frac{dy}{d\tau} &= g(x, y + \tilde{g}(x, w), u_f^*, w, \varepsilon) - \varepsilon \frac{\partial \tilde{g}}{\partial w}\dot{w} - \varepsilon \frac{\partial \tilde{g}}{\partial x} f(x, y + \tilde{g}(x, w), u_s^*, w, \varepsilon)\end{aligned} \tag{7.39}$$

Setting $\varepsilon = 0$, the following closed-loop fast subsystem is obtained:

$$\frac{d\bar{y}}{d\tau} = g(x, \bar{y} + \tilde{g}(x, w), u_f^*, w, 0) \tag{7.40}$$

Now, we focus on the singularly perturbed system of Eq. 7.39. Considering the fast subsystem state $y(t)$ of Eq. 7.39 and assuming that $x(t)$ is bounded in Ω_{ρ^s} (which will be proved later), it can be obtained using a Lyapunov argument that there exist positive constants δ_{x_1} and ε_1 such that if $\max\{|x(0)|, |y(0)|, \|w\|, \|\dot{w}\|\} \leq \delta_{x_1}$ and $\varepsilon \in (0, \varepsilon_1]$, there exists a positive constant k_1 such that:

$$|z - \tilde{g}(x, w)| = |y(t)| \leq \beta_y\left(\delta_{x_1}, \frac{t}{\varepsilon}\right) + \gamma_y(\rho_f^*) + k_1 \tag{7.41}$$

for all $t \geq 0$. We consider $t \in (0, \Delta_s]$ and $t \geq \Delta_s$ separately and prove that if the conditions stated in Theorem 7.1 are satisfied, the boundedness of the states of the system of Eq. 7.39 is ensured. When $x(0) \in B_{\delta_{x_2}} \subset \Omega_{\rho_e} \subset \Omega_{\rho^s}$, where δ_{x_2} is a positive real number, considering the closed-loop slow subsystem of Eq. 7.39 state trajectory:

$$\dot{x}(t) = f(x, y + \tilde{g}(x, w), u_s^*, w, \varepsilon),\ \forall\, t \in (0, \Delta_s]$$

and considering the facts that

$$\begin{aligned}|f(x, z, u_s^*, w, \varepsilon)| &\leq |f_s(x, u_s^*, w)| + |f(x, z, u_s^*, w, \varepsilon) - f_s(x, u_s^*, w)|,\\ |f_s(x, u_s^*, w)| &\leq M_s,\\ |y(t)| &\leq \beta_y(\delta_{x_2}, 0) + \gamma_y(\rho_f^*) + k_1 < M_y\end{aligned}$$

where M_y is a positive constant such that

$$\begin{aligned}|f(x, z, u_s^*, w, \varepsilon) - f_s(x, u_s^*, w)| &= |f(x, z, u_s^*, w, \varepsilon) - f(x, \tilde{g}(x, w), u_s^*, w, 0)|\\ &\leq L_z|z - \tilde{g}(x, w)| + L_\varepsilon\varepsilon \leq L_z M_y + L_\varepsilon\varepsilon\end{aligned} \tag{7.42}$$

and

$$\begin{aligned}V_s(x(t)) &= V_s(x(0)) + \int_0^t \dot{V}_s(x(s))\, ds\\ &= V_s(x(0)) + \int_0^t \frac{\partial V_s(x(s))}{\partial x}\dot{x}(s)\, ds\end{aligned}$$

$$\leq \rho_e + (M_s + L_z M_y + L_\varepsilon \varepsilon)\Delta_s \alpha_{s_4}(\alpha_{s_1}^{-1}(\rho^s)) \tag{7.43}$$

Thus, there exists Δ_1 and ε_2 such that if $\Delta_s \in (0, \Delta_1]$ and $\varepsilon \in (0, \varepsilon_2]$, Eq. 7.33 holds and

$$V_s(x(t)) < \rho^s, \ \forall\, t \in (0, \Delta_s] \tag{7.44}$$

Picking $\varepsilon_3 = \min\{\varepsilon_1, \varepsilon_2\}$ ensures that $x(t) \in \Omega_{\rho^s}$ and $y(t) \in \Omega_{\rho^f}$ for all $t \in [0, \Delta_s)$.

For $t \geq \Delta_s$, considering Eq. 7.41, there exists a positive real number $\bar{M}_y$ such that

$$|y(t)| \leq \beta_y\left(\delta_{x_2}, \frac{\Delta_s}{\varepsilon}\right) + \gamma_y(\rho_f^*) + k_1 \leq \bar{M}_y \tag{7.45}$$

and we can write the time derivative of the Lyapunov function $V_s(\cdot)$ along the closed-loop system state of Eq. 7.1 under the LEMPC of Eq. 7.22 for $t \in [t_{k_s}, t_{k_s+1})$ (assuming without loss of generality that $t_{k_s} = \Delta_s$) as follows

$$\dot{V}_s(x(t)) = \frac{\partial V_s(x(t))}{\partial x} f(x(t), z(t), u_s^*(t_{k_s}), w(t), \varepsilon) \tag{7.46}$$

Adding/subtracting the terms:

$$\frac{\partial V_s(x(t_{k_s}))}{\partial x} f_s(x(t_{k_s}), u_s(t_{k_s}), 0) \text{ and } \frac{\partial V_s(x(t))}{\partial x} f_s(x(t), u_s(t_{k_s}), w(t))$$

to/from the above inequality and taking advantage of Eqs. 7.7 and 7.22f and the fact that

$$\begin{aligned}
&|f(x(t), z(t), u_s^*(t_{k_s}), w(t), \varepsilon)| \\
&\quad \leq |f_s(x(t), u_s^*(t_{k_s}), 0)| \\
&\qquad + |f_s(x(t), u_s^*(t_{k_s}), w(t)) - f_s(x(t_{k_s}), u_s^*(t_{k_s}), 0)| \\
&\qquad + |f(x(t), z(t), u_s^*(t_{k_s}), w(t), \varepsilon) - f_s(x(t), u_s^*(t_{k_s}), w(t))|
\end{aligned}$$

and considering

$$\left|\frac{\partial V_s(x(t))}{\partial x} f_s(x, u_s^*, w)\right| \leq \alpha_{s_4}(\alpha_{s_1}^{-1}(\rho^s))M_s, \tag{7.47}$$

$$\begin{aligned}
&\left|\frac{\partial V_s(x(t))}{\partial x} f(x(t), z(t), u_s^*(t_{k_s}), w(t), \varepsilon) - \frac{\partial V_s(x(t))}{\partial x} f_s(x(t), u_s^*(t_{k_s}), w(t))\right| \\
&\quad \leq L_{\tilde{z}}|z - \tilde{g}(x, w)| + L_{\tilde{\varepsilon}}\varepsilon,
\end{aligned} \tag{7.48}$$

$$|z - \tilde{g}(x, w)| \leq \bar{M}_y, \tag{7.49}$$

$$\left| \frac{\partial V_s(x(t))}{\partial x} f_s(x(t), u_s^*(t_{k_s}), w(t)) - \frac{\partial V_s(x(t_{k_s}))}{\partial x} f_s(x(t_{k_s}), u_s^*(t_{k_s}), 0) \right| \le L_{\bar{x}} |x(t) - x(t_{k_s})| + L_{w_s}\theta, \tag{7.50}$$

and

$$|x(t) - x(t_{k_s})| \le M_s \Delta_s, \tag{7.51}$$

we can obtain

$$\dot{V}_s(x(t)) \le -\alpha_{s_3}(\alpha_{s_1}^{-1}(\rho_s^s)) + d_1 \tag{7.52}$$

where

$$d_1 = \alpha_{s_4}(\alpha_{s_1}^{-1}(\rho^s))M_s + L_{\bar{z}}\bar{M}_y + L_{\bar{\varepsilon}}\varepsilon + L_{\bar{x}} M_s \Delta_s + L_{w_s}\|w\| \tag{7.53}$$

where d_1 is a positive constant. Picking δ_{x_2}, ε_4 and Δ_2 such that for any $\varepsilon \in (0, \varepsilon_4]$, $\max\{|x(0)|, |y(0)|, \|w\|, \|\dot{w}\|\} \le \delta_{x_2}$ and for all $\Delta_s \in (0, \Delta_2]$, Eq. 7.34 is satisfied, the closed-loop system state $x(t)$ is bounded in Ω_{ρ^s} for all $t \ge \Delta_s$. Finally, using similar arguments to the proof of Theorem 1 in [11], there exist a class $\mathscr{KL}$ function β_x and a class $\mathscr{K}$ function γ_x, positive real numbers (δ_x, d_x) where $\delta_x < \min\{\delta_{x_1}, \delta_{x_2}\}$, and $0 < \varepsilon^* < \min\{\varepsilon_1, \varepsilon_2, \varepsilon_3, \varepsilon_4\}$ and $0 < \Delta^* < \min\{\Delta_1, \Delta_2\}$ such that if $\max\{|x(0)|, |y(0)|, \|w\|, \|\dot{w}\|\} \le \delta_x$, $\varepsilon \in (0, \varepsilon^*]$ and $\Delta_s \in (0, \Delta^*]$, then, the bound of Eq. 7.35 holds for all $t \ge t^* > t_s$.

Remark 7.4 It should be emphasized that in Theorem 7.1, it has been indicated that for operation periods corresponding to LEMPC mode 1 operation, both of fast and slow reduced order subsystem states are bounded in the invariant sets $\Omega_{\tilde{\rho}^s}$ and $\Omega_{\tilde{\rho}^f}$ to ensure that the actual states of the system are bounded in certain stability regions, i.e., Ω_{ρ^s} and Ω_{ρ^f} through restricting their corresponding initial states. On the other hand, for operation periods corresponding to LEMPC mode 2 operation, both of system states are asymptotically bounded in a small invariant set containing the origin.

Remark 7.5 While the present work focuses on nonlinear singularly perturbed systems and general economic stage cost functions, the results of this work are novel and apply to the case of linear singularly perturbed systems; however, in the linear case, the verification of the assumption that there is an isolated equilibrium manifold for the fast dynamics (Assumption 1), the construction of the explicit control laws for the slow and fast subsystems imposed in Sect. 2.4 and the computation of the associated closed-loop stability regions, and the solution of the LEMPC and LMPC optimization problems when convex economic cost functions are used simplify significantly, given the availability of robust and efficient tools for matrix calculations and convex optimization.

7.2.5 Application to a Chemical Process Example

Consider a well-mixed, non-isothermal continuous stirred tank reactor (CSTR) where an irreversible, second-order, endothermic reaction $A \rightarrow B$ takes place, where A is the reactant and B is the desired product. The feed to the reactor consists of the reactant A and an inert gas at flow rate F, temperature T_0 and molar concentration C_{A0}. Due to the non-isothermal nature of the reactor, a jacket is used to provide heat to the reactor. The dynamic equations describing the behavior of the reactor, obtained through material and energy balances under standard modeling assumptions, are given below:

$$\frac{dC_A}{dt} = \frac{F}{V_R}(C_{A0} - C_A) - k_0 e^{-E/RT} C_A^2 \tag{7.54a}$$

$$\rho_R C_p \frac{dT}{dt} = \frac{F \rho_R C_p}{V_R}(T_0 - T) - \Delta H k_0 e^{-E/RT} C_A^2 + \frac{Q}{V_R} \tag{7.54b}$$

where C_A denotes the concentration of the reactant A, T denotes the temperature of the reactor, Q denotes the rate of heat supply to the reactor, V_R represents the volume of the reactor, ΔH, k_0, and E denote the enthalpy, pre-exponential constant and activation energy of the reaction, respectively, and C_p and ρ_R denote the heat capacity and the density of the fluid in the reactor, respectively. The values of the process parameters used in the simulations are shown in Table 7.1. The process model of Eq. 7.54 is numerically simulated using an explicit Euler integration method with integration step $h_c = 1.0 \times 1.0^{-6}$ h.

The process model has one unstable steady-state in the operating range of interest. The control objective is to optimize the process operation in a region around the unstable steady-state (C_{As}, T_s) to maximize the average production rate of B through manipulation of the concentration of A in the inlet to the reactor. The steady-state input value associated with the steady-state point is denoted by C_{A0s}. Defining $\varepsilon = \rho_R c_P$, the following nonlinear state-space model can be obtained

$$\dot{x}(t) = f(x(t), z(t), u_s(t), 0, \varepsilon)$$
$$\varepsilon \dot{z}(t) = g(x(t), z(t), u_f(t), 0, \varepsilon)$$

Table 7.1 Parameter values

$T_0 = 300$	K	$F = 5$	$\mathrm{m^3\,h^{-1}}$
$V_R = 1.0$	$\mathrm{m^3}$	$E = 5 \times 10^4$	$\mathrm{kJ\,kmol^{-1}}$
$k_0 = 8.46 \times 10^6$	$\mathrm{m^3\,kmol^{-1}\,h^{-1}}$	$\Delta H = -19.91$	$\mathrm{kJ\,kmol^{-1}}$
$C_p = 0.02$	$\mathrm{kJ\,kg^{-1}\,K^{-1}}$	$R = 8.314$	$\mathrm{kJ\,kmol^{-1}\,K^{-1}}$
$\rho_R = 20$	$\mathrm{kg\,m^{-3}}$	$C_{As} = 1.95$	$\mathrm{kmol\,m^{-3}}$
$T_s = 401.87$	K	$C_{A0s} = 4$	$\mathrm{kmol\,m^{-3}}$
$Q_s = 0$	$\mathrm{kJ\,h^{-1}}$		

where $x = C_A - C_{As}$ and $z = T - T_s$ are the states, $u_s = C_{A0} - C_{A0s}$ and $u_f = Q - Q_s$ are the inputs and f and g are scalar functions. The inputs are subject to constraints as follows: $|u_s| \leq 3.5\,\text{kmol}\,\text{m}^{-3}$ and $|u_f| \leq 5.0 \times 10^5\,\text{kJ}\,\text{h}^{-1}$. The economic measure considered in this example is as follows:

$$\frac{1}{t_N} \int_0^{t_N} k_0 e^{-E/RT(\tau)} C_A^2(\tau)\, d\tau \tag{7.55}$$

where $t_N = 1.0\,\text{h}$ is the time duration of a reactor operating period. This economic objective function highlights the maximization of the average production rate over a process operation period for $t_N = 1.0\,\text{h}$ (of course, different, yet finite, values of t_N can be chosen). A limitation on the amount of reactant material that may be used over the period t_N is considered. Specifically, the control input trajectory of u_s should satisfy the following constraint:

$$\frac{1}{t_N} \int_0^{t_N} u_s(\tau)\, d\tau = 1.0\,\text{kmol}\,\text{m}^{-3}. \tag{7.56}$$

This constraint means that the available amount of reactant material over one period is fixed. For the sake of simplicity, the constraint of Eq. 7.56 is referred for the material constraint for the remainder.

In terms of the Lyapunov-based controllers, feedback linearization techniques are utilized for the design of explicit controllers for the fast and slow reduced-order subsystems subject to input constraints and quadratic Lyapunov functions $V_s(x) = x^2$ and $V_f(y) = y^2$ are used to compute the stability regions. Through feedback linearization and evaluating $\dot{V}_s(\cdot)$ subject to the input constraint, $\dot{V}_s(x) \leq 0$ when $x \in \Omega_{\rho^s}$ and $\rho^s = 4$. Furthermore, to guarantee that $C_A > 0$ and $T \leq 480$ K, the corresponding stability Ω_{ρ^s} is defined as $\Omega_{\rho^s} = \{x | -1.15 \leq x \leq 3.95\}$.

In this example, a slow LEMPC is designed to regulate the slow subsystem state, which maximizes the average production rate of the desired product B, and a fast feedback linearizing controller is designed to stabilize the fast subsystem state. With respect to the fast feedback linearizing controller, the deviation variable $y(t)$ is defined as $z(t) - \bar{z}(t)$ where $\bar{z}(t)$ is the unique root of the algebraic equation $g(x(t), \bar{z}(t), 0, 0, 0) = 0$ given $x(t)$. For the purpose of simulation, this unique root has been approximated through a tenth-order polynomial. Furthermore, we assume that the state measurements are available every $\Delta_f = 1.0 \times 1.0^{-6}\,\text{h}$ and the manipulated input u_f is obtained every Δ_f such that

$$g(x, y + \tilde{g}(x, 0), 0, u_f, 0) = -\mu y \tag{7.57}$$

where $\mu = 100$. Regarding the slow dynamics, the LEMPC obtains its manipulated input trajectory u_s every $\Delta_s = 1.0 \times 10^{-2}\,\text{h}$ by optimizing the objective function of Eq. 7.55 using the one dimensional slow subsystem which is independent of ε. As a result, the slow subsystem used in the LEMPC is well conditioned and is integrated with time step $1.0 \times 10^{-3}\,\text{h}$ resulting in a nearly-three order of magnitude improvement in the computational time needed to compute the control action for u_s. The LEMPC operates only under mode 1 operation to highlight the effect of economic optimization. Considering the material constraint which needs to be satisfied through one period of process operation, a shrinking prediction horizon sequence is employed in the LEMPC in the sense that $N_0, \ldots, N_{99}$ where $N_i = 100 - i$ denotes the prediction horizon at time step i and $i = 0, \ldots, 99$.

Figures 7.1, 7.2, 7.3, 7.4 and 7.5 illustrate closed-loop state and manipulated input trajectories of the chemical process of Eq. 7.54 under the mode one operation of the LEMPC design of Eq. 7.22 and feedback linearization of Eq. 7.57 for an initial condition of $(C_A(0), T(0)) = (3.0\,\text{kmol m}^{-3}, 400\,\text{K})$. From these figures, the economic cost of Eq. 7.55 is optimized by dictating time-varying operation by the LEMPC when considering the material constraint of Eq. 7.56. Furthermore, u_f through feedback linearization ensures that the fast subsystem state $y(t)$ converges to zero. We point out that either the open-loop or closed-loop dynamics can evolve on different time-scales. In this example, feedback linearization is used with a gain chosen to drive the deviation variable y to zero fast relative to the slow state C_A as observed in Fig. 7.3. Therefore, this illustrative example possesses two-time-scale behavior.

A set of simulations is performed to compare the economic closed-loop performance of the method versus the case that the input material is fed to the reactor uniformly in time, i.e., $u_s(t) = 1.0\,\text{kmol m}^{-3}$ for all $t \in [0, 1.0\,\text{h}]$. To carry out this comparison, the total cost of each scenario is computed based on the index of the following form:

$$J = \frac{1}{t_M} \sum_{i=0}^{M-1} \left(k_0 e^{-E/RT(t_i)} C_A^2(t_i) \right)$$

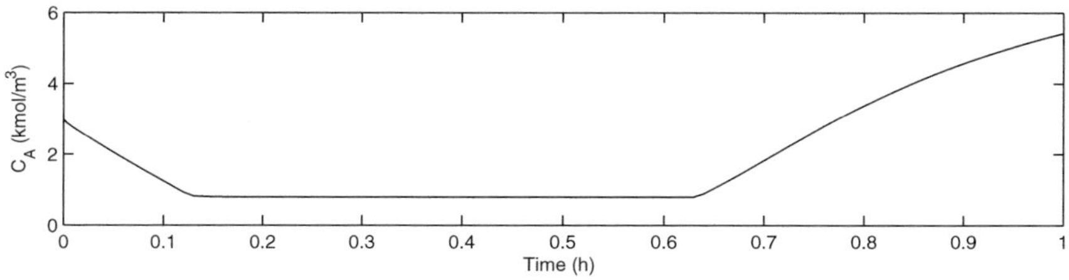

Fig. 7.1 The closed-loop reactant concentration profile under the composite control structure

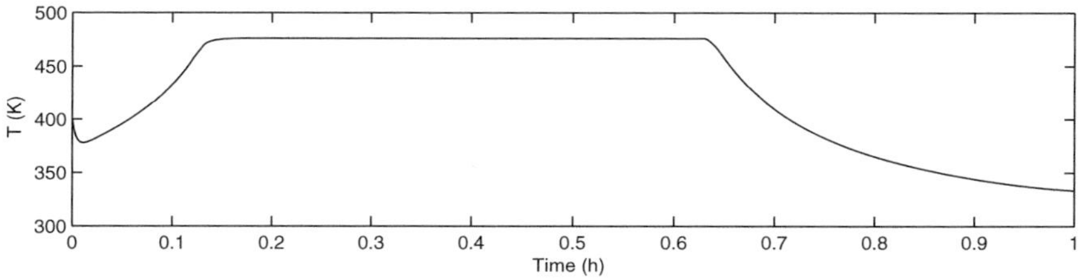

Fig. 7.2 The closed-loop temperature profile under the composite control structure

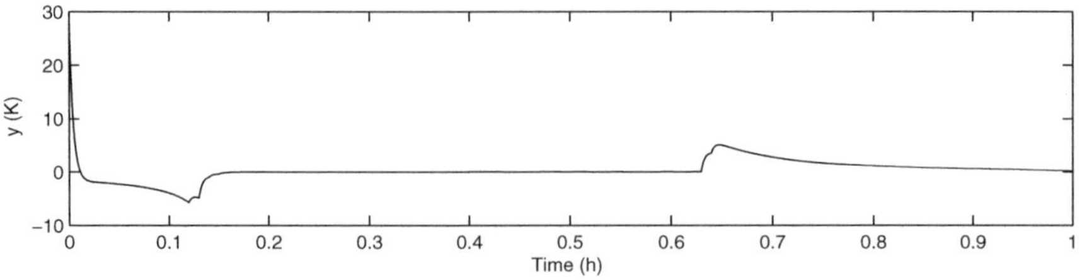

Fig. 7.3 The closed-loop profile of $y = z - \tilde{g}(x, 0)$

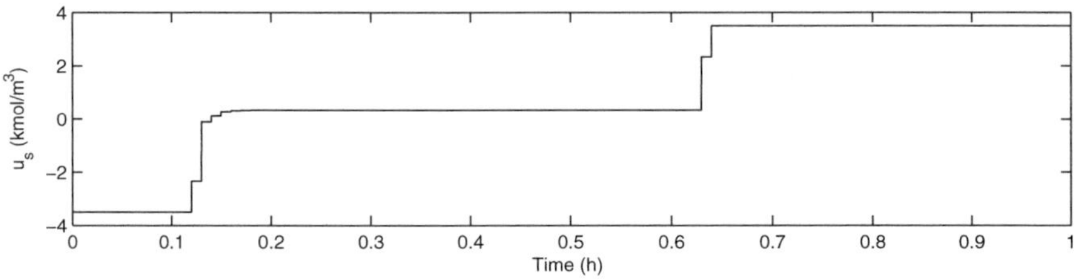

Fig. 7.4 The manipulated input u_s profile under the slow LEMPC, which optimizes the production rate of the desired product B

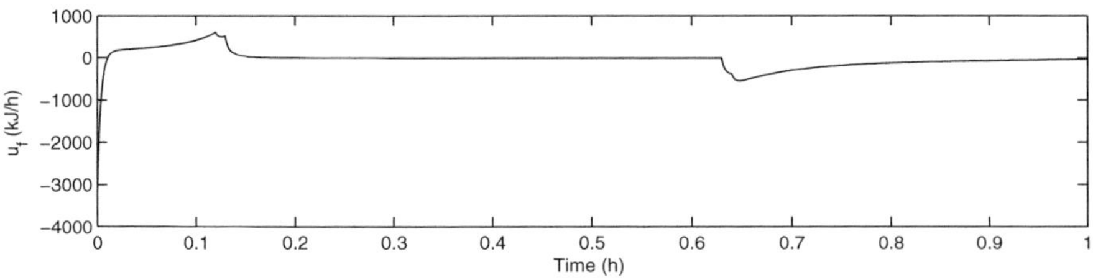

Fig. 7.5 The manipulated input u_f profile under the feedback linearizing controller of Eq. 7.57

where $t_0 = 0.0\,\text{h}$, $t_i = i0.01\,\text{h}$ and $M = 100$. By comparing the performance index J for these two cases, the LEMPC through a time-varying operation achieves a greater cost value, i.e., larger average production rate, compared to the case that the reactant material is fed to the reactor uniformly in time (13.12 versus 5.92, respectively).

7.3 Distributed EMPC: Evaluation of Sequential and Iterative Architectures

It is possible that significant computation delay may result, which may affect closed-loop stability and performance, when computing control actions for large-scale process systems with many states and inputs. In the context of control of large-scale nonlinear chemical process networks, an alternative is to employ a distributed

MPC (DMPC) architecture (see, for example, the review [12] and references contained therein). DMPC has the ability to control large-scale process systems subject to input and state constraints while remaining computationally feasible to be implemented on-line through a distributed implementation of the computations. Numerous formulations, implementation strategies, and theoretical results have been developed within the context of tracking DMPC, e.g., [13, 14]; see, also, the reviews of [12, 15] and the references therein. In the context of distributed EMPC (DEMPC), some work has been completed including DEMPC for linear systems [16, 17] and for nonlinear systems [18, 19].

In this section, sequential and iterative DEMPC strategies are developed and applied to the benchmark catalytic reactor to produce ethylene oxide from ethylene, which was first presented in Sect. 1.3.1. Recall, in Sect. 3.2, the application of EMPC to the catalytic reactor resulted in improved average yield of ethylene oxide compared to the yield of steady-state operation and to the yield achieved under an open-loop optimal periodic switching of the inputs considered in [20]. Here, several EMPC implementation strategies (centralized and distributed) are applied to the catalytic reactor. A description of the DEMPC implementation strategies is provided. Several closed-loop simulations are performed to evaluate the approaches. Two key performance metrics are considered in the evaluation: the closed-loop economic performance under the various DEMPC strategies and the on-line computation time required to solve the EMPC optimization problems.

Regarding the implementation details of the EMPC systems below, a sampling period of $\Delta = 1.0$ (dimensionless time units) was used. The optimization problems were solved using the interior point solver Ipopt [21]. To account for real-time computation considerations, the solver was forced to terminate after 100 iterations and/or after 100 seconds of computation time. The tolerance of the solver was set to 10^{-5}. To satisfy the constraint on the amount of ethylene that may be fed to the reactor, this constraint was enforced over operating windows of length $t_p = 47$, that is the average molar flow rate of ethylene must be equal to 0.175 at the end of each operating window (refer to Sect. 1.3.1 for more details regarding this average material constraint). A shrinking horizon approach was used within EMPC: at the beginning of the jth operating window, the prediction horizon was set to $N_k := t_p/\Delta$ and the horizon was decreased by one at every subsequent sampling time ($N_k = N_{k-1} - 1$ at the sampling instance t_k). At the beginning of the $(j+1)$th operating window, the prediction horizon was set to t_p/Δ.

The closed-loop simulations below were programmed using C++ on a desktop computer with an Ubuntu Linux operating system and an Intel® Core™ i7 3.4 GHz processor. To recursively solve the catalytic reactor dynamic model, the explicit Euler method was used. A step size of 0.00001 was used to simulate the closed-loop dynamics of the reactor, while a step size of 0.005 was used to solve the model within the EMPC problem; both led to stable numerical integration.

7.3.1 Centralized EMPC

For this computational study, a centralized EMPC (C-EMPC) strategy is considered to compare against the two distributed implementation strategies. Recall, the catalytic reactor that produces ethylene oxide from ethylene has three inputs: the inlet flow rate, u_1, the ethylene concentration in the inlet, u_2, and the coolant temperature in the reactor jacket. Also, the reactor model has the form of:

$$\dot{x} = f(x, u_1, u_2, u_3) \tag{7.58}$$

where the state vector $x \in \mathbb{R}^4$ includes the reactor content density, x_1, the reactor ethylene concentration, x_2, the reactor ethylene oxide concentration, x_3, and the reactor temperature, x_4. The C-EMPC formulation with an economic stage cost function that maximizes the yield of ethylene oxide is given by:

$$\max_{u_1,u_2,u_3 \in S(\Delta)} \int_{t_k}^{t_k+N_k\Delta} u_1(\tau)\tilde{x}_4(\tau)\tilde{x}_3(\tau)\, d\tau \tag{7.59a}$$

$$\text{s.t.} \quad \dot{\tilde{x}}(t) = f(\tilde{x}(t), u_1(t), u_2(t), u_3(t)) \tag{7.59b}$$

$$u_i(t) \in \mathbb{U}_i, \ \forall\, t \in [t_k, t_k + N_k\Delta),\ i = 1, 2, 3 \tag{7.59c}$$

$$\frac{1}{t_p}\int_{t_k}^{t_k+N_k\Delta} u_1(t)u_2(t)\, dt = 0.175 - \frac{1}{t_p}\int_{t_0+jt_p}^{t_k} u_1^*(t)u_2^*(t)\, dt \tag{7.59d}$$

where $\mathbb{U}_i$ denotes the set of admissible values of the ith input (refer to Sect. 1.3.1 for more details) and u_1^* and u_2^* denote the optimal control actions applied to the reactor from the beginning of the current operating window to current sampling time, t_k. The EMPC problem of Eq. 7.59 maximizes the yield of ethylene oxide (or more precisely, the numerator of the yield) over the prediction horizon (Eq. 7.59a) subject to the dynamic process model to predict the future behavior of the reactor (Eq. 7.59b), the inputs constraints (Eq. 7.59c), and the average constraint on amount of ethylene that may be fed to the reactor (Eq. 7.59d).

Figures 7.6 and 7.7 depict the closed-loop state and input trajectories under the C-EMPC scheme over ten operating windows. Similar to the results of [22], the C-EMPC distributes the ethylene in a non-uniform fashion with respect to time to optimize the yield of ethylene oxide. The average yield of ethylene oxide of the reactor under the C-EMPC is 10.22 yield of ethylene oxide of the reactor over the same length of operation under constant steady-state input values is 6.38 and the average yield under EMPC is 60 percent better than that achieved under steady-state operation.

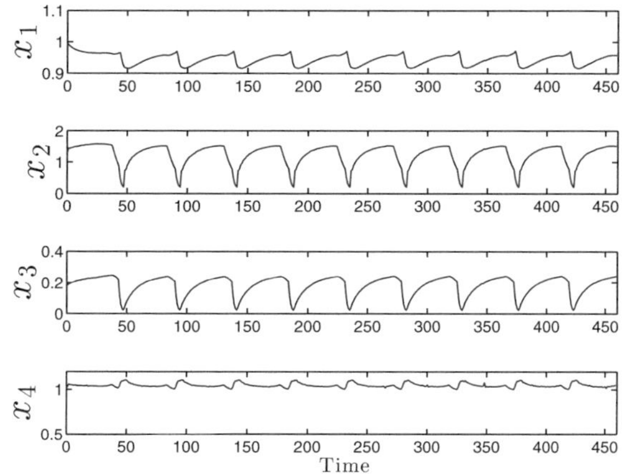

Fig. 7.6 State trajectories under C-EMPC

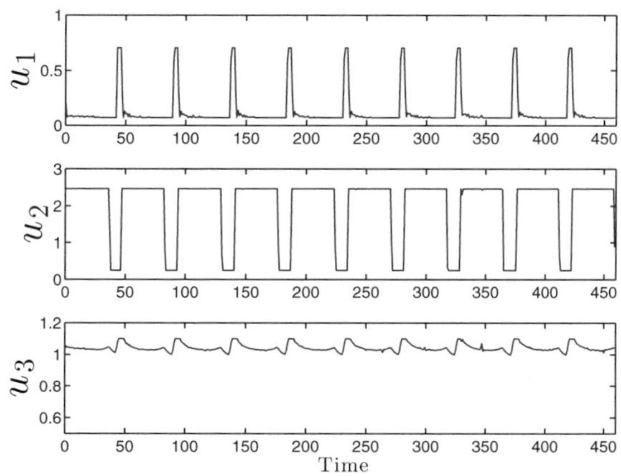

Fig. 7.7 Input trajectories computed by the C-EMPC

7.3.2 Sequential DEMPC

A sequential DEMPC implementation strategy computes the control actions by sequentially solving a series of DEMPC problems. One-way communication is used between controllers to send the computed input trajectories from one EMPC to the next EMPC. The next EMPC also receives the input trajectory from all other previously solved EMPCs. Once all the input trajectories are received, the EMPC is solved utilizing this information. The resulting trajectories are then sent to the subsequent EMPC. The process is repeated until all EMPCs are solved and the control actions for all inputs computed by the sequential DEMPC approach are obtained.

For the catalytic reactor example, which has three inputs, a reasonable choice of input grouping can be made as a consequence of the integral input constraint. The inputs u_1 and u_2 should be computed by the same EMPC, while it is worth investigating if the input u_3 can be placed on another EMPC system. This input pairing will be used in all the DEMPC schemes below. The formulation of the EMPC problem that computes control actions for u_1 and u_2, which is denoted as EMPC-1, is given by

$$\max_{u_1,u_2 \in S(\Delta)} \int_{t_k}^{t_k+N_k\Delta} u_1(\tau)\tilde{x}_4(\tau)\tilde{x}_3(\tau)\,d\tau \tag{7.60a}$$

$$\text{s.t.} \quad \dot{\tilde{x}}(t) = f(\tilde{x}(t), u_1(t), u_2(t), \hat{u}_3(t)) \tag{7.60b}$$

$$u_i(t) \in \mathbb{U}_i,\ \forall\, t \in [t_k, t_k + N_k\Delta),\ i = 1, 2 \tag{7.60c}$$

$$\frac{1}{t_p}\int_{t_k}^{t_k+N_k\Delta} u_1(t)u_2(t)\,dt = 0.175 - \frac{1}{t_p}\int_{t_0+jt_p}^{t_k} u_1^*(t)u_2^*(t)\,dt \tag{7.60d}$$

and the formulation of the EMPC that computes control actions for u_3, which is denoted as EMPC-2 is given by:

$$\max_{u_3 \in S(\Delta)} \int_{t_k}^{t_k+N_k\Delta} u_1(\tau)\tilde{x}_4(\tau)\tilde{x}_3(\tau)\,d\tau \tag{7.61a}$$

$$\text{s.t.} \quad \dot{\tilde{x}}(t) = f(\tilde{x}(t), \hat{u}_1(t), \hat{u}_2(t), u_3(t)) \tag{7.61b}$$

$$u_3(t) \in \mathbb{U}_3,\ \forall\, t \in [t_k, t_k + N_k\Delta) \tag{7.61c}$$

In the problems of Eqs. 7.60–7.61, the input trajectories denoted by $\hat{u}_i$ must be provided before the problems may be solved. The input trajectory $\hat{u}_3$ must be assumed if EMPC-1 is solved first. In general, the assumed input trajectory may be a constant input trajectory, an input trajectory computed by an explicit controller, or the input trajectory of EMPC-2 computed at the previous sampling time. Similarly, if EMPC-2 is solved first in the sequential DEMPC architecture, the input trajectories $\hat{u}_1$ and $\hat{u}_2$ must be assumed before solving EMPC-2.

7.3.2.1 Sequential DEMPC 1-2

The first configuration considered, which is referred to as the sequential DEMPC 1-2 and abbreviated to S-DEMPC 1-2, first solves the EMPC-1 problem for the optimal input trajectories $u_1^*(t|t_k)$ and $u_2^*(t|t_k)$ for $t \in [t_k, t_{k+N})$. Then, the EMPC-2 problem is solved to compute the input trajectory $u_3^*(t|t_k)$ after receiving $u_1^*(t|t_k)$ and $u_2^*(t|t_k)$ from EMPC-1. A block diagram of the resulting control architecture showing the communication between the controllers is given in Fig. 7.8. Since the input trajectory $\hat{u}_3(t)$ for $t \in [t_k, t_{k+N})$ has not been determined when the EMPC-1 problem is solved, it is set to be the resulting input trajectory under a proportional-integral (PI) controller implemented in a sample-and-hold fashion over the prediction horizon (other methods for the assumed profile of $\hat{u}_3(t)$ within EMPC-1 could be considered).

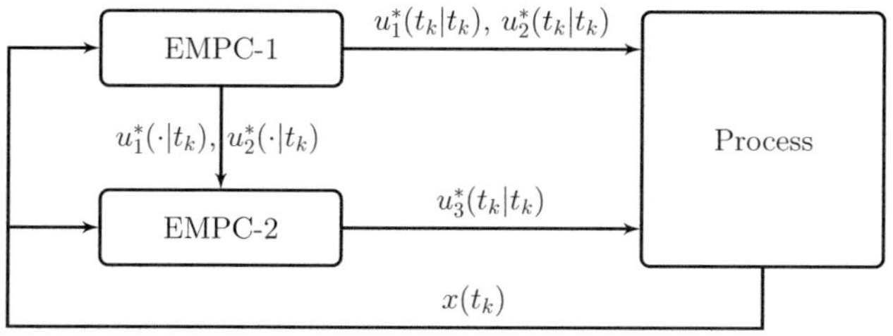

Fig. 7.8 A block diagram of the S-DEMPC 1-2 scheme

Fig. 7.9 Closed-loop state trajectories of the catalytic reactor under the S-DEMPC 1-2

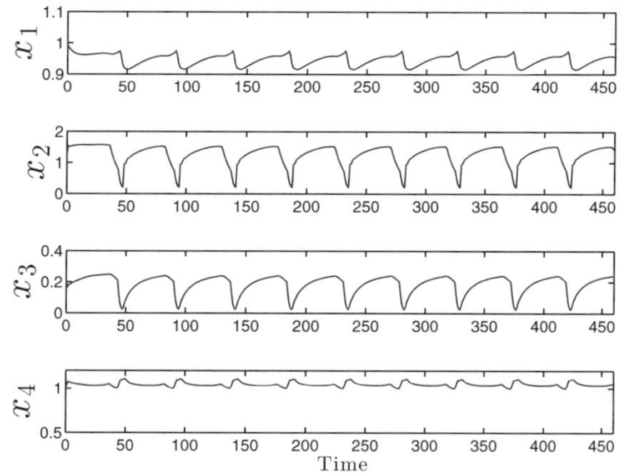

The input constraints are accounted for in the computed PI input trajectory, e.g., if the PI controller computes a control action greater than the upper bound on u_3, it is set to $u_{3,\max}$. For the input trajectories $\hat{u}_1$ and $\hat{u}_2$ in the EMPC-2 problem, the optimal input trajectories computed by the EMPC-1 problem are used.

Figures 7.9 and 7.10 show the closed-loop state and input trajectories under the S-DEMPC 1-2, respectively. The trajectories are similar to those under the C-EMPC (Figs. 7.6 and 7.7). For the closed-loop simulation, the average yield was 10.20 (recall, the average yield under the C-EMPC was 10.22 differences in the state trajectories are observed from Figs. 7.6 and 7.9, e.g., in the x_1 and x_4 trajectories. It is important to note that given the nonlinear nature of the process considered, there is no guarantee, in general, that the centralized EMPC and sequential EMPC scheme will lead to the same or even similar optimal input trajectories.

7.3.2.2 Sequential DEMPC 2-1

Another sequential distributed implementation of EMPC-1 and EMPC-2 may be considered by reversing the execution of EMPC-1 and EMPC-2. In this DEMPC approach, which is shown in Fig. 7.11, EMPC-2 computes its optimal input trajectory

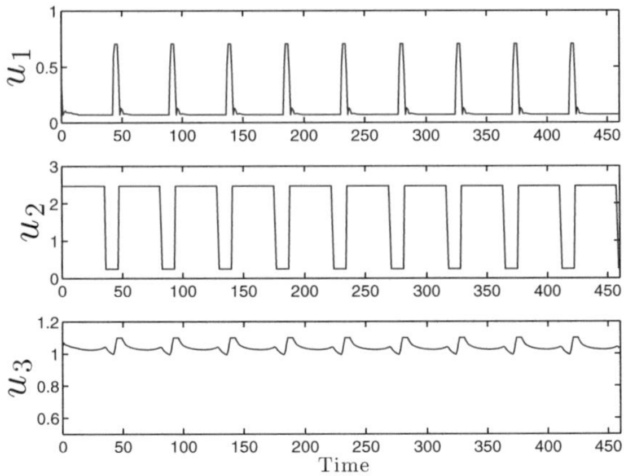

Fig. 7.10 Closed-loop input trajectories computed by the S-DEMPC 1-2

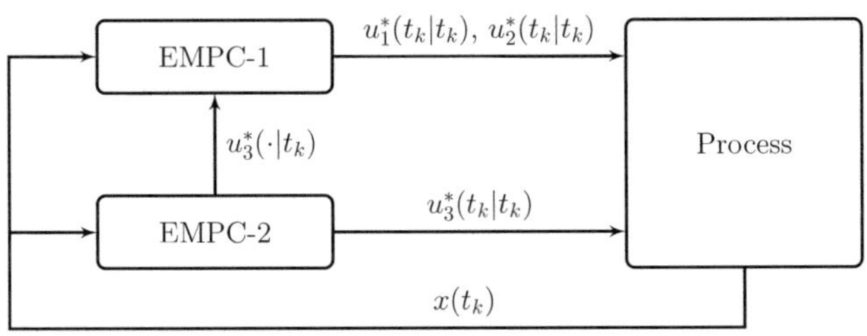

Fig. 7.11 A block diagram of the S-DEMPC 1-2 scheme

$u_3^*(t|t_k)$ for $t \in [t_k, t_{k+N})$ first. The sequential DEMPC approach is referred to as the sequential DEMPC 2-1 (S-DEMPC 2-1). To solve EMPC-2, the trajectories $\hat{u}_1(t)$ and $\hat{u}_2(t)$ for $t \in [t_k, t_{k+N})$ are set to the input trajectories resulting from two PI controllers implemented in sample-and-hold fashion. While the bounds on admissible input values are accounted for in the PI input trajectories, the input average constraint is not accounted for in the PI input trajectories. Figures 7.12 and 7.13 give the closed-loop state and input trajectories under the S-DEMPC 2-1 approach. Noticeable differences are observed between the closed-loop trajectories under the S-DEMPC 2-1 approach and those under the C-EMPC approach (Figs. 7.6 and 7.7).

7.3.3 Iterative DEMPC

Instead of sequential computation of the distributed EMPC problems, parallel computation may be employed in the sense that each problem may be solved simultaneously. Given the control actions are computed without the knowledge of the control actions computed by the other distributed EMPC schemes, an iterative approach

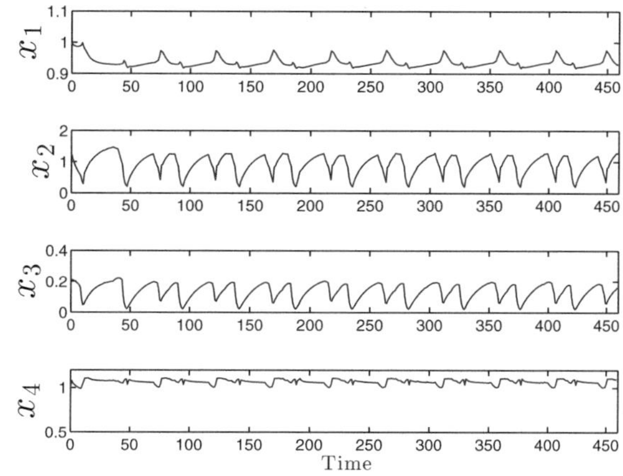

Fig. 7.12 Closed-loop state trajectories of the catalytic reactor under the S-DEMPC 2-1

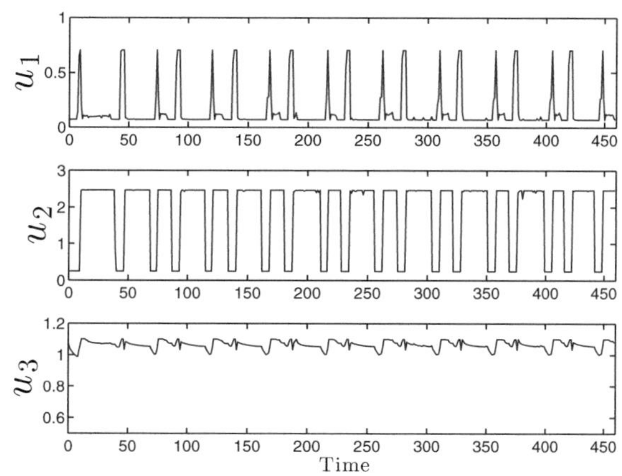

Fig. 7.13 Closed-loop input trajectories of computed by the S-DEMPC 2-1

may be used to (ideally) compute control actions closer to the centralized solution. Again, it is important to emphasize that given the nonlinearity and non-convexity of the optimization problems, it is difficult, in general, to guarantee that an iterative DEMPC strategy will converge to the centralized solution even after infinite iterations. Moreover, there is no guarantee that the input solution computed at each iteration improves upon the closed-loop performance over the previous iteration.

An iterative DEMPC (I-DEMPC) scheme is designed for the catalytic reactor and a block diagram of the I-DEMPC control architecture is given in Fig. 7.14. The computed input trajectories at each iteration of the I-DEMPC is denoted as $u_i^{*,c}(t|t_k)$ for $t \in [t_k, t_{k+N})$, $i = 1,\ 2,\ 3$ where c is the iteration number. At the first iteration, the input trajectory $\hat{u}_3(t)$ for $t \in [t_k, t_{k+N})$ in EMPC-1 is initialized with the sample-and-hold input trajectory computed from the same PI controller used in the S-DEMPC 1-2 scheme, and similarly, the input trajectories $\hat{u}_2(t)$ and $\hat{u}_3(t)$ for $t \in [t_k, t_{k+N})$ in EMPC-2 are initialized with the input trajectories computed from

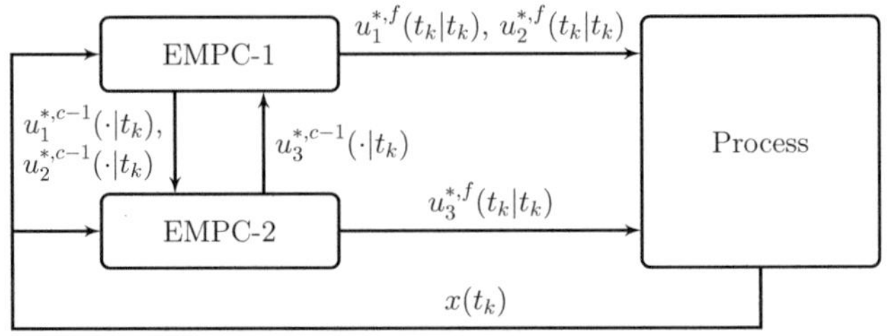

Fig. 7.14 A block diagram of the I-DEMPC scheme

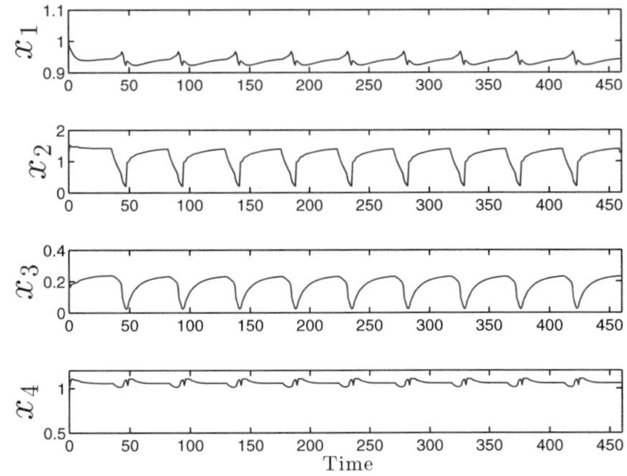

Fig. 7.15 Closed-loop state trajectories of the catalytic reactor under the I-DEMPC (1 iteration)

the PI controllers of the S-DEMPC 2-1 scheme. The control action applied to the reactor is denoted as $u_i^{*,f}(t_k|t_k)$ for $i = 1, 2, 3$ where f is the total number of iterations of the iterative DEMPC scheme (f is a design parameter of the scheme). When $f = 1$, the I-EMPC scheme is a decentralized DEMPC approach in the sense that there is no communication between EMPC-1 and EMPC-2 and each problem are solved independently of each other.

For this example, no closed-loop performance benefit was observed after iterating more than once through the I-DEMPC scheme. In fact, using the previous iterate solution to compute the next iterative gave worse closed-loop performance than applying the first computed iteration to the process. One method considered to compensate for this problem was to use the best computed input solution over all iterations to compute the next iteration. However, minimal closed-loop performance benefit was observed with this method. Thus, $f = 1$, which corresponds to a decentralized DEMPC approach, was selected for this case given that using more than one iteration did not improve the closed-loop performance. The resulting closed-loop trajectories are given in Figs. 7.15 and 7.16. The trajectories have similar characteristics as the centralized case.

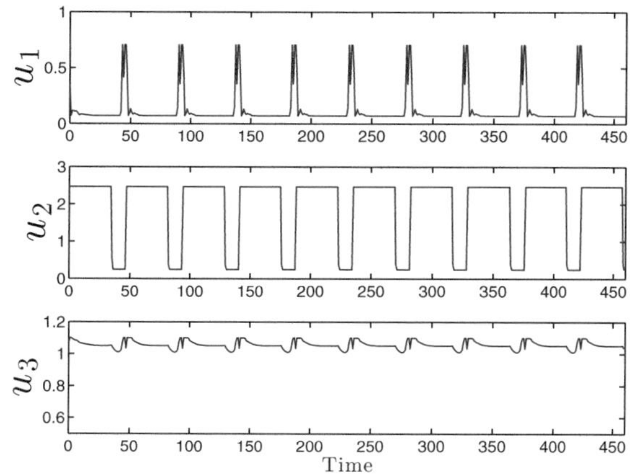

Fig. 7.16 Closed-loop input trajectories of computed by the I-DEMPC (1 iteration)

7.3.4 Evaluation of DEMPC Approaches

The average yield and average computation time required to solve the optimization problem at each sampling time over the entire simulation were considered for all the cases. The sequential DEMPC computation time is computed as the sum of the computation time of EMPC-1 and EMPC-2 at each sampling time because the sequential DEMPC schemes are computed sequentially. The iterative DEMPC computation time is the maximum computation time over all EMPCs at each sampling time (recall only one iteration was used). The average yield and average computation time for all the cases is given in Table 7.2. The closed-loop performance under the centralized EMPC, the sequential DEMPC 1-2, and the iterative DEMPC schemes was similar. The sequential DEMPC 1-2 and iterative DEMPC resulted in approximately a 70 reduction in computation time over the centralized EMPC. The sequential DEMPC 2-1 scheme not only had the worst performance of all the strategies considered (albeit still better than steady-state operation), but also, required a comparable amount of time to solve the optimization problems as the centralized case, thereby implying a strong dependence of closed-loop performance on controller calculation sequence. DEMPC was able to yield comparable closed-loop performance while substantially reducing the on-line computation time. This demonstrates that a distributed implementation may allow EMPC to be used on processes where centralized control is not feasible due to the solve time.

This example illustrates another key point within the context of DEMPC. Specifically, the inclusion of integral constraint in EMPC may be an important consideration for input selection in DEMPC. From the sequential DEMPC results, the computed u_3 profile is impacted by the assumed input profiles $\bar{u}_1$ and $\bar{u}_2$ (Fig. 7.13), while u_1 and u_2 are not affected as much by the assumed profile $\bar{u}_3$ (Fig. 7.10) compared to the centralized EMPC case (Fig. 7.7). This behavior may be due to the enforcement of the integral input constraint, and for this example, there may only be one method

Table 7.2 The average yield and computation time under the EMPC strategies

Strategy	Yield (%)	Comp. time (s)
Sequential DEMPC 1-2	10.20	1.039
Sequential DEMPC 2-1	9.92	2.969
Iterative DEMPC ($f = 1$)	10.05	0.832
Centralized EMPC	10.22	4.244

to distribute a fixed amount of ethylene to the reactor that maximizes the yield that is independent of u_3.

7.4 Real-Time Economic Model Predictive Control of Nonlinear Process Systems

Besides designing EMPC strategies that improves the computational efficiency such as the use of two-layer EMPC or distributed EMPC implementations, it may also be important to consider an EMPC implementation strategy that explicitly addresses potential computational delay. Some of the early work addressing computational delay within tracking MPC includes developing an implementation strategy of solving the MPC problem intermittently to account for the computational delay [23] and predicting the future state after an assumed constant computational delay to compute an input trajectory to be implemented after the optimization problem is solved [24, 25]. Nominal feasibility and stability has been proved for tracking MPC subject to computational delay formulated with a positive definite stage cost (with respect to the set-point or steady-state), a terminal cost, and a terminal region constraint [24, 25]. Another option to handle computational delay would be to force the optimization solver to terminate after a pre-specified time to ensure that the solver returns a solution by the time needed to ensure closed-loop stability. This concept is typically referred to as suboptimal MPC [26] because the returned solution will likely be suboptimal. It was shown that when the returned solution of the MPC with a terminal constraint is any feasible solution, the operating steady-state of the closed-loop system is asymptotically stable [26].

More recently, more advanced strategies have been proposed. Particularly, nonlinear programming (NLP) sensitivity analysis has demonstrated to be a useful tool to handle computational delay by splitting the MPC optimization problem into two parts: (1) solving a computationally intensive nonlinear optimization problem which is completed before feedback is received and (2) performing a fast on-line update of the precomputed input trajectories using NLP sensitivities (when the active-set does not change) after the current state measurement is obtained, e.g., [27, 28]; see, also, the review [29]. If the active-set changes, various methods have been proposed to cope with changing active-sets, e.g., solving a quadratic program like that proposed in [30]. In this direction, the advanced-step MPC [28] has been proposed which computes the solution of the optimization problem one sampling period in advance using a prediction of the state at the next sampling period. At the next sampling period (when the precomputed control action will be applied), the optimal solution is updated employing NLP sensitivities after state feedback is received. The advanced-step (tracking) MPC has been extended to handle computation spanning multiple sampling periods [31] and to EMPC [32]. Another related approach involves a hierarchical control structure [33, 34]. The upper layer is the full optimization problem which is solved infrequently. In the lower layer, NLP sensitivities are used to update the control actions at each sampling period that are applied to the system. The aforementioned schemes solve an optimization problem to (local) optimality using a prediction of the state at the sampling time the control action is to be applied to the system.

As another way, the so-called real-time nonlinear MPC (NMPC) scheme [35] only takes one Newton-step of the NLP solver instead of solving the optimization problem to optimality at each sampling period. To accomplish this, the structure of the resulting dynamic optimization program, which is solved using a direct multiple shooting method, is exploited to divide the program into a preparation phase and a feedback phase. In the preparation phase, the computationally expensive calculations are completed before feedback is received. In the feedback phase, a measurement is received and the remaining fast computations of the Newton-step are completed on-line to compute the control action to apply to the system. The advantage of such a strategy is that the on-line computation after a feedback measurement is obtained is insignificant compared to solving the optimization problem to optimality. The disadvantage is one would expect to sacrifice at least some closed-loop performance as a result of not solving the problem to optimality.

Clearly, the available computing power has significantly increased since the early work on computational delay of MPC and if this trend continues, one may expect a significant increase in computing power over the next decade. Moreover, more efficient solution strategies for nonlinear dynamic optimization problems continue to be developed (see, for example, the overview paper [36] and the book [37] for results in this direction). However, the ability to guarantee that a solver will converge within the time needed for closed-loop stability remains an open problem especially for nonlinear, non-convex dynamic optimization problems and systems with fast dynamics. Additionally, EMPC is generally more computationally intensive compared to tracking MPC given the additional possible nonlinearities in the stage cost of EMPC.

In this section, a real-time implementation strategy for LEMPC, referred to as real-time LEMPC, is developed to account for possibly unknown and time-varying computational delay. The underlying implementation strategy is inspired by event-triggered control concepts [38] since the LEMPC is only recomputed when stability conditions dictate that it must recompute a new input trajectory. If the precomputed control action satisfies the stability conditions, the control action is applied to the closed-loop system. If not, a back-up explicit controller, which has negligible computation time, is used to compute the control action for the system at the current sampling instance. This type of implementation strategy has the advantage of being easy to implement and the strategy avoids potential complications of active-set changes because the re-computation condition is only formulated to account for closed-loop stability considerations. Closed-loop stability under the real-time LEMPC scheme is analyzed and specific stability conditions are derived. The real-time LEMPC scheme is applied to an illustrative chemical process network to demonstrate closed-loop stability under the control scheme. The example also demonstrates that real-time LEMPC improves closed-loop economic performance compared to operation at the economically optimal steady-state.

7.4.1 Class of Systems

The class of nonlinear systems considered has the following state-space form:

$$\dot{x}(t) = f(x(t), u(t), w(t)) \tag{7.62}$$

where $x(t) \in \mathbb{R}^n$ is the state vector, $u(t) \in \mathbb{U} \subset \mathbb{R}^m$ is the manipulated input vector, $\mathbb{U}$ is a compact set, $w(t) \in \mathbb{W} \subset \mathbb{R}^l$ is the disturbance vector, and f is a locally Lipschitz vector function. The disturbance vector is bounded in the following set:

$$\mathbb{W} := \{w \in \mathbb{R}^l : |w| \leq \theta\} \tag{7.63}$$

where $\theta > 0$ bounds the norm of the disturbance vector. Without loss of generality, the origin of the unforced system is assumed to be the equilibrium point of Eq. 7.62, i.e., $f(0, 0, 0) = 0$.

The following stabilizability assumption further qualifies the class of systems considered and is similar to the assumption that the pair (A, B) is stabilizable in linear systems.

Assumption 7.3 There exists a feedback controller $h(x) \in \mathbb{U}$ with $h(0) = 0$ that renders the origin of the closed-loop system of Eq. 7.62 with $u(t) = h(x(t))$ and $w \equiv 0$ asymptotically stable for all $x \in D_0$ where D_0 is an open neighborhood of the origin.

Applying converse theorems [7, 9], Assumption 7.3 implies that there exists a continuously differentiable Lyapunov function, $V : D \to \mathbb{R}_+$, for the closed-loop system of Eq. 7.62 with $u = h(x) \in \mathbb{U}$ and $w \equiv 0$ such that the following inequalities hold:

$$\alpha_1(|x|) \leq V(x) \leq \alpha_2(|x|), \tag{7.64a}$$

$$\frac{\partial V(x)}{\partial x} f(x, h(x), 0) \leq -\alpha_3(|x|), \tag{7.64b}$$

$$\left|\frac{\partial V(x)}{\partial x}\right| \leq \alpha_4(|x|) \tag{7.64c}$$

for all $x \in D$ where D is an open neighborhood of the origin and α_i, $i = 1, 2, 3, 4$ are functions of class $\mathcal{K}$. A level set of the Lyapunov function Ω_ρ, which defines a subset of D (ideally the largest subset contained in D), is taken to be the stability region of the closed-loop system under the controller $h(x)$.

Measurements of the state vector of Eq. 7.62 are assumed to be available synchronously at sampling instances denoted as $t_k := k\Delta$ where $\Delta > 0$ is the sampling period and $k = 0, 1, \ldots$. As described below, the EMPC computes sample-and-hold control actions and thus, the resulting closed-loop system, which consists of the continuous-time system of Eq. 7.62 under a sample-and-hold controller, is a sampled-data system. If the controller $h(x)$ is implemented in a sample-and-hold fashion, it possesses a certain degree of robustness to uncertainty in the sense that the origin of the closed-loop system is rendered practically stable when a sufficiently small sampling period is used and the bound θ on the disturbance vector is sufficiently small; see, for example, [10] for more discussion on this point.

7.4.2 Real-Time LEMPC Formulation

The overall objective of the real-time LEMPC is to account for the real-time computation time required to solve the optimization problem for a (local) solution. Particularly, the case when the average computation time, which is denoted as $\bar{t}_s$, is greater than one sampling period is considered, i.e., $N_s = \lceil \bar{t}_s / \Delta \rceil \geq 1$ where N_s is the average number of sampling periods required to solve the optimization problem. During the time the solver is solving the optimization problem, the control actions computed at a previous sampling period are applied to the system if there are precomputed control actions available and if the stability conditions described below are satisfied. If no precomputed control actions are available or the stability conditions are violated, the explicit controller $h(x)$ is used to compute and apply control actions during the time that the real-time LEMPC is computing. In this fashion, the LEMPC is used to compute control actions to improve the economic performance when possible.

Specifically, when the closed-loop state is in the subset of the stability region $\Omega_{\rho_e} \subset \Omega_\rho$, the control actions of the precomputed LEMPC problem may be applied to the system. When the state is outside the subset, the explicit controller is used because maintaining the closed-loop state in Ω_ρ is required for guaranteeing the existence of a feasible input trajectory that maintains closed-loop stability (in the sense that the closed-loop state trajectory is always bounded in Ω_ρ). To force the state back to the subset of the stability region Ω_{ρ_e}, the Lyapunov function must decrease over each sampling period in the presence of uncertainty. This requires the incorporation of feedback, i.e., recomputing the control action at each sampling period using a measurement of the current state. Owing to the computational burden of solving the LEMPC optimization problem, it may not be possible to achieve convergence of the optimization solver within one sampling period. Hence, the controller $h(x)$ is used when the state is outside of Ω_{ρ_e}.

For real-time implementation, only mode 1 of the LEMPC of Eq. 4.3 is used and the LEMPC is solved infrequently (not every sampling period) which will be made clear when the implementation strategy is discussed. The real-time LEMPC is formulated as follows:

$$\min_{u \in S(\Delta)} \int_{t_{j+1}}^{t_{j+N}} l_e(\tilde{x}(t), u(t))\, dt \tag{7.65a}$$

$$\text{s.t.} \quad \dot{\tilde{x}}(t) = f(\tilde{x}(t), u(t), 0) \tag{7.65b}$$

$$\tilde{x}(t_j) = x(t_j) \tag{7.65c}$$

$$u(t) = \tilde{u}(t_j),\ \forall\, t \in [t_j, t_{j+1}) \tag{7.65d}$$

$$u(t) \in \mathbb{U},\ \forall\, t \in [t_{j+1}, t_{j+N}) \tag{7.65e}$$

$$V(\tilde{x}(t)) \le \rho_e,\ \forall\, t \in [t_{j+1}, t_{j+N}) \tag{7.65f}$$

where the notation and constraints are similar to that used in LEMPC of Eq. 4.3 except for an additional constraint of Eq. 7.65d. This additional constraint is used because a predetermined control action is applied to the system over the first sampling period of the prediction horizon. The predetermined control action is either the control action computed by the LEMPC at a previous sampling period or the control action from the explicit controller $h(x)$, i.e., the input trajectory over the first sampling period of the prediction horizon is not a degree of freedom in the optimization problem. The LEMPC of Eq. 7.65 may dictate a time-varying operating policy to optimize the economic cost as long as the predicted evolution is maintained in the level set $\Omega_{\rho_e} \subset \Omega_\rho$. The notation t_j denotes the sampling time at which the LEMPC problem is initialized with a state measurement and the solver begins solving the resulting optimization problem. The optimal solution of the LEMPC is denoted as $u^*(t|t_j)$ and is defined for $t \in [t_{j+1}, t_{j+N})$. Feasibility of the optimization problem is considered in Sect. 7.4.4. However, it is important to point out that $x(t_j) \in \Omega_{\rho_e}$ and $\tilde{x}(t_{j+1}) \in \Omega_{\rho_e}$ owing to the real-time implementation strategy, and thus, the real-time LEMPC has a feasible solution (refer to the proof of Theorem 7.2).

7.4.3 Implementation Strategy

Before the implementation strategy is presented, the following discrete-time signals are defined to simplify the presentation of the implementation strategy. The first signal is used to keep track of whether the solver is currently solving an LEMPC optimization problem:

$$s_1(k) = \begin{cases} 1, & \text{solving the LEMPC} \\ 0, & \text{not solving the LEMPC} \end{cases} \tag{7.66}$$

where k denotes the k-th sampling period, i.e., t_k. The second signal keeps track if there is a previously computed input trajectory currently stored in memory:

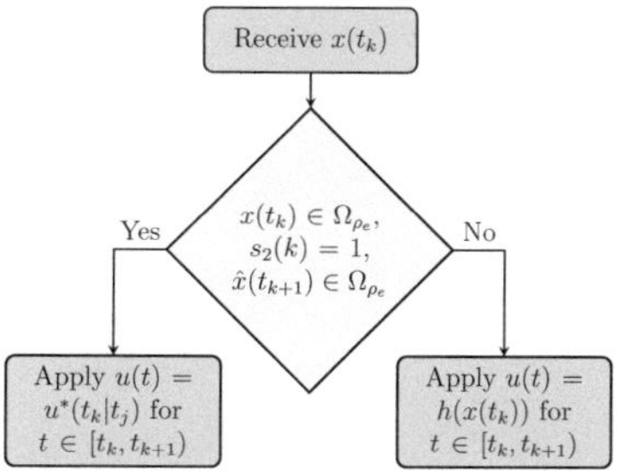

Fig. 7.17 Implementation strategy for determining the control action at each sampling period. The notation $u^*(t_k|t_j)$ is used to denote the control action to be applied over the sampling period t_k to t_{k+1} from the precomputed input solution of the real-time LEMPC of Eq. 7.65 solved at time step t_j

$$s_2(k) = \begin{cases} 1, & \text{previous input solution stored} \\ 0, & \text{no previous input solution stored} \end{cases} \tag{7.67}$$

At each sampling period, a state measurement $x(t_k)$ is received from the sensors and three conditions are used to determine if a precomputed control action from LEMPC or if the control action from the explicit controller $h(x)$ is applied to the system. If the following three conditions are satisfied the control action applied to the system in a sample-and-hold fashion is the precomputed control action from the LEMPC: (1) the current state must be in Ω_{ρ_e} $(x(t_k) \in \Omega_{\rho_e})$, (2) there must be a precomputed control action available for the sampling instance t_k, i.e., $s_2(k) = 1$, and (3) the predicted state under the precomputed control action must satisfy: $\hat{x}(t_{k+1}) \in \Omega_{\rho_e}$ where $\hat{x}(t_{k+1})$ denotes the predicted state. To obtain a prediction of the state at the next sampling period, the nominal model of Eq. 7.62 with $w \equiv 0$ is recursively solved with the input $u(t) = u^*(t_k|t_j)$ for $t \in [t_k, t_{k+1})$ (the on-line computation time to accomplish this step is assumed to be negligible). The control action decision at a given sampling instance t_k is summarized by the flow chart of Fig. 7.17.

A series of decisions are made at each sampling period to determine if the LEMPC should begin resolving, continue solving, or terminate solving the optimization problem and is illustrated in the flow chart of Fig. 7.18. The computation strategy is summarized in the following algorithm. To initialize the algorithm at $t_0 = 0$, get the state measurement $x(0) \in \Omega_\rho$. If $x(0) \in \Omega_{\rho_e}$, begin solving the LEMPC problem with $k = j = 0$ and $x(0)$. Set $s_1(0) = 1$, $s_2(0) = 0$, and $\tilde{u}(t_j) = h(x(0))$. Go to Step 8. Else, set $s_1(0) = s_1(1) = s_2(0) = s_2(1) = 0$ and go to Step 9.

1. Receive a measurement of the current state $x(t_k)$ from the sensors; go to Step 2.
2. If $x(t_k) \in \Omega_{\rho_e}$, then go to Step 2.1. Else, go to Step 2.2.

   2.1 If $s_2(k) = 1$, go to Step 3. Else, go to Step 6.

   2.2 Terminate solver if $s_1(k) = 1$, set $s_1(k+1) = 0$ and $s_2(k+1) = 0$, and go to Step 9.

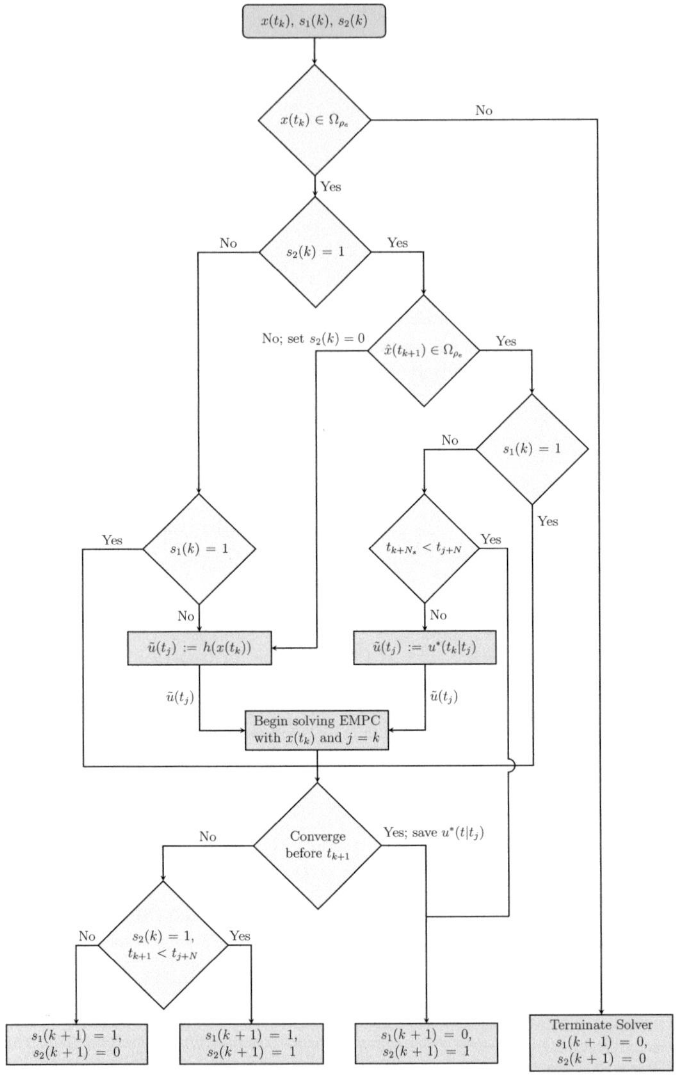

Fig. 7.18 Computation strategy for the real-time LEMPC scheme

3. If $\hat{x}(t_{k+1}) \in \Omega_{\rho_e}$, go to Step 4. Else, set $s_2(k) = 0$ and $\tilde{u}(t_j) = h(x(t_k))$; go to Step 7.
4. If $s_1(k) = 1$, go to Step 8. Else, go to Step 5.
5. If $t_{k+N_s} < t_{j+N}$, set $s_1(k+1) = 0$ and $s_2(k+1) = 1$, and go to Step 9. Else, set $\tilde{u}(t_j) = u^*(t_k|t_j)$; go to Step 7.
6. If $s_1(k) = 1$, go to Step 8. Else, set $\tilde{u}(t_j) = h(x(t_k))$; go to Step 7.

7. If the solver is currently solving a problem ($s_1(k) = 1$), terminate the solver. Begin solving the LEMPC problem with $j = k$ and $x(t_j) = x(t_k)$. Go to Step 8.
8. If the solver converges before t_{k+1}, then go to Step 8.1. Else, go to Step 8.2.

   8.1 Save $u^*(t|t_j)$ for $t \in [t_k, t_{j+N})$. Set $s_1(k+1) = 0$ and $s_2(k+1) = 1$. Go to Step 9.

   8.2 Set $s_1(k+1) = 1$. If $s_2(k) = 1$ and $t_{k+1} < t_{j+N}$, the go to Step 8.2.1. Else, go to Step 8.2.2.

      8.2.1 Set $s_2(k+1) = 1$. Go to Step 9.

      8.2.2 Set $s_2(k+1) = 0$. Go to Step 9.

9. Go to Step 1 ($k \leftarrow k+1$).

In practice, N_s may be unknown or possibly time varying. If N_s is unknown, then one may specify the number of sampling periods that the real-time LEMPC may apply a precomputed input trajectory before it must start re-computing a new input trajectory as a design parameter. This condition may be used instead of Step 5 of the algorithm above. Additionally, it may be beneficial from a closed-loop performance perspective to force the LEMPC to recompute its solution more often than prescribed by the implementation strategy described above.

A possible input trajectory resulting under the real-time LEMPC scheme is given in Fig. 7.19. In the illustration, the solver begins to solve an LEMPC optimization problem at t_0 and returns a solution at t_5. It is assumed that the closed-loop state is maintained in Ω_{ρ_e} from t_0 to t_5 so that the solver is not terminated. Over the time the solver is solving, the explicit controller is applied to the system since a precomputed LEMPC input trajectory is not available. The precomputed LEMPC solution is applied from t_5 to t_{13}. At t_{10}, the solver begins to solve a new LEMPC problem. The solver returns a solution at t_{13}. At t_{16}, the stability conditions are

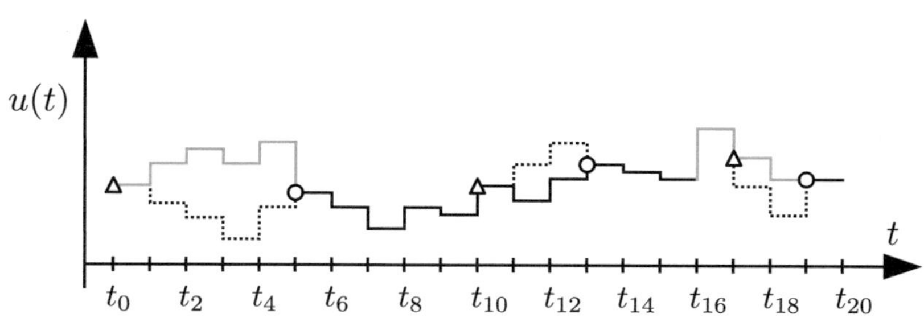

Fig. 7.19 An illustration of an example input trajectory resulting under the real-time LEMPC scheme. The *triangles* are used to denote the time instances when the LEMPC begins to solve the optimization problem, while the *circles* are used to denote when the solver converges to a solution. The *solid black* trajectory represents the control actions computed by the LEMPC which are applied to the system, the *dotted* trajectory represents the computed input trajectory by the LEMPC (not applied to the system), and the *solid gray* trajectory is the input trajectory of the explicit controller which is applied to the system

not satisfied for the precomputed LEMPC input trajectory, so the explicit controller computes a control action and applies it to the system.

7.4.4 Stability Analysis

In this section, sufficient conditions such that the closed-loop state under the real-time LEMPC is bounded in Ω_ρ are presented which make use of the following properties. Since f is a locally Lipschitz vector function of its arguments and the Lyapunov function V is a continuously differentiable function, there exist positive constants L_x, L_w, L'_x, and L'_w such that the following bounds hold:

$$|f(x_a, u, w) - f(x_b, u, 0)| \leq L_x |x_a - x_b| + L_w |w| \tag{7.68}$$

$$\left| \frac{\partial V(x_a)}{\partial x} f(x_a, u, w) - \frac{\partial V(x_b)}{\partial x} f(x_b, u, 0) \right| \leq L'_x |x_a - x_b| + L'_w |w| \tag{7.69}$$

for all $x_a, x_b \in \Omega_\rho$, $u \in \mathbb{U}$ and $w \in \mathbb{W}$. Furthermore, there exists $M > 0$ such that

$$|f(x, u, w)| \leq M \tag{7.70}$$

for all $x \in \Omega_\rho$, $u \in \mathbb{U}$ and $w \in \mathbb{W}$ owing to the compactness of the sets Ω_ρ, $\mathbb{U}$, and $\mathbb{W}$ and the locally Lipschitz property of the vector field.

The following proposition bounds the difference between the actual state trajectory of the system of Eq. 7.62 ($w \not\equiv 0$) and the nominal state trajectory ($w \equiv 0$).

Proposition 7.3 (Proposition 4.1) *Consider the state trajectories $x(t)$ and $\hat{x}(t)$ with dynamics:*

$$\dot{x}(t) = f(x(t), u(t), w(t)), \tag{7.71}$$

$$\dot{\hat{x}}(t) = f(\hat{x}(t), u(t), 0), \tag{7.72}$$

input trajectory $u(t) \in \mathbb{U}$, $w(t) \in \mathbb{W}$, and initial condition $x(0) = \hat{x}(0) \in \Omega_\rho$. If $x(t)$, $\hat{x}(t) \in \Omega_\rho$ for all $t \in [0, T]$ where $T \geq 0$, then the difference between $x(T)$ and $\hat{x}(T)$ is bounded by the function $\gamma_e(\cdot)$:

$$\left|x(T) - \hat{x}(T)\right| \leq \gamma_e(T) := \frac{L_w \theta}{L_x} \left(e^{L_x T} - 1\right). \tag{7.73}$$

Owing to the compactness of the set Ω_ρ, the difference in Lyapunov function values for any two points in Ω_ρ may be bounded by a quadratic function which is stated in the following proposition.

Proposition 7.4 (Proposition 4.2) *Consider the Lyapunov function V of the closed-loop system of Eq. 7.62 under the controller $h(x)$. There exists a scalar-valued quadratic function $f_V(\cdot)$ such that*

$$V(x_a) \leq V(x_b) + f_V(|x_a - x_b|) \tag{7.74}$$

for all $x_a, x_b \in \Omega_\rho$ where

$$f_V(s) := \alpha_4(\alpha_1^{-1}(\rho))s + \beta s^2 \tag{7.75}$$

and β is a positive constant.

Theorem 7.2 provides sufficient conditions such that the real-time LEMPC renders the closed-loop state trajectory bounded in Ω_ρ for all times. The conditions such that the closed-loop state trajectory is maintained in Ω_ρ are independent of the computation time required to solve the LEMPC optimization problem. From the perspective of closed-loop stability, computational delay of arbitrary size may be handled with the real-time LEMPC methodology. In the case where the computational delay is always greater than the prediction horizon, the real-time LEMPC scheme would return the input trajectory under the explicit controller applied in a sample-and-hold fashion.

Theorem 7.2 *Consider the system of Eq. 7.62 in closed-loop under the real-time LEMPC of Eq. 7.65 based on a controller $h(x)$ that satisfies the conditions of Eq. 7.64 that is implemented according to the implementation strategy of Fig. 7.17. Let $\varepsilon_w > 0$, $\Delta > 0$ and $\rho > \rho_e \geq \rho_{\min} > \rho_s > 0$ satisfy*

$$-\alpha_3(\alpha_2^{-1}(\rho_s)) + L'_x M\Delta + L'_w\theta \leq -\varepsilon_w/\Delta, \tag{7.76}$$

$$\rho_{\min} = \max\{V(x(t+\Delta)) \mid V(x(t)) \leq \rho_s\}, \tag{7.77}$$

and

$$\rho_e < \rho - f_V(\gamma_e(\Delta)). \tag{7.78}$$

If $x(t_0) \in \Omega_\rho$ and $N \geq 1$, then the state trajectory $x(t)$ of the closed-loop system is always bounded in Ω_ρ for $t \geq t_0$.

Proof If the real-time LEMPC is implemented according to the implementation strategy of Fig. 7.17, the control action to be applied over the sampling period has been (pre)computed by the LEMPC or the explicit controller $h(x)$. To prove that the closed-loop state is bounded in Ω_ρ, we will show that when the control action is computed from the explicit controller and $x(t_k) \in \Omega_\rho$, then the state at the next sampling period will be contained in Ω_ρ. If the control action comes from a precomputed LEMPC solution, we will show that if $x(t_k) \in \Omega_{\rho_e}$, then $x(t_{k+1}) \in \Omega_\rho$ owing to the stability conditions imposed on applying the precomputed LEMPC solution. The proof consists of two parts. In the first part, the closed-loop properties when the control action is computed by the explicit controller $h(x)$ are analyzed. This part of the proof is based on the proof of [10] which considers the stability properties of an explicit controller of the form assumed for $h(x)$ implemented in a sample-and-hold fashion. In the second part, the closed-loop stability properties of the precomputed control actions by the LEMPC are considered. In both cases, the closed-loop state trajectory is shown to be maintained in Ω_ρ for $t \geq t_0$ when $x(t_0) \in \Omega_\rho$.

Part 1: First, consider the properties of the control action computed by the explicit controller $h(x)$ applied to the system of Eq. 7.62 in a sample-and-hold fashion. Let $x(t_k) \in \Omega_\rho \setminus \Omega_{\rho_s}$ for some $\rho_s > 0$ such that the conditions of Theorem 7.2 are satisfied, i.e., Eq. 7.76. The explicit controller $h(x)$ computes a control action that has the following property (from condition of Eq. 7.64):

$$\frac{\partial V(x(t_k))}{\partial x} f(x(t_k), h(x(t_k)), 0) \leq -\alpha_3(|x(t_k)|) \leq -\alpha_3(\alpha_2^{-1}(\rho_s)) \tag{7.79}$$

for any $x(t_k) \in \Omega_\rho \setminus \Omega_{\rho_s}$. Over the sampling period, the time-derivative of the Lyapunov function is:

$$\begin{aligned}\dot{V}(x(t)) = &\frac{\partial V(x(t_k))}{\partial x} f(x(t_k), h(x(t_k)), 0) + \frac{\partial V(x(t))}{\partial x} f(x(t), h(x(t_k)), w(t)) \\ &- \frac{\partial V(x(t_k))}{\partial x} f(x(t_k), h(x(t_k)), 0)\end{aligned} \tag{7.80}$$

for all $t \in [t_k, t_{k+1})$. From the bound on the time-derivative of Lyapunov function of Eq. 7.79, the Lipschitz bound of Eq. 7.69, and the bound on the norm of the disturbance vector, the time-derivative of the Lyapunov function is bounded for $t \in [t_k, t_{k+1})$ as follows:

$$\begin{aligned}\dot{V}(x(t)) &\leq -\alpha_3(\alpha_2^{-1}(\rho_s)) \\ &\quad + \left| \frac{\partial V(x(t))}{\partial x} f(x(t), h(x(t_k)), w(t)) - \frac{\partial V(x(t_k))}{\partial x} f(x(t_k), h(x(t_k)), 0) \right| \\ &\leq -\alpha_3(\alpha_2^{-1}(\rho_s)) + L'_x |x(t) - x(t_k)| + L'_w |w(t)| \\ &\leq -\alpha_3(\alpha_2^{-1}(\rho_s)) + L'_x |x(t) - x(t_k)| + L'_w \theta\end{aligned} \tag{7.81}$$

for all $t \in [t_k, t_{k+1})$. Taking into account of Eq. 7.70 and the continuity of $x(t)$, the following bound may be written for all $t \in [t_k, t_{k+1})$:

$$|x(t) - x(t_k)| \leq M\Delta. \tag{7.82}$$

From Eqs. 7.81 and 7.82, the bound below follows:

$$\dot{V}(x(t)) \leq -\alpha_3(\alpha_2^{-1}(\rho_s)) + L'_x M\Delta + L'_w \theta \tag{7.83}$$

for all $t \in [t_k, t_{k+1})$. If the condition of Eq. 7.76 is satisfied, i.e., Δ and θ is sufficiently small, then there exists $\varepsilon_w > 0$ such that:

$$\dot{V}(x(t)) \leq -\varepsilon_w/\Delta \tag{7.84}$$

for all $t \in [t_k, t_{k+1})$. Integrating the above bound, yields:

$$V(x(t)) \leq V(x(t_k)), \quad \forall\, t \in [t_k, t_{k+1}), \tag{7.85}$$

$$V(x(t_{k+1})) \leq V(x(t_k)) - \varepsilon_w. \tag{7.86}$$

For any state $x(t_k) \in \Omega_\rho \backslash \Omega_{\rho_s}$, the state at the next sampling period will be in a smaller level set when the control action $u(t) = h(x(t_k))$ is applied for $t \in [t_k, t_{k+1})$. Also, the state will not come out of Ω_ρ over the sampling period owing to Eq. 7.84. Once the closed-loop state under the explicit controller $h(x)$ implemented in a sample-and-hold fashion has converged to Ω_{ρ_s}, the closed-loop state trajectory will be maintained in $\Omega_{\rho_{\min}}$ if $\rho_{\min} \leq \rho$ and $\rho_{\min}$ is defined according to Eq. 7.77. Thus, the sets Ω_ρ and $\Omega_{\rho_{\min}}$ are forward invariant sets under the controller $h(x)$ and if $x(t_k) \in \Omega_\rho$, then $x(t_{k+1}) \in \Omega_\rho$ under the explicit controller $h(x)$.

Part 2: In this part, the closed-loop stability properties of the input precomputed by the LEMPC for the sampling period t_k to t_{k+1} are considered. For clarity of presentation, the notation $\hat{x}(t)$ denotes the prediction of closed-loop state at time t, i.e., this prediction used in the implementation strategy to determine which control action to apply to the system, while the notation $\tilde{x}(t)$ will be reserved to denote the predicted state in the LEMPC of Eq. 7.65. The predicted state in the LEMPC of Eq. 7.65 at t_{j+1}, which is denoted as $\tilde{x}(t_{j+1})$, satisfies $\hat{x}(t_{j+1}) = \tilde{x}(t_{j+1})$ because both predicted states use the nominal model with the same initial condition and same piecewise constant input applied from t_j to t_{j+1}.

First, feasibility of the optimization problem is considered. Owing to the formulation of the LEMPC of Eq. 7.65, the optimization problem is always feasible if ρ_e satisfies: $\rho > \rho_e \geq \rho_{\min}$. Recall, the input over the sampling period t_j to t_{j+1} is not a degree of freedom in the optimization problem. If this control action is precomputed from a previous LEMPC solution, it must have the property that $\hat{x}(t_{j+1}) = \tilde{x}(t_{j+1}) \in \Omega_{\rho_e}$ which is imposed as a condition of the implementation strategy of Fig. 7.17. If the control action is computed by the explicit controller, the control action over the sampling period t_j to t_{j+1} will maintain $\tilde{x}(t_{j+1}) \in \Omega_{\rho_e}$. Thus, $\tilde{x}(t_{j+1}) \in \Omega_{\rho_e}$ in the LEMPC of Eq. 7.65. Feasibility of the optimization problem follows from the fact that the input trajectory obtained from the explicit controller $h(x)$ over the prediction horizon is a feasible solution, that is $u(t) = h(\hat{x}(t_i))$ for $t \in [t_i, t_{i+1}), i = j+1,\ j+2,\ \ldots,\ j+N-1$ where $\hat{x}(t)$ is obtained by recursively solving the model:

$$\dot{\hat{x}}(t) = f(\hat{x}(t), h(\hat{x}(t_i)), 0) \tag{7.87}$$

for $t \in [t_i, t_{i+1})$ and $i = j+1,\ j+1\ \ldots,\ j+N-1$ with the initial condition $\hat{x}(t_{j+1}) = \tilde{x}(t_{j+1})$. Furthermore, the set Ω_{ρ_e} is forward invariant under the controller

$h(x)$ (the proof is analogous to Part 1 where the set Ω_{ρ_e} is used instead of Ω_ρ). Thus, the LEMPC of Eq. 7.65 is always feasible for any $x(t_j) \in \Omega_{\rho_e}$.

If the LEMPC is implemented according to the implementation strategy of Fig. 7.17, then the precomputed input for t_k by the LEMPC is only used when $x(t_k) \in \Omega_{\rho_e}$ and the predicted state at the next sampling period $\hat{x}(t_{k+1}) \in \Omega_{\rho_e}$. When $x(t) \in \Omega_\rho$ for $t \in [t_k, t_{k+1})$, i.e., a sufficiently small sampling period is used, the following bound on the Lyapunov function value at the next sampling period t_{k+1} may be derived from Propositions 7.3–7.4:

$$V(x(t_{k+1})) \leq V(\hat{x}(t_{k+1})) + f_V(\gamma_e(\Delta)). \tag{7.88}$$

Since $\hat{x}(t_{k+1}) \in \Omega_{\rho_e}$ and if the condition of Eq. 7.78 is satisfied, $x(t_{k+1}) \in \Omega_\rho$.

To summarize, if the control action to be applied over the sampling period t_k to t_{k+1} is $u(t_k) = h(x(t_k))$, the state at the next sampling period will be in Ω_ρ ($x(t_{k+1}) \in \Omega_\rho$). If the control action to be applied over the sampling period t_k to t_{k+1} is from a precomputed LEMPC input, the state at the next sampling period will also be contained in Ω_ρ which completes the proof of boundedness of the closed-loop state trajectory $x(t) \in \Omega_\rho$ under the real-time LEMPC for $t \geq t_0$.

Remark 7.6 No closed-loop performance guarantees may be made because performance constraints, e.g., terminal constraints, are not imposed on the LEMPC and the closed-loop performance may be adversely affected with greater computation time. The latter point is associated with the fact that the LEMPC problem allows for the input trajectory from t_{j+1} to t_{j+N_s}, i.e., the time the solver converges, to be degrees of freedom in the optimization problem. However, the actual closed-loop input trajectory applied over this period may be different from that computed by the LEMPC over the same time period. Potentially, one may also employ sensitivity-based corrections to the precomputed control actions after receiving state feedback like that employed in [27, 28] to improve closed-loop performance. However, active set changes must be handled appropriately which may introduce additional on-line computation. It is important to point out that the computed solution of the LEMPC may dictate a time-varying operating policy to optimize the process economics. Even in the presence of uncertainty, the closed-loop performance under the real-time LEMPC may be substantially better (with respect to the economic cost) than traditional control methods, which is the case for the chemical process network considered in Sect. 7.4.5.

Remark 7.7 In the current section, unknown and possibly time-varying computational delay is considered for operation affected by unknown bounded disturbance. If, instead of the computation algorithm described above, a hard cap was placed on the solver to terminate and return a (suboptimal) solution by a certain number of sampling times, one could account for the control actions that are applied to the system over the computation time by setting the input trajectory in the LEMPC problem over the specified number of sampling periods of the prediction horizon be equal to a predetermined input trajectory. This potential strategy, however, does not account for the fact that the solver may return a solution before the end of specified number of sampling periods.

Remark 7.8 From the proof of Theorem 7.2, recursive feasibility of the LEMPC in the presence of bounded uncertainty is guaranteed if the initial state is in Ω_ρ. It is difficult in general to characterize the feasible set under EMPC formulated with a terminal constraint, i.e., the set of points where recursive feasibility is maintained in the presence of uncertainty. Thus, it may be difficult to ensure that the closed-loop state is maintained in the feasible set under EMPC with a terminal constraint in the presence of uncertainty and computational delay. In this respect, LEMPC has a unique advantage for real-time implementation compared to EMPC with a terminal constraint in that LEMPC maintains the closed-loop state inside Ω_ρ where recursive feasibility is guaranteed.

Remark 7.9 The number of times that the explicit controller is applied to the closed-loop system may be a factor in the closed-loop economic performance. Whether the control action is from a precomputed LEMPC problem or the explicit controller is mainly influenced by how close the state measurement is to the boundary of Ω_{ρ_e}. To decrease the number of times that the explicit controller is applied to the system, one could potentially add penalization terms to the stage cost of the LEMPC to penalize the closeness of the state to the boundary of Ω_{ρ_e}.

7.4.5 Application to a Chemical Process Network

Consider a chemical process network consisting of two continuous stirred-tank reactors (CSTRs) in series followed by a flash separator shown in Fig. 7.20. In each of the reactors, the reactant A is converted to the desired product B through an exothermic and irreversible reaction of the form $A \rightarrow B$. A fresh feedstock containing a dilute solution of the reactant A in an inert solvent D is fed to each reactor. The reaction rate is second-order in the reactant concentration. The CSTRs are denoted as CSTR-1 and CSTR-2, respectively. A flash separator, which is denoted as SEP-1, is used to recover some unreacted A. The overhead vapor from the flash tank is condensed and recycled back to CSTR-1. The bottom stream is the product stream of the process network which contains the desired product B. In the separator, a negligible amount of A is assumed to be converted to B through the reaction. The two reactors have both heating and cooling capabilities and the rate of heat supplied to or removed from the reactors is denoted as Q_j, $j = 1,\ 2$. While the heat supplied to or removed from the vessel contents is modeled with one variable, two different actuators may be used in practice for supplying heat to and removing heat from each vessel. To vaporize some contents of the separator, heat is supplied to the separator at a rate of Q_3. The liquid holdup of each vessel is assumed to be constant and the liquid density throughout the process network is also assumed to be constant.

Applying first principles, a dynamic model of the process network may be obtained (neglecting the dynamics of the condenser and the solvent) and is given by the following ordinary differential equations (ODEs) (see Table 7.3 for parameter notation and values):

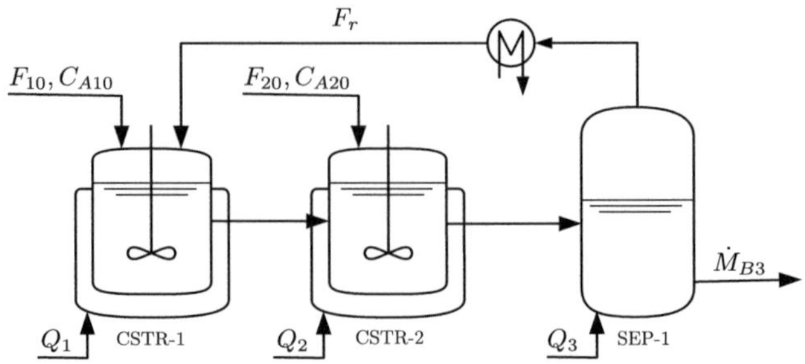

Fig. 7.20 Process flow diagram of the reactor and separator process network

Table 7.3 Process parameters of the reactor and separator process network

Symbol/Value	Description	Symbol/Value	Description
$T_{10} = 300\,\text{K}$	Temp.: CSTR-1 inlet	$k_0 = 1.9 \times 10^9\,\text{m}^3\,\text{kmol}^{-1}\,\text{h}^{-1}$	Pre-exponential factor
$T_{20} = 300\,\text{K}$	Temp.: CSTR-2 inlet	$E = 7.1 \times 10^4\,\text{kJ}\,\text{kmol}^{-1}$	Activation energy
$F_{10} = 5.0\,\text{m}^3\,\text{h}^{-1}$	Flow: CSTR-1 inlet	$\Delta H = -7.8 \times 10^3\,\text{kJ}\,\text{kmol}^{-1}$	Heat of reaction
$F_{20} = 5.0\,\text{m}^3\,\text{h}^{-1}$	Flow: CSTR-2 inlet	$\Delta H_{\text{vap}} = 4.02 \times 10^4\,\text{kJ}\,\text{kmol}^{-1}$	Heat of vaporization
$F_r = 2.0\,\text{m}^3\,\text{h}^{-1}$	Flow: SEP-1 vapor	$C_p = 0.231\,\text{kJ}\,\text{kg}^{-1}\,K^{-1}$	Heat capacity
$V_1 = 5.0\,\text{m}^3$	Volume: CSTR-1	$R = 8.314\,\text{kJ}\,\text{kmol}^{-1}\,\text{K}^{-1}$	Gas constant
$V_2 = 5.0\,\text{m}^3$	Volume: CSTR-2	$\rho_L = 1000\,\text{kg}\,\text{m}^{-3}$	Liquid solution density
$V_3 = 3.0\,\text{m}^3$	Volume: SEP-1	$MW_A = 18\,\text{kg}\,\text{kmol}^{-1}$	Molecular weight: *A*
$\alpha_A = 3.0$	Relative volatility: *A*	$MW_B = 18\,\text{kg}\,\text{kmol}^{-1}$	Molecular weight: *B*
$\alpha_B = 0.8$	Relative volatility: *B*	$MW_D = 40.0\,\text{kg}\,\text{kmol}^{-1}$	Molecular weight: *D*
$\alpha_D = 1.0$	Relative volatility: *D*		

$$\frac{dT_1}{dt} = \frac{F_{10}}{V_1}T_{10} + \frac{F_r}{V_1}T_3 - \frac{F_1}{V_1}T_1 - \frac{\Delta H k_0}{\rho_L C_p}e^{-E/RT_1}C_{A1}^2 + \frac{Q_1}{\rho_L C_p V_1} \tag{7.89a}$$

$$\frac{dC_{A1}}{dt} = \frac{F_{10}}{V_1}C_{A10} + \frac{F_r}{V_1}C_{Ar} - \frac{F_1}{V_1}C_{A1} - k_0 e^{-E/RT_1}C_{A1}^2 \tag{7.89b}$$

$$\frac{dC_{B1}}{dt} = \frac{F_r}{V_1}C_{Br} - \frac{F_1}{V_1}C_{B1} + k_0 e^{-E/RT_1}C_{A1}^2 \tag{7.89c}$$

$$\frac{dT_2}{dt} = \frac{F_{20}}{V_2}T_{20} + \frac{F_1}{V_2}T_1 - \frac{F_2}{V_2}T_2 - \frac{\Delta H k_0}{\rho_L C_p}e^{-E/RT_2}C_{A2}^2 + \frac{Q_2}{\rho_L C_p V_2} \tag{7.89d}$$

$$\frac{dC_{A2}}{dt} = \frac{F_{20}}{V_2}C_{A20} + \frac{F_1}{V_2}C_{A1} - \frac{F_2}{V_2}C_{A2} - k_0 e^{-E/RT_2}C_{A2}^2 \tag{7.89e}$$

$$\frac{dC_{B2}}{dt} = \frac{F_1}{V_2}C_{B1} - \frac{F_2}{V_2}C_{B2} + k_0 e^{-E/RT_2}C_{A2}^2 \tag{7.89f}$$

$$\frac{dT_3}{dt} = \frac{F_2}{V_3}(T_2 - T_3) - \frac{\Delta H_{\text{vap}} \dot{M}_r}{\rho_L C_p V_3} + \frac{Q_3}{\rho_L C_p V_3} \tag{7.89g}$$

$$\frac{dC_{A3}}{dt} = \frac{F_2}{V_3} C_{A2} - \frac{F_r}{V_3} C_{Ar} - \frac{F_3}{V_3} C_{A3} \tag{7.89h}$$

$$\frac{dC_{B3}}{dt} = \frac{F_2}{V_3} C_{B2} - \frac{F_r}{V_3} C_{Br} - \frac{F_3}{V_3} C_{B3} \tag{7.89i}$$

where T_j denotes the temperature of the j-th vessel ($j = 1$ denotes CSTR-1, $j = 2$ denotes CSTR-2, and $j = 3$ denotes SEP-1), C_{ij} denotes the concentration of the i-th species ($i = A,\ B$) in the j-th vessel, and $\dot{M}_r$ denotes the molar flow rate of the recycle stream.

The relative volatility of each species is assumed to be constant within the operating temperature range of the flash tank. The following algebraic equations are used to model the composition of the recycle stream:

$$C_{D3} = (\rho_L - C_{A3} MW_A - C_{B3} MW_B) / MW_D \tag{7.90a}$$

$$C_{ir} = \frac{\alpha_i \rho_L C_{i3}}{\sum_{j \in \{A,B,D\}} \alpha_j C_{j3} MW_j}, \quad i = A,\ B,\ D \tag{7.90b}$$

$$\dot{M}_r = F_r \left(C_{Ar} + C_{Br} + C_{Dr} \right) \tag{7.90c}$$

where C_{ir} is the overhead vapor concentration of the separator. Given the assumption of constant liquid hold-up and constant liquid density, the volumetric flow rates are given by the following equations:

$$F_1 = F_r + F_{10} \tag{7.91a}$$

$$F_2 = F_1 + F_{20} \tag{7.91b}$$

$$F_3 = F_2 - F_r \tag{7.91c}$$

where F_j is the volumetric flow rate of the outlet stream of the j-th vessel.

The process network has five manipulated inputs: the three heat rates Q_j, $j = 1,\ 2,\ 3$ and the inlet concentration of the reactant A in the feedstock to each reactor (C_{A10} and C_{A20}). The bounds on the available control action are $Q_j \in [-1.0,\ 1.0] \times 10^5\ \text{kJ}\,\text{h}^{-1}$ for $j = 1,\ 2$, $Q_3 \in [2.2,\ 2.5] \times 10^6\ \text{kJ}\,\text{h}^{-1}$, and $C_{Aj0} \in [0.5,\ 7.5]\ \text{kmol}\,\text{m}^{-3}$ $j = 1,\ 2$. In addition to the input constraints, the reactions take place within the temperature range from 370.0 to 395.0 K and thus, the reactors are to be operated within this temperature range. The separation occurs at 390.0 K.

The real-time economics of the process network are assumed to be described by the molar flow rate of desired product B leaving the process network which is denoted as $\dot{M}_{B3}$. The time-averaged amount of reactant that may be fed to each reactor is constrained to an average amount of 20.0 kmol h^{-1} which gives rise to the following two input average constraints:

$$\frac{1}{t_f - t_0} \int_{t_0}^{t_f} F_{j0} C_{Aj0}(t)\, dt = 20.0\,\text{kmol}\,\text{h}^{-1} \tag{7.92}$$

for $j = 1, 2$ where t_0 and t_f are the initial and final time of the operation of the process network. Since the inlet flow rates F_{10} and F_{20} are constant, the average input constraint may be written in terms of the inlet concentration of A only such that the time-averaged value of C_{Aj0} must be equal to 4.0 kmol m^{-3}.

The economically optimal steady-state (which is simply referred to as the optimal steady-state for the remainder) will be used in the design of a real-time LEMPC, i.e., the stability region for the optimal steady-state will be used in the LEMPC formulation. Since the reaction rate is maximized at high temperature, computing the optimal steady-state with the exact acceptable temperature operating range will give an optimal steady-state with the greatest acceptable reactor operating temperature. Much like current practice, the optimal steady-state is computed with a degree of conservativeness or "back-off" introduced in the acceptable operating temperature range, so that the reactor temperature is maintained within the acceptable operating range over the length of operation in the presence of uncertainty and disturbances (see [39] and the references therein, for instance, for more details on the back-off methodology). Thus, the optimal steady-state must satisfy a restricted temperature range of $T_{js} \in [370.0, 380.0]$ K for $j = 1, 2$. The steady-state optimization problem is given by:

$$\begin{aligned}
\max_{x_s, u_s} \quad & F_3 C_{B3s} \\
\text{s.t.} \quad & f(x_s, u_s) = 0 \\
& 370.0\,\text{K} \le T_{1s} \le 380.0\,\text{K} \\
& 370.0\,\text{K} \le T_{2s} \le 380.0\,\text{K} \\
& T_{3s} = 390.0\,\text{K} \\
& -1.0 \times 10^5\,\text{kJh}^{-1} \le Q_{1s} \le 1.0 \times 10^5\,\text{kJh}^{-1} \\
& -1.0 \times 10^5\,\text{kJh}^{-1} \le Q_{2s} \le 1.0 \times 10^5\,\text{kJh}^{-1} \\
& 2.2 \times 10^6\,\text{kJh}^{-1} \le Q_{3s} \le 2.5 \times 10^6\,\text{kJh}^{-1} \\
& C_{A10s} = C_{A20s} = 4.0\,\text{kmol}\,\text{m}^{-3}
\end{aligned} \tag{7.93}$$

where $f(x_s, u_s) = 0$ represents the steady-state model. The optimal steady-state vector (omitting units) is:

$$\begin{aligned}
x_s^* &= \left[T_{1s}^* \; C_{A1s}^* \; C_{B1s}^* \; T_{2s}^* \; C_{A2s}^* \; C_{B2s}^* \; T_{3s}^* \; C_{A3s}^* \; C_{B3s}^* \right]^T \\
&= \left[380.0 \; 2.67 \; 2.15 \; 380.0 \; 2.42 \; 2.06 \; 390.0 \; 1.85 \; 2.15 \right]^T,
\end{aligned} \tag{7.94}$$

and the optimal steady-state input vector is

$$u_s^* = \left[Q_{1s}^* \; Q_{2s}^* \; Q_{3s}^* \; C_{A10s}^* \; C_{A20s}^* \right]^T$$
$$= \left[-4.21 \times 10^3 \; 1.70 \times 10^4 \; 2.34 \times 10^6 \; 4.0 \; 4.0 \right]^T . \tag{7.95}$$

The optimal steady-state is open-loop unstable.

The control objective of the process network is to optimize the economics through real-time operation while maintaining the closed-loop state trajectory inside a well-defined state-space set. To accomplish this objective, the real-time LEMPC scheme is applied to the process network. In stark contrast to traditional tracking control that forces the closed-loop state to converge to the (optimal) steady-state, applying LEMPC to the process network is not expected to achieve convergence to the optimal steady-state. Instead, LEMPC may force the process network to operate in a consistently transient manner to achieve better closed-loop performance compared to the closed-loop performance at the optimal steady-state.

For the implementation of the LEMPC, the acceptable temperature range is not treated as a hard constraint. Instead, the acceptable temperature range is accounted for by imposing quadratic penalty terms in the stage cost of the LEMPC. Thus, the stage cost used in the objective function of the LEMPC is

$$l_e(x, u) = -F_3 C_{B3} + \sum_{i=1}^{3} Q_{c,i} (T_i - T_{is}^*)^2 \tag{7.96}$$

where T_{is}^*, $i = 1, 2, 3$ are the optimal steady-state temperatures. The stage cost of Eq. 7.96 includes the economics and the quadratic penalty terms for the temperature. The weight coefficients are $Q_{c,1} = 0.018$, $Q_{c,2} = 0.022$, and $Q_{c,3} = 0.01$ and have been tuned such that the closed-loop temperatures are maintained near the optimal steady-state temperature. Since no hard or soft constraints are imposed on the temperature in the LEMPC, it is emphasized that there is no guarantee that the temperatures are maintained within the acceptable temperature range described above ($T_j \in [370.0, 395.0]$ K for $j = 1, 2$ and $T_3 \approx 390.0$ K). In this example, small violations over a short period are considered acceptable. If maintaining the operation within the acceptable operating temperature range is considered critical, one may add various modifications to the LEMPC to achieve this objective such as decreasing the size of Ω_{ρ_e}, adding hard or soft constraints on the temperature in the LEMPC, or adding a contractive constraint on the temperature ODEs.

An explicit stabilizing controller is designed using feedback linearization techniques to make the dynamics of the temperature ODEs linear (in a state-space region where the input constraints are satisfied) under the explicit controller. Specifically, the temperature ODEs are input-affine in the heat rate input and have the form:

$$\dot{T}_j = f_j(x) + b_j Q_j \tag{7.97}$$

where $f_j(x)$ is a nonlinear scalar-valued function, b_j is constant and $j = 1, 2, 3$. The controller that makes the closed-loop temperature dynamics linear is:

$$Q_j = -\frac{1}{b_j}\left(f_j(x) + K_j(T_j - T_{js}^*)\right) \tag{7.98}$$

where K_j denotes the controller gain. In this case, the controller gains are $K_1 = 5$, $K_2 = 5$, and $K_3 = 7$, respectively. The inlet concentration input values are fixed to the average values (4.0 kmol m^{-3}). Through extensive closed-loop simulations under the state feedback controller, a quadratic Lyapunov function for the process network under the feedback controller $h(x)$ is determined. An estimate of the stability region of the process network under the feedback controller was characterized by computing the state-space points where $\dot{V} < 0$ and taking the stability region to be a level set of the Lyapunov function containing only state-space points where the time-derivative of the Lyapunov function is negative. The quadratic Lyapunov function has the form:

$$V(x) = (x - x_s^*)^T P (x - x_s^*) \tag{7.99}$$

where P is the following positive definite matrix:

$$P = \text{diag}\left[0.001\ 1.5\ 0.05\ 0.001\ 1.5\ 0.05\ 0.001\ 1.5\ 0.05\right]. \tag{7.100}$$

The estimated stability region Ω_ρ is the level set of the Lyapunov function where $V(x) \leq 11.0$, i.e., $\rho = 11.0$. The subset of the stability region which defines the mode 1 constraint of the LEMPC is $\rho_e = 10.0$ and has been determined through extensive closed-loop simulation under LEMPC as the subset of the stability region Ω_ρ where the closed-loop state under LEMPC is maintained in Ω_ρ.

The input average constraint is imposed over successive, finite-length operating periods. Specifically, the average constraint must be satisfied over each operating period $t_M = M\Delta$ where M is the number of sampling periods in the operating period. This ensures that over the entire length of operation the average constraint will be satisfied. For this example, the operating period was chosen to be $t_M =$ 2.4 h which leads to better asymptotic average economic performance under LEMPC (assuming no computational delay) than the asymptotic average performance at the economically optimal steady-state.

To solve the dynamic optimization problem of the LEMPC, orthogonal collocation with three Radau collocation points per sampling period is employed for discretization of the ODEs (see, for instance, [37] for details on solving a dynamic optimization problem using a simultaneous approach). The open-source nonlinear optimization solver Ipopt [21] was employed owing to its ability to exploit the high degree of sparsity of the resulting optimization problem. Analytical first and second-order derivative information was provided to the solver. The closed-loop simulations were coded in C++ and performed on an Intel® Core™ 2 Quad 2.66 GHz processor running an Ubuntu Linux operating system. The sampling period of the LEMPC used in the simulations below is $\Delta = 0.01$ h. To simulate forward in time the closed-loop process network, the fourth-order Runge-Kutta method was used with a time step of 0.0001 h.

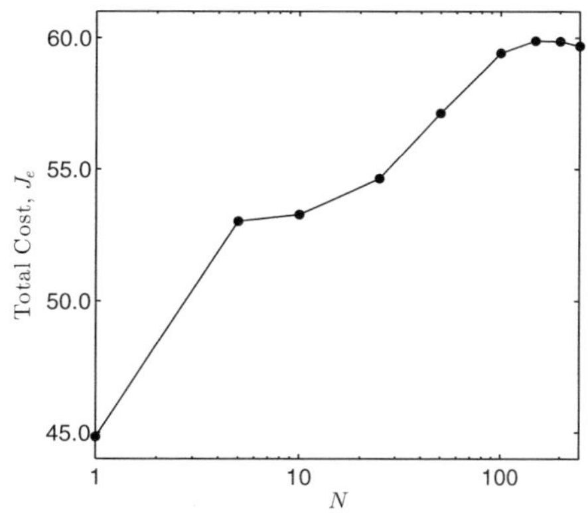

Fig. 7.21 The total economic cost J_e over one operating window length of operation (2.4 h) of the process network under LEMPC with the prediction horizon length

In the first set of simulations, nominal operation of the process network under LEMPC implemented in a typical receding horizon fashion is considered under ideal computation, i.e., assuming no computational delay. The closed-loop economic performance under LEMPC is assessed using the economic performance index which is defined as:

$$J_e = \int_0^{t_f} F_3 C_{B3}\, dt. \tag{7.101}$$

Since the LEMPC does not directly optimize the molar flow rate of product out of the process network, the stage cost index will also be considered as a measure of the closed-loop performance and is given by:

$$L_e = -\int_0^{t_f} l_e(x, u)\, dt. \tag{7.102}$$

First, the effect of the prediction horizon on the closed-loop economic performance over one operating period (2.4 h) is considered. The closed-loop performance index of Eq. 7.101 plotted against the prediction horizon length is given in Fig. 7.21. A significant increase in closed-loop performance is observed initially with increasing prediction horizon length until the closed-loop performance becomes approximately constant. Owing to this fact, a prediction horizon of $N = 200$ is used in all subsequent simulations. A simulation over many operating periods such that the effect of the initial condition on closed-loop performance becomes negligible is performed (with $N = 200$). The asymptotic average closed-loop economic performance, which is the time-averaged economic cost after the effect of the initial condition becomes negligible, is determined from this simulation to be 25.0 kmol h^{-1} (in this case, the time-averaged production rate over each operating window becomes constant after a sufficiently long length of operation). The optimal steady-state production rate of B is 21.5 kmol h^{-1}. Thus, the asymptotic production rate of the process network under the LEMPC is 16.3 % better than the production rate at the economically optimal steady-state.

The effect of computational delay is considered in the next set of simulations, and two scenarios are considered: (1) the closed-loop process network under LEMPC implemented in a typical receding horizon fashion where the control action is subject to real-time computational delay (for the sake of simplicity, this case will be referred to as the closed-loop process network under LEMPC for the remainder) and (2) the closed-loop process network under the real-time LEMPC scheme (also, subject to real-time computational delay). For the former scenario, the LEMPC begins to compute a control action at each sampling instance after receiving a state measurement. Owing to the computational delay, the control action applied to the process network is the most up-to-date control action. For example, if it takes 0.002 h to compute the control action at the sampling instance t_k, then $u(t_{k-1})$ is applied to the process network from t_k to $t_k + 0.002$h (assuming $u(t_{k-1})$ is available at t_k) and applies $u(t_k)$ to the process network from $t_k + 0.002$ h to $t_{k+1} = t_k + \Delta$. For each scenario, a 12.0 h length of closed-loop operation is simulated. For the real-time LEMPC case, the LEMPC is forced to recompute a new solution after three sampling periods have elapsed since the last time an LEMPC solution was computed, i.e., the solver starts computing a new solution at the beginning of the fourth sampling period.

The average computation time required to solve the LEMPC (of scenario (1)) at each sampling time was 11.2 s (31.2 % of the sampling period) with a standard deviation of 7.42 s. The maximum computation time over the simulation is 61.9s which is almost double the sampling period. The computation time exceeds the sampling period ten out of the 1,200 sampling periods in the simulation. Over the course of both simulations, the closed-loop state is maintained in Ω_ρ. The closed-loop trajectories under the real-time LEMPC scheme are given in Fig. 7.22 (the closed-loop behavior under the LEMPC subject to real-time computational delay was similar). The difference between the performance index of the two cases is less

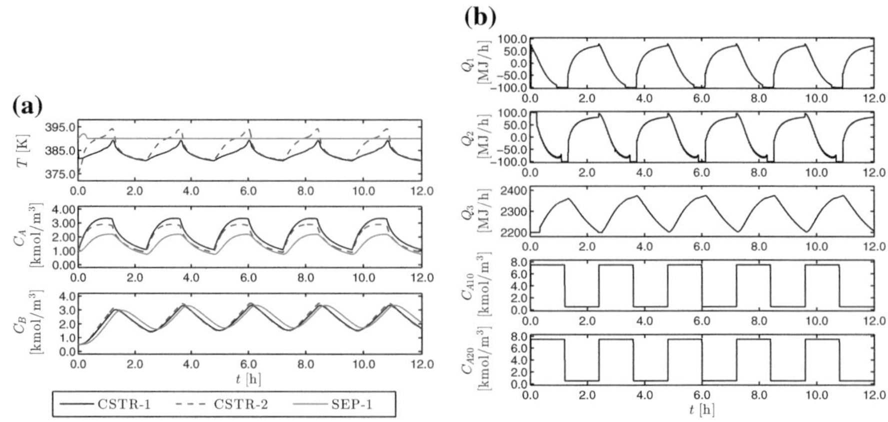

Fig. 7.22 The closed-loop **a** state and **b** input trajectories of the nominally operated process network under the real-time LEMPC scheme

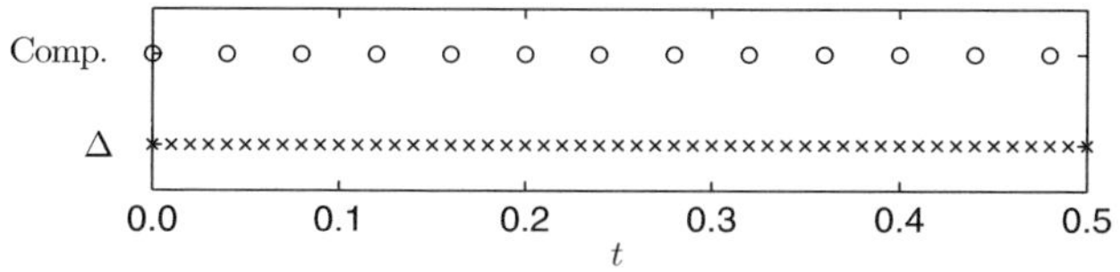

Fig. 7.23 The number of times the LEMPC problem was solved (Comp.) as dictated by the real-time implementation strategy compared to the sampling period (Δ) over the first 0.5 h of operation

than 0.5 % (the performance indexes for case (1) and case (2) were 284.3 and 283.3, respectively).

While little difference between the two cases in terms of closed-loop performance is observed, it is important to note that an a priori guarantee on closed-loop stability under the real-time LEMPC may be made. Also, the total on-line computation time to solve the LEMPC over the two simulations is 3.74 and 0.94 h, respectively. The real-time LEMPC reduces the total on-line computation requirement by 75 % compared to LEMPC implemented in a receding horizon fashion because the real-time LEMPC does not recompute a control action at each sampling instance, while LEMPC, implemented in a receding horizon fashion, recomputes a control action at each sampling instance. To better illustrate this point, Fig. 7.23 shows the frequency the LEMPC problem is solved under the real-time implementation strategy with respect to the sampling period over the first 0.5 h of operation. Over this time, the LEMPC optimization problem is solved at a rate of one out of every four sampling periods. This trend continues over the remainder of the 12.0 h length of operation and hence, the 75 % reduction in total computational time.

Since the computational delay depends on many factors, e.g., model dimension, prediction horizon, solution strategy used to solve the dynamic optimization problem, the nonlinear optimization solver used, and computer hardware, it is also important to consider computational delay greater than one sampling period to demonstrate that the real-time LEMPC scheme may handle computation delay of arbitrary length. Therefore, another set of simulations is considered where longer computational delay is simulated. The computation delay is modeled as a bounded uniformly-distributed random number and the maximum computational delay is assumed to be less than 10 sampling periods. Both the LEMPC (receding horizon implementation) and the real-time LEMPC scheme are considered. To make the comparison as consistent as possible, the computational delay, at the time steps the real-time LEMPC is solved, is simulated to be the same as the computation delay to solve the LEMPC at the same time step (recall the real-time LEMPC is not solved at each sampling period). Given the computational delay is much greater for this set of simulations than in the previous set of simulations, the real-time LEMPC is forced to recompute a new solution after 15 sampling periods have elapsed since the last time it computed a solution.

Several simulations are performed, each starting at a different initial condition, and the performance indexes of these simulations are given in Table 7.4. Applying

Table 7.4 The performance indices of the process network under the back-up explicit controller, under the LEMPC subject to computational delay, and under the real-time LEMPC for several simulations

Sim.	Back-up controller		LEMPC		Real-time LEMPC	
	J_e	L_e	J_e	L_e	J_e	L_e
1	225.5	225.4	277.0	245.0	295.1	216.5
2	254.2	254.1	318.7	278.6	307.3	279.6
3	260.5	260.4	319.9	286.3	318.1	294.7
4	232.7	230.6	290.7	255.7	299.2	266.4
5	250.0	250.0	308.7	276.9	322.8	282.9

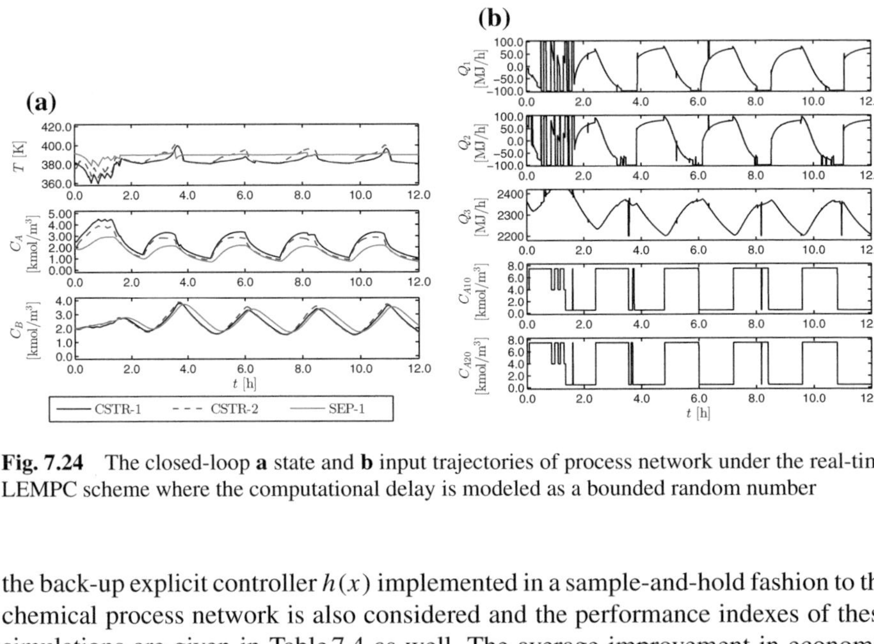

Fig. 7.24 The closed-loop **a** state and **b** input trajectories of process network under the real-time LEMPC scheme where the computational delay is modeled as a bounded random number

the back-up explicit controller $h(x)$ implemented in a sample-and-hold fashion to the chemical process network is also considered and the performance indexes of these simulations are given in Table 7.4 as well. The average improvement in economic performance compared to the process network under the back-up controller is 26.1 % under the real-time LEMPC scheme and 23.9 % under the LEMPC (implemented in a receding horizon). Thus, a substantial economic benefit is achieved by applying LEMPC to the process network. While the real-time LEMPC does not always achieve better performance (either measured in terms of the economic performance index or stage cost index) compared to the performance under LEMPC, the closed-loop trajectories between the two cases are significantly different. Figures 7.24 and 7.25 give the closed-loop trajectories of simulation 2 (as labeled in Table 7.4). The input trajectory computed by the real-time LEMPC has chattering initially over the first operating period because of the effect of the initial condition, but after the first operating period when the effect of the initial condition dissipates, the computed input trajectory is significantly smoother. On the other hand, chattering in the input profiles

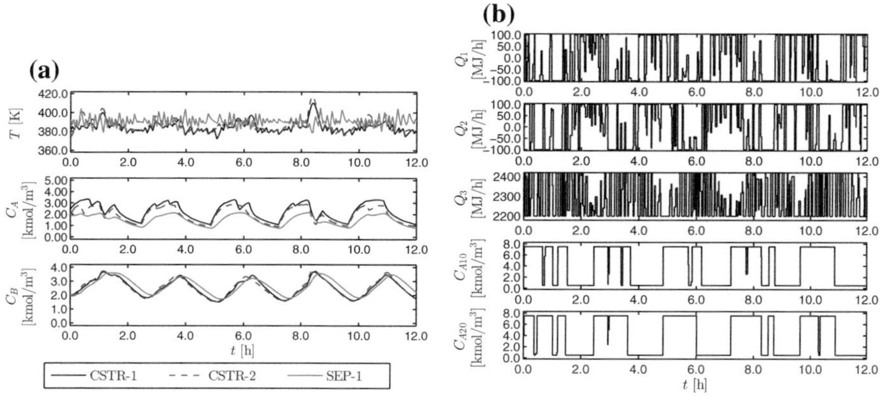

Fig. 7.25 The closed-loop **a** state and **b** input trajectories of process network under LEMPC subject to computational delay where the computational delay is modeled as a bounded random number

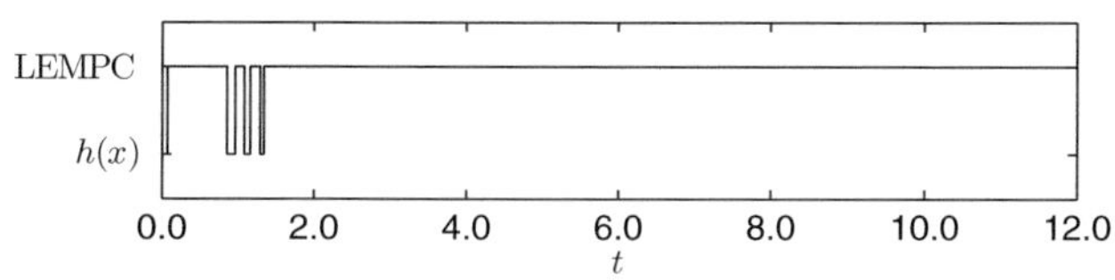

Fig. 7.26 A discrete trajectory depicting when the control action applied to the process network over each sampling period is from a precomputed LEMPC solution or from the back-up controller for the closed-loop simulation of Fig. 7.24

is observed throughout the entire simulation under the LEMPC. If we compare the performance index of operation from $t = 2.4$ h to $t = 12.0$ h (after the first operating period) for simulation 2, the indexes are $J_e = 249.8$ and $L_e = 227.9$ for operation under the real-time LEMPC and $J_e = 248.5$ and $L_e = 217.4$ for operation under the LEMPC; the performance under the real-time LEMPC is better over this period than under LEMPC.

Over the five simulations under the real-time LEMPC strategy, the explicit controller was applied on average 19 out of 1200 sampling periods. For the simulation of Fig. 7.24, a discrete trajectory showing when the control action applied to the process network under the real-time LEMPC strategy is from a precomputed LEMPC solution or from the back-up controller is given in Fig. 7.26. For this case, the back-up controller is used 31 out of 1200 sampling periods (2.7 % of the sampling periods). From Fig. 7.26, the back-up controller is only applied over the first operating period and is not used in any subsequent sampling period. Thus, the source of performance degradation for this case (Sim. 2 in Table 7.4) is due to applying the explicit back-up controller to maintain the closed-loop state in Ω_ρ. Again, it is emphasized that there is no a priori guarantee that the LEMPC implemented in a receding horizon fashion subject to computational delay may maintain the closed-loop state inside Ω_ρ.

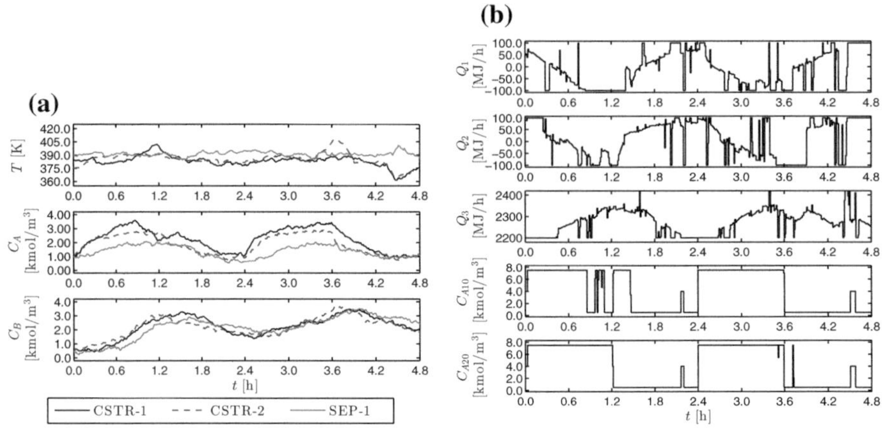

Fig. 7.27 The closed-loop **a** state and **b** input trajectories of process network under the real-time LEMPC scheme with bounded process noise

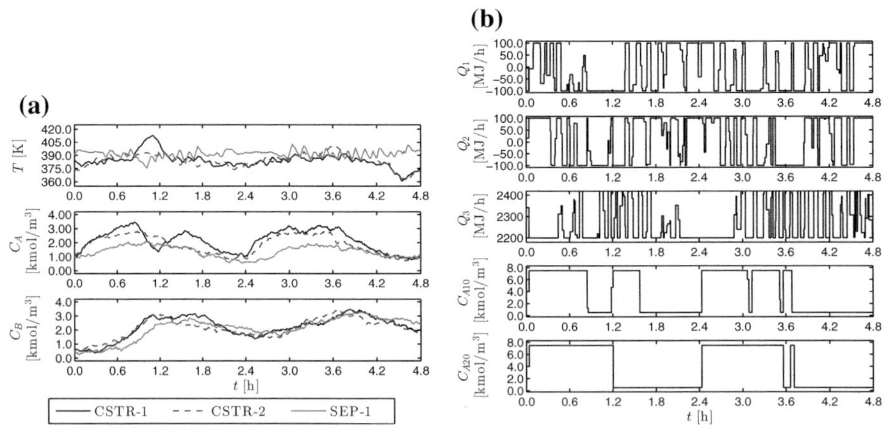

Fig. 7.28 The closed-loop **a** state and **b** input trajectories of process network under LEMPC subject to computational delay with bounded process noise

In the last set of simulations, significant bounded Gaussian process noise with zero mean is added to the model states. The standard deviations of the noise added to the temperature and concentration states are 5.0 and 0.5, respectively and the bounds on the noise are 2.0 and 15.0, respectively. Two closed-loop simulations over 12.0 h length of operation are completed with the same realization of the process noise. In the first simulation, the process network is controlled by the real-time LEMPC and the closed-loop trajectories are given in Fig. 7.27 over the first two operating periods. For this case, the back-up controller is applied 69 out of 1200 sampling periods (5.8 % of the sampling periods). In the second simulation, the process network is controlled by LEMPC subject to computation delay (trajectories shown in Fig. 7.28).

From Fig. 7.28, a significant degree of chattering and bang-bang type actuation in the input trajectory is observed. This behavior tries to combat the effect of the added process noise and is due to not penalizing control actions in the stage cost and not imposing rate of change constraints on the control actions. In practice, one could add one or both of these elements to the LEMPC if the computed input trajectory is not implementable. On the other hand, the real-time LEMPC implements a much smoother input trajectory (Fig. 7.27) because the precomputed input trajectory of the real-time LEMPC has a degree of smoothness like the closed-loop trajectory of the nominally operated process network (Fig. 7.22). If the precomputed input trajectory satisfies the stability conditions, it will be applied to the closed-loop process network with disturbances. The closed-loop system under the real-time LEMPC has guaranteed stability properties, but is not recomputed at each sampling period like the receding horizon implementation of LEMPC which will try to combat the effect of the disturbance on performance. In both cases, the state is maintained in Ω_ρ. The performance indexes of the two cases are 301.6 under the real-time LEMPC and 295.5 under the LEMPC; the closed-loop performance under the real-time LEMPC scheme is 2.0 % better than applying LEMPC without accounting for the computational delay. Moreover, the back-up controller is also applied to the process network subject to the same realization of the process noise. The economic performance index for this case is 242.3. For operation with process noise, the economic performance improvement over the process network under the back-up controller is 24.4 % under the real-time LEMPC strategy and 21.9 % under the receding horizon LEMPC for the same initial condition.

7.5 Conclusions

In this chapter, three EMPC designs were considered. The first section focused on the design of economic MPC for a class of nonlinear singularly perturbed systems. Under appropriate stabilizability assumptions, fast sampling of the fast states and slow sampling of the slow states, the presented composite control system featuring an EMPC may dictate a possible time-varying operation to address economic considerations while guaranteeing closed-loop stability. Closed-loop stability was addressed through singular perturbation arguments.

In the second section, an application study of several distributed EMPC strategies was presented. Two important performance metrics were considered to assess the EMPC strategies including the closed-loop performance and the computational time required to solve the optimization problem(s) at each sampling time. From the closed-loop simulation results of application study, a distributed EMPC strategy was capable of delivering similar closed-loop performance as a centralized EMPC approach while reducing the on-line computation time required to solve the optimization problems.

In the final section, a strategy for implementing Lyapunov-based economic model predictive control (LEMPC) in real-time with computation delay was developed. The implementation strategy uses a triggering condition to precompute an input trajectory from LEMPC over a finite-time horizon. At each sampling period, if a certain stability (triggering) condition is satisfied, then the precomputed control action by LEMPC is applied to the closed-loop system. If the stability condition is violated, then a backup explicit stabilizing controller is used to compute the control action for the sampling period. In this fashion, the LEMPC is used when possible to optimize the economics of the process. Conditions such that the closed-loop state under the real-time LEMPC is always bounded in a compact set were derived.

References

1. Kokotovic P, Khalil HK, O'Reilly J (1999) Singular perturbation methods in control: analysis and design, vol. 25. SIAM
2. Christofides PD, Daoutidis P (1996) Feedback control of two-time-scale nonlinear systems. Int J Control 63:965–994
3. Kumar A, Daoutidis P (2002) Nonlinear dynamics and control of process systems with recycle. J Process Control 12:475–484
4. Baldea M, Daoutidis P (2012) Control of integrated chemical process systems using underlying DAE models. In: Biegler LT, Campbell SL, Mehrmann V (eds) Control and optimization with differential-algebraic constraints. SIAM, pp 273–291
5. Chen X, Heidarinejad M, Liu J, Christofides PD (2012) Composite fast-slow MPC design for nonlinear singularly perturbed systems. AIChE J 58:1802–1811
6. Chen X, Heidarinejad M, Liu J, Muñoz de la Peña D, Christofides PD (2011) Model predictive control of nonlinear singularly perturbed systems: application to a large-scale process network. J Process Control 21:1296–1305
7. Massera JL (1956) Contributions to stability theory. Ann Math 64:182–206
8. Lin Y, Sontag E, Wang Y (1996) A smooth converse Lyapunov theorem for robust stability. SIAM J Control Optim 34:124–160
9. Khalil HK (2002) Nonlinear Systems, 3rd edn. Prentice Hall, Upper Saddle River, NJ
10. Muñoz de la Peña D, Christofides PD (2008) Lyapunov-based model predictive control of nonlinear systems subject to data losses. IEEE Trans Autom Control 53:2076–2089
11. Christofides P, Teel A (1996) Singular perturbations and input-to-state stability. IEEE Trans Autom Control 41:1645–1650
12. Christofides PD, Scattolini R, Muñoz de la Peña D, Liu J (2013) Distributed model predictive control: a tutorial review and future research directions. Comput Chem Eng 51:21–41
13. Liu J, Muñoz de la Peña D, Christofides PD (2009) Distributed model predictive control of nonlinear process systems. AIChE J 55:1171–1184
14. Liu J, Chen X, Muñoz de la Peña D, Christofides PD (2010) Sequential and iterative architectures for distributed model predictive control of nonlinear process systems. AIChE J 56:2137–2149
15. Scattolini R (2009) Architectures for distributed and hierarchical model predictive control—a review. J Process Control 19:723–731
16. Müller MA, Allgöwer F (2014) Distributed economic MPC: a framework for cooperative control problems. In: Proceedings of the 19th world congress of the international federation of automatic control. Cape Town, South Africa, pp 1029–1034
17. Driessen PAA, Hermans RM, van den Bosch PPJ (2012) Distributed economic model predictive control of networks in competitive environments. In: Proceedings of the 51st IEEE conference on decision and control, Maui, HI, pp 266–271
18. Chen X, Heidarinejad M, Liu J, Christofides PD (2012) Distributed economic MPC: application to a nonlinear chemical process network. J Process Control 22:689–699

19. Lee J, Angeli D (2012) Distributed cooperative nonlinear economic MPC. In: Proceedings of the 20th international symposium on mathematical theory of networks and systems. Melbourne, Australia
20. Özgülşen F, Adomaitis RA, Çinar A (1992) A numerical method for determining optimal parameter values in forced periodic operation. Chem Eng Sci 47:605–613
21. Wächter A, Biegler LT (2006) On the implementation of an interior-point filter line-search algorithm for large-scale nonlinear programming. Math Program 106:25–57
22. Ellis M, Christofides PD (2014) Optimal time-varying operation of nonlinear process systems with economic model predictive control. Ind Eng Chem Res 53:4991–5001
23. Ronco E, Arsan T, Gawthrop PJ (1999) Open-loop intermittent feedback control: practical continuous-time GPC. IEE Proc—Control Theory Appl 146:426–434
24. Chen WH, Ballance DJ, O'Reilly J (2000) Model predictive control of nonlinear systems: computational burden and stability. IEE Proc—Control Theory Appl 147:387–394
25. Findeisen R, Allgöwer F (2004) Computational delay in nonlinear model predictive control. In: Proceedings of the IFAC international symposium of advanced control of chemical processes, Hong Kong, pp 427–432
26. Scokaert POM, Mayne DQ, Rawlings JB (1999) Suboptimal model predictive control (feasibility implies stability). IEEE Trans Autom Control 44:648–654
27. Würth L, Hannemann R, Marquardt W (2009) Neighboring-extremal updates for nonlinear model-predictive control and dynamic real-time optimization. J Process Control 19:1277–1288
28. Zavala VM, Biegler LT (2009) The advanced-step NMPC controller: optimality, stability and robustness. Automatica 45:86–93
29. Biegler LT, Yang X, Fischer GAG (2015) Advances in sensitivity-based nonlinear model predictive control and dynamic real-time optimization. J Process Control 30:104–116
30. Ganesh N, Biegler LT (1987) A reduced Hessian strategy for sensitivity analysis of optimal flowsheets. AIChE J 33:282–296
31. Yang X, Biegler LT (2013) Advanced-multi-step nonlinear model predictive control. J Process Control 23:1116–1128
32. Jäschke J, Yang X, Biegler LT (2014) Fast economic model predictive control based on NLP-sensitivities. J Process Control 24:1260–1272
33. Würth L, Hannemann R, Marquardt W (2011) A two-layer architecture for economically optimal process control and operation. J Process Control 21:311–321
34. Wolf IJ, Muñoz DA, Marquardt W (2014) Consistent hierarchical economic NMPC for a class of hybrid systems using neighboring-extremal updates. J Process Control 24:389–398
35. Diehl M, Bock H, Schlöder J (2005) A real-time iteration scheme for nonlinear optimization in optimal feedback control. SIAM J Control Optim 43:1714–1736
36. Diehl M, Ferreau HJ, Haverbeke N (2009) Efficient numerical methods for nonlinear MPC and moving horizon estimation. In: Magni L, Raimondo DM, Allgöwer F (eds) Nonlinear model predictive control, vol 384., Lecture Notes in Control and Information SciencesSpringer, Berlin Heidelberg, pp 391–417
37. Biegler LT (2010) Nonlinear programming: concepts, algorithms, and applications to chemical processes. SIAM, Philadelphia, PA
38. Tabuada P (2007) Event-triggered real-time scheduling of stabilizing control tasks. IEEE Trans Autom Control 52:1680–1685
39. Kookos IK, Perkins JD (2002) An algorithmic method for the selection of multivariable process control structures. J Process Control 12:85–99

Index

M. Ellis et al., *Economic Model Predictive Control*,
Advances in Industrial Control, DOI 10.1007/978-3-319-41108-8

MIX
Papier aus verantwortungsvollen Quellen
Paper from responsible sources
FSC® C105338

If you have any concerns about our products,
you can contact us on
ProductSafety@springernature.com

In case Publisher is established outside the EU,
the EU authorized representative is:
Springer Nature Customer Service Center GmbH
Europaplatz 3, 69115 Heidelberg, Germany

Printed by Libri Plureos GmbH
in Hamburg, Germany